An Introduction to Mining Seismology

This is Volume 55 in the
INTERNATIONAL GEOPHYSICS SERIES
A series of monographs and textbooks
Edited by RENATA DMOWSKA and JAMES R. HOLTON

A complete list of the books in this series appears at the end of this volume.

An Introduction to Mining Seismology

Slawomir Jerzy Gibowicz and Andrzej Kijko
INSTITUTE OF GEOPHYSICS
POLISH ACADEMY OF SCIENCES
WARSAW, POLAND

ACADEMIC PRESS
A Division of Harcourt Brace & Company
San Diego   New York   Boston
London   Sydney   Tokyo   Toronto

Cover photograph   Rockburst damage in a tunnel from a seismic event in one of the deep gold mines in South Africa. Photo courtesy of Rommel Grobbelaar.

This book is printed on acid-free paper.

Academic Press, Inc.
525 B Street, Suite 1900, San Diego, California 92101-4495

United Kingdom Edition published by
Academic Press Limited
24–28 Oval Road, London NW1 7DX

Library of Congress Cataloging-in-Publication Data

Gibowicz, Slawomir J., Date.
    An introduction to mining seismology / S.J. Gibowicz, A. Kijko.
        p.   cm. -- (international geophysics ; v. 55)
    Includes bibliographical references (p   ) and index.
    ISBN 0-12-282120-3
    1. Rock bursts.   2. Induced seismicity.   I. Kijko, Andrzej.
    II. Title.   III. Series
    TN317.G52   1993
    622' .28--dc20   93-20660
                                                                  CIP
PRINTED IN THE UNITED STATES OF AMERICA
      94   95   96   97   98   QW   9   8   7   6   5   4   3   2   1

Contents

Preface ix

Chapter 1 Introduction

1.1 Seismicity in Underground Mines 2
1.2 Seismicity Induced by Surface Mining 10
1.3 The Scope of the Book 12

Chapter 2 Principal Types of Mine Tremors

2.1 Seismic Events at Stope Faces 15
2.2 Geologic Discontinuities and Seismicity 19
2.3 Bimodal Distribution of Mine Tremors: An Introduction 22

Chapter 3 Seismic Waves at Short Distances

3.1 Displacement and Strain 24
3.2 Stress 26
3.3 Stress–Strain Relations 29
3.4 Equations of Motion 31
3.5 Plane Seismic Waves 34
3.6 Spherical Waves 37
3.7 Plane Seismic Waves in a Layered Medium 38

Chapter 4 Location of Seismic Events in Mines

4.1 Classic Approach and Its Computational Aspects 49
4.2 Bayesian Approach 58
4.3 Fedorov's Generalization of Least-Squares Procedure: Location
    with Approximate Velocity Models 63
4.4 Relative Location Technique 65
4.5 Simultaneous Hypocenter and Velocity Determination 69
4.6 Other Location Methods 74

Chapter 5   Optimal Planning of Seismic Networks in Mines

5.1   Theoretical Background 79
5.2   Special Cases and Applications 82

Chapter 6   Selected Topics from Seismic Tomography in Mines

6.1   Mathematical Principles 94
6.2   Effects of Seismic Ray Bending 97
6.3   Inversion Techniques 99

Chapter 7   Stress-Induced Anisotropy and the Propagation of Seismic Waves

7.1   Stress–Strain Relations 106
7.2   Equations of Motion 107
7.3   Anisotropic Symmetry Systems 109
7.4   Wave Propagation in Cracked Solids 111
7.5   Shear-Wave Splitting Induced by Anisotropy 115
7.6   Shear-Wave Splitting in Isotropic Media 123

Chapter 8   Attenuation and Scattering of Seismic Waves

8.1   Anelastic Effects 129
8.2   Viscoelastic Constitutive Relations 130
8.3   Intrinsic Attenuation and Its Quality Factor Q 137
8.4   Attenuation in Anisotropic Media 144
8.5   Scattering Effects and Coda Waves 146
8.6   Single-Scattering Models of Coda Generation 153
8.7   Multiple-Scattering Models 157
8.8   Frequency Dependence of Q and Backscattering Coefficient 163
8.9   Methods of Calculation of Quality Factor Q 167

Chapter 9   Focal Mechanism of Mine Tremors

9.1   Single Forces in Homogeneous Media 177
9.2   Concentrated Force Couples 185
9.3   Double-Couple Sources 189
9.4   Determination of Fault-Plane Solutions 198
9.5   Moment-Tensor Inversion 209
9.6   Non-Double-Couple Seismic Events 220

Chapter 10   Seismic Source Modeling

10.1   Recent Developments in Fracture Mechanics 224
10.2   Dislocation and Crack Source Models 232
10.3   Complex Source Models: Asperities and Barriers 242

10 4 Earthquake Sequences. 249
10.5 Synthetic Seismograms. 255
10.6 Faults and Fractals. 259

Chapter 11 Seismic Spectra and Source Parameters

11.1 Spectral and Time-Domain Parameters 265
11 2 Seismic Moment 277
11.3 Magnitude. 278
11.4 Seismic Energy 281
11.5 Source Dimensions. 284
11.6 Stress-Release Estimates 291
11.7 Scaling Relations. 294

Chapter 12 Statistical Assessment of Seismic Hazard in Mines: Statistical Prediction

12.1 The Gutenberg and Richter Frequency–Magnitude Distribution. 301
12.2 Extreme Distributions of Magnitudes 310
12.3 Bimodal Distributions of Mine Tremors. 314
12.4 Seismicity and Rock Extraction. 322
12.5 The Number and Energy of Seismic Events in the Stope Area 327
12.6 Statistics with Incomplete and Uncertain Data 329

Chapter 13 Prediction and Prevention of Large Seismic Events

13.1 Precursory Phenomena 337
13.2 Dilatency Models. 342
13.3 Prevention. 345

References 347
Selected Bibliography. 386
Index 389
International Geophysics Series 397

Preface

The safety and productivity of underground mining can be severely impacted by seismic activity. In particular, the occurrence of rockbursts is the most serious and least understood problem in deep mining operations throughout the world—a problem that becomes progressively more severe as the average depth and extent of mining operations increases. Therefore, it is not surprising that the monitoring of seismicity induced by mining has a long history, going back to the beginning of this century. However, except for a few book collections of papers from symposia, no comprehensive book on mining seismology has appeared until now.

The methods and techniques employed to study seismicity in mines have been transferred directly from earthquake seismology. Although no systematic differences between large mine tremors and natural earthquakes have been found thus far, specific differences between these two types of seismicity are apparent, especially in terms of their observation. The seismic networks in mines are three-dimensional, and the observations are made in the source area. They are often run by the industry itself and professional seismologists are not always available. Most often either mining or electrical engineers are responsible for seismic monitoring in mines, and their experience in interpretation and methodology is, naturally, limited.

In the spring of 1989, Dr. Renata Dmowska, coeditor of *Advances in Geophysics*, invited me to contribute an article entitled "Seismicity Induced by Mining," which was published in the 1990 volume. This review was meant to be the most up-to-date comprehensive account of current knowledge on seismicity in mines, with an emphasis on the results obtained, rather than on the methods employed. At the same time, the review could readily become a nucleus for something much bigger—a book where the methods could also be described in some detail. I therefore accepted a second invitation from Dr. Dmowska to contribute a volume to the International Geophysics Series published by Academic Press, entitled *An Introduction to Mining Seismology*. Seismological methods are so extensive that a reasonably comprehensive introduction confined to one volume demands a strong selection of topics considered the most important in mining practice. I invited, in turn, Professor Andrzej Kijko, my colleague from the Institute of Geophysics of the Polish Academy of Sciences, now with the Integrated Seismic System (ISS)

International Limited in South Africa,[1] to join me as coauthor of this book. He is a well-known expert in mathematical and statistical methods in seismology and his contribution to the book (Location of Seismic Events in Mines, Optimal Planning of Seismic Networks in Mines, Selected Topics from Seismic Tomography in Mines, and Statistical Assessment of Seismic Hazard in Mines: Statistical Prediction) is substantial.

This book introduces consecutive topics of interest in several chapters, forming a series of logical steps from a concise theory of the propagation of seismic waves described in one of the first chapters to the efforts undertaken in the field of rockburst prediction and prevention presented in the last chapter. Following the introduction of seismic waves, a series of topics related to their travel times are considered: location of mine tremors, optimum planning of three-dimensional seismic networks in underground mines, seismic tomography, and S-wave splitting as a result of stress-induced anisotropy. Displacement field and amplitude observations form the basis for the subsequent topics introduced in the book: attenuation and scattering of seismic waves, focal mechanisms of mine tremors, seismic source modeling, and seismic spectra and source parameters. The important problem of the assessment of seismic hazards in mines is considered in detail at the end of the book.

An Introduction to Mining Seismology is intended as a handbook for seismologists and engineers at the mine who make everyday observations of seismic events induced by mining and as a handbook and reference book for university level researchers not only in the field of seismicity in mines but in the field of local natural seismicity as well. Practical aspects of the subject are emphasized, and each topic is illustrated, whenever possible, with the key case histories taken from various mining districts worldwide. A comprehensive list of some 900 references is included for those interested in following specific topics in more details. A selected bibliography, listing contemporary books relevant to the subject, is also added.

We are grateful to the management of the Institute of Geophysics in Warsaw and of the ISS International Limited in Welkom for all the facilities provided during the writing of this book and during the final preparation of the manuscript.

Slawomir Jerzy Gibowicz

---

[1] Present address: ISS International, P.O. Box 2083, Welkom 9460, Republic of South Africa

Chapter 1 | Introduction

An increase in seismicity in seismic areas and the generation of seismicity in aseismic areas have been observed as a result of deep underground mining and large-scale surface quarrying, the filling of reservoirs behind high dams, the injection of fluids in rocks at depth, the removal of fluids from subsurface formations, and the detonation of large underground explosions. This type of seismicity is usually called *induced seismicity* to indicate the triggering nature of engineering activities in the release of preexisting stresses of tectonic origin. The primary requirement for inducing seismicity appears to be human activity where the rocks are in a highly prestressed condition (e.g., Kisslinger, 1976; Simpson, 1986).

Seismicity associated with underground mining is probably the most adverse phenomenon, among the different types of triggered earthquakes, in relation to the safety and productivity of mining. Rockbursts are very often the major cause of fatalities in mines. The problem becomes progressively more severe as the average depth and the extent of mining operations increase.

The distinction between a seismic event in a given mine, often called a *mine tremor*, and a rockburst should be noted. Rockbursts are violent failures of rock that result in damage to excavations (e.g., Cook, 1976; Ortlepp, 1984). Thus only those seismic events that cause damage in accessible areas of the mine are called *rockbursts*. Although the distinction is arbitrary, introduced and used by mining engineers, it is of considerable practical importance. Out of several thousand seismic events recorded annually by seismic networks at some mines, only a few become rockbursts. Rockbursts form only a small subset within a large set of seismic events induced by mining, and there are no clear-cut lines that could imply that a specific seismic event in a given situation would become a rockburst (e.g., Salamon, 1983).

Seismic monitoring provides a powerful means for the detection and evaluation of seismic events occurring around underground openings. It is not surprising, therefore, that the monitoring of seismicity induced by mining has a long history. The first seismological observatory was established for this purpose in Bochum, in the Ruhr coal basin in Germany, by Mintrop as early as 1908 (Mintrop, 1909). The station was equipped with the famous horizontal seismograph of Wiechert, which was in continuous operation until World

War II, when it was destroyed by bombing. In the Witwatersrand region of South Africa, mine tremors associated with deep gold mining operations were first noticed in 1908 (Gane *et al.*, 1946). In 1910 a horizontal seismograph of Wiechert was installed a few kilometers from the mining area.

The first seismic network to monitor seismicity in mines was established in the then German part of the Upper Silesia coal basin, now Poland, by Mainka at the end of the 1920s. The network comprised four stations, with one station placed at depth at the Rozbark coal mine, equipped with the horizontal seismograph of Mainka. The network, extended and updated, was in operation for some years after World War II (e.g., Gibowicz, 1963) and in the mid-1960s was replaced by modern seismic stations operated at the surface and underground.

In South Africa five mechanical seismographs were designed and deployed in 1939 in an array at the surface, specifically to locate the tremors (Gane *et al.*, 1946). Although the association of seismicity with mining operations has been apparent in many areas shortly after the commencement of mining, the direct relationship between seismicity and deep gold mining in the Witwatersrand area was first described by Gane *et al.* (1946).

The largest mine tremor ever observed occurred on March 13, 1989 in the potash mining district along the river Werra, in southern Germany, with magnitude $M_L = 5.6$ and $m_b = 5.5$ (Knoll, 1990). Another outstanding event of similar size occurred in the same district on June 23, 1975, with magnitude $M_L = 5.2$ and $m_b = 5.4$ (Hurtig *et al.*, 1982). The largest mine tremor in South Africa occurred in the Klerksdorp gold mining district on April 7, 1977, with magnitude $M_L = 5.2$ and $m_b = 5.5$ (Fernandez and van der Heever, 1984). The Lubin copper mining district in Poland is another area where large seismic events are generated, with magnitude M_L up to 4.5, like that of March 24, 1977 (Gibowicz *et al.*, 1979).

No systematic differences have been found between mine tremors and natural earthquakes, and the methods and techniques employed at mines for seismic monitoring are those directly transferred from earthquake seismology. They will be described in some detail later on. A short description of seismicity observed so far in mines in various parts of the world seems to be relevant at this introductory stage of the book. The description, modified and extended, is taken from the review of seismicity induced by mining recently published by Academic Press in the *Advances in Geophysics* serial publication (Gibowicz, 1990b).

1.1   Seismicity in Underground Mines

Rock failure and seismic activity are often unavoidable phenomena in extensive mining deep below the Earth's surface. Seismicity induced by mining is usually defined as the appearance of seismic events caused by rock failures as a result of changes in the stress field in the rockmass near mining excavations (e.g., Cook, 1976). The total state of stress around a mine excavation is the sum of the ambient stress state in the rockmass and the stresses induced by mining. The ambient stress state tends to be lithostatic, corresponding to the weight of overburden.

Seismicity induced by underground mining is observed in numerous mining districts throughout the world. The literature is extensive and only more recent results, published in accessible journals and books, are briefly reported here.

The most comprehensive studies of seismicity in deep mines have been carried out for a long time in South Africa. The gold-bearing reefs of the Witwatersrand system are mined by stoping at depths down to 3.5 km below the surface. This creates flat voids in the quartzitic strata extending horizontally up to several kilometers with an initial excavated thickness of a meter (Cook, 1976). A close spatial relationship of seismicity to mining is observed at several mines, notably at the East Rand Proprietary Mines (ERPM) (e.g., Cook, 1963; McGarr et al., 1975). A similar relationship is observed at the Vaal Reefs mine in the Klerksdorp district, where the mining is typically 2.3 km deep (Gay et al., 1984), and at the Blyvooruitzicht mine in the Carletonville area (e.g., Spottiswoode, 1984). In the mines in the Orange Free State district also some of the seismicity appears to be closely related to active mining, with some events located above and below the mining horizons at depths from 400 to 2300 m (e.g., Lawrence, 1984). The source parameters of seismic events in gold mines have been estimated (e.g., Spottiswoode and McGarr, 1975) and the differences in seismicity in major gold mining districts evaluated (e.g., Dempster et al., 1983; McGarr et al., 1989). Rockburst hazard and strong ground-motion studies are also being conducted intensively (e.g., McGarr et al., 1981; Salamon, 1983; Ortlepp, 1984). A comprehensive review of recent seismic and rockburst research in South Africa has been published recently (Spottiswoode, 1989).

Seismicity induced by underground mining is a well-known long-studied phenomenon in Poland, observed in the Upper Silesia coal basin (e.g., Gibowicz, 1963, 1979, 1984; Kijko, 1975, 1978; Droste and Teisseyre, 1976; Gibowicz et al., 1977; Ostrihansky and Gerlach, 1982; Goszcz, 1986; Syrek and Kijko, 1988) and in the Lower Silesia coal basin (e.g., Gibowicz and Cichowicz, 1986), where mining has been carried out for many decades, and in the Lubin copper district in Lower Silesia (e.g., Gibowicz et al., 1979, 1989; Stopinski and Dmowska, 1984; Gibowicz, 1985; Kazimierczyk et al., 1988), where mining was started only over 20 years ago. In both the major mining areas, the Upper Silesia basin and the Lubin district, the rockbursts are severe. Several underground seismic networks have been operated by the mining industry in Upper Silesia since the mid-1960s and in the Lubin district since the mid-1970s, at depths from about 600 to 1100 m, corresponding to the mining horizons. Several thousand mine tremors are recorded annually, but only about a dozen of them reach a local magnitude value higher than 3. Very seldom does an excessively large tremor, with magnitude exceeding 4, occur (Gibowicz, 1984).

Mine-induced seismicity in Czechoslovakia is experienced in four mining districts (e.g., Rudajev and Bucha, 1988). Rockbursts are the most severe in the Ostrava-Karvina coal district, where eight mines are in operation at a depth of about 800 m. The first rockburst was reported in 1917 and the strongest and most disastrous rockburst occurred on April 27, 1983, with seismic energy of 10^{10} J (e.g., Holub et al., 1988). Seismic monitoring in mines was started in 1977 and a regional seismic network is in operation since 1988 (Konecny, 1989). The uranium mines in the Pribram district are about 1700 m deep. A seismic network is in operation there to monitor seismicity, and various methods are used to improve the safety of mining operations (Rudajev and Bucha, 1988). In the Kladno coal district a single and almost horizontal coal seam is excavated at a depth of about 450 m. The first seismic event was felt in 1872, and since 1961 the tremors have been recorded by a seismic station (Skala and Roček, 1985). In the mines in the North Bohemian lignite basin, mostly pillar bumps are observed (Rudajev and Bucha, 1988).

In the former Federal Republic of Germany (West Germany) seismicity induced by mining is best recognized in the Ruhr coal basin, where the maximum depth of excavation is about 1100 m (e.g., Casten and Cete, 1980; Hinzen, 1982). To monitor such seismicity, a modern digital seismic network was established in 1982 by Ruhr University. About 1000 mine tremors, with local magnitude of up to 3.0, are recorded each year by the network consisting of a local small array and two remote stations in mines (at depths of 410 and 890 m). These records provide an opportunity for investigating the focal mechanism and source parameters of mine tremors in the area (Gibowicz et al., 1990).

Induced seismicity in the territory of the former German Democratic Republic (East Germany) is best known in the potash mining district along the river Werra near the town of Sünna, and is characterized by the occurrence of infrequent but very large seismic events (Hurtig et al., 1982). A seismic network composed of 25 stations is in operation at two mines of the Kalibetrieb Werra Company, at two underground levels at a depth between 500 and 1000 m and at the surface (Knoll et al., 1989). Several studies of source parameters of mine tremors and their relation to mining and geologic factors have been undertaken (e.g., Knoll and Kuhnt, 1990).

In the United States, the U.S. Bureau of Mines continues to be the major research organization involved in studies of mine-induced seismicity (e.g., Bolstad, 1990). Since the mid-1960s the Bureau of Mines has been studying rockburst and methods of providing warning of their occurrence and devising means of their control (e.g., Leighton, 1984). The focal mechanism and source parameters of tremors associated with underground coal mining in eastern Utah have been studied and spectral analysis methods applied to mine tremors for the first time (Smith *et al.*, 1974). The mechanism of mine-related seismic events in the Wasatch Plateau, Utah, has been studied in detail, providing some evidence for non-double-couple events (Williams and Arabasz, 1989; Wong *et al.*, 1989). Occasional studies of mine-induced seismicity in other parts of the United States have been reported (e.g., Wong, 1984; Bollinger, 1989; Sprenke *et al.*, 1991).

An overview of seismicity induced by mining in Canada was published recently by Hasegawa *et al.* (1989). Seismic events and rockburst are observed in metalliferous, potash, and coal mines. Mine-induced tremors were first noticed in the Sudbury and Kirkland Lake deep metalliferous mines in Ontario during the early 1930s. In recent years there has been a growing rockburst problem in northern Ontario hard-rock mines operating at depths down to 2 km (e.g., Cook and Bruce, 1983; Morrison, 1989). In response, seismic monitoring systems have been extended and installed in the four mining districts (Red Lake, Elliot Lake, Sudbury, and Kirkland Lake) experiencing rockbursts. The preliminary results from the study of the focal mechanism and source parameters of seismic events at Strathcona mine, Sudbury, have been published (Young *et al.*, 1989a). Induced seismicity in the potash district of Saskatchewan is a new and unexpected phenomenon. The potash mines operate at a depth of about 1000 m. Mining on a commercial scale started in 1962 and by 1973 10 potash mines were in operation. Since 1976 four mines have been generating seismic events with magnitudes in the range from 2.3 to 3.6, large enough to be felt on the surface (Hasegawa *et al.*, 1989). The failure mechanism of potash mine-induced tremors seems to be quite different from that observed in many hard-rock mines because no surface faulting or rockbursts have ever been observed in a potash mine. The failure is thought to be confined to the competent limestones some 40 m above the mine and caused by subsidence (Gendzwill, 1984).

Rockburst hazard in British coal mines is not great. In the North Staffordshire coalfield, however, underlying the densely populated area of Stoke-on-Trent, mining operations are associated with seismic events that, since the mid-1970s, have reached magnitudes of up to 3.5. The seismicity is attributed directly to mining and is believed not to be associated with seismic movements on faults (Kusznir *et al.*, 1980, 1984; Westbrook *et al.*, 1980). Tremors induced by coal mining are often seen on the widely spaced seismic stations operated by the British Geological Survey. These events account for about 25% of the earthquakes recorded by the network. The South Wales, Staffordshire, Nottinghamshire, and Midlothian coalfields are particularly clearly marked. Many of these events are recorded only instrumentally, but a significant number are felt and a few, with magnitudes approaching $M_L = 3$, reportedly caused damage (Redmayne, 1988).

In the Provence coalfields in France a single seam deposit is excavated at a depth of about 700 m. The deposit has been subject to rockbursts for many years. A seismoacoustic monitoring system was developed and used (Dechelette *et al.*, 1984), and later supplemented by a seismic system (Revalor *et al.*, 1990), for the analysis and prediction of rockburst phenomena. Preliminary results of the study of seismic events associated with mining in the Lorraine coal basin were published recently (Hoang-Trong *et al.*, 1988). A Freyming–Merlebach digital seismic network, composed of six surface and two underground (at a depth of 950 and 1250 m) stations, has been in operation since November 1986. The largest observed event was of magnitude 3.3. The focal mechanism of most events shows thrust faulting, but the largest events seem to be implosive.

In India rockbursts have been known for a long time in the Kolar Gold Fields situated in the Karnataka State in the southern part of India. Mining operations are carried out there for more than a century, now at depths down to 3.2 km, and related rock mechanics studies are almost classic (Murthy and Gupta, 1983). The rockbursts were reported as early as in 1902 (Behera, 1990) and in 1912 a Wiechert seismograph was installed to monitor seismicity. At present seismicity is monitored by a seismic network, established in 1978 and composed of 14 stations situated on the surface and underground (Krishnamurthy and Shringarputale, 1990). Induced seismicity has also been reported from a number of mines in the Eartern Coalfields (Chouhan, 1986).

Rockbursts are one of the severe problems in Japanese coal mines. The Miike coal mine is an undersea mine at a depth of 650 m, in which rockbursts have occurred at the longwall faces. The seismicity in the mine is attributed to the nonuniformity of the stress field and to the presence of geologic discontinuities (Kaneko *et al.*, 1990). Since 1980 microseismicity associated with longwall mining has been monitored by a mine-wide seismic network at Horonai coal mine in the Ishikari coal basin, the deepest coal mine in Japan. The seismic moment-tensor inversion technique has been used to analyze the focal mechanism of a few seismic events that occurred in February 1984 near the longwall face at a depth of 1100 m (Fujii and Sato, 1990).

Unfortunately, little is known about seismicity in deep mines on the territory of the former Soviet Union, an area where major mining operations are conducted. In the 1950s extensive application of the seismoacoustic technique to underground mining was under way in the Soviet Union (e.g., Antsyferov, 1966). Recent developments, however, have not appeared in international literature, and we were able to trace only a few studies that have been published in the Russian journals that were accessible to us.

The All-Union Research Institute of Geomechanics and Mine Surveying (VNIMI) in Leningrad has published at least three catalogs of rockbursts in the USSR mines; they appeared in 1967, 1973, and 1981. The catalog published in 1981 (Anonymous, 1981) provides detailed and systematic description, with corresponding mining sketches, of 73 rockbursts that occurred between 1973 and 1980 in mines situated in nine coal mining districts: Kizelovsk basin, Kuznetsk basin, Vorkuta deposits, Partizan deposits, Bukatchatchinsk and Lipovetsk deposits, Tkibuli-Shaorsk deposits, Shurabsk deposits, Donbass central district, and Tchelabinsk basin.

In the Northern Ural bauxite basin first rockbursts occurred at the beginning of 1970s, at a depth of 350 m. Seismic observations are carried out there from 1980 (Lomakin et al., 1989). The seismic network extends over an area of about 10 km^2 and is composed of 15 three-component stations. About 1000 seismic events are recorded each year; half of them occur within a 50-m-wide zone around the stoping areas. About 3 percent of seismic events become rockbursts. The largest tremors are characterized by seismic energy of the order of 10^7–10^8 J, and some of them occur at depths distinctly greater than the depth of mining (Voinov et al., 1987).

Similarly, in the Tkibuli-Shaorsk coal basin in the Black Sea area, some seismic events, often referred to by the Russian authors as the "mining–tectonic events," occur at much greater depths than the depth of excavations not exceeding 1100 m (Petukhov et al., 1980). Complex studies of induced seismicity in the Tkibuli-Shaorsk basin were undertaken in 1967 and a regional network composed of six seismic stations is in operation.

The mining of the Khibiny massif apatite deposits in the Kola Peninsula was started in 1929. The first seismic event with magnitude 4 was felt in 1948. At present a regional network, composed of four stations, is in operation. Since 1986 about 100 events with magnitude from 2.2 to 4.2 were recorded. The strongest tremor with magnitude 4.2 occurred on April 16, 1989 at a depth of about 1 km, causing extensive damage at Kirovsk mine and minor damage at Kirovsk town. The maximum measured displacement was 15–20 cm and was traced on the surface for 1200 m and observed at depth of at least 220 m. The main tremor was followed by several hundred aftershocks during the next 2 months. The event occurred simultaneously with a 240-ton explosion in one of the Kirovsk mines, which implies that the blast triggered the tremor. A similar pair of events occured in 1982 and 1991 (E. O. Kremenetskaya and V. M. Trijpizin, 1990, written communication; Kremenetskaya, 1991).

As in Russia, little is known about seismicity in deep mines in China, another country involved in major mining operations. It is known, however, that seismic monitoring systems are employed in a number of Chinese mines. The State Seismological Bureau of China and the China National Coal

Corporation have chosen the mines from the Beijing (Peking) Coal Mining Corporation as test sites to conduct effective cooperation in studies of induced seismicity (S.-Q. Zhang and M.-Y. Yang, 1990, written communication). Two large underground networks are in routine operation at Men-Tou-Gou and Fang-Shan mines. At Men-Tou-Gou mine, active for several decades, the largest seismic event had magnitude $M_1 = 3.9$ and was associated with severe rockburst. Although the Fang-Shan mine is active during the last few years only, the rockburst hazard there is very high. The excavation is carried out at depths from 650 to 900 m under extremely hazardous conditions, with coal seams steeply dipping and embedded in strong quartzite. Between 1985 and 1989 three tragic rockbursts occurred and the largest seismic event had magnitude $M_L = 3.2$.

An interesting study on the performance of underground coal mines during the destructive Tangshan earthquake of 1976 was published (Lee, 1987). There are eight coal mines in the Tangshan area, forming the largest underground coal mining operation in China. Approximately 10,000 miners were working underground at the time of the earthquake. With few exceptions they survived and returned safely to the surface. It was found that seismic damage in the underground mines was far less severe than that at the surface.

Specific studies on seismicity in underground mines are occasionally reported in other countries. A relatively large seismic event with $M_L = 3.2$ occurred on August 30, 1974 at the Grängesberg iron ore mine in central Sweden, followed by a long sequence of tremors with properties intermediate between those of an aftershock sequence and an earthquake swarm (Båth, 1984). The Mount Charlotte gold mine near Kalgoorlie, Western Australia, operating at a depth of 650 m, has been experiencing seismicity for a long time. A seismic event of magnitude 3.0 was associated with widespread shear displacement on a thin rough fault. Good correlation was achieved between observations and predicted ground behavior from numerical modeling (Lee et al., 1990).

It follows from studies of seismicity induced by underground mining that mine tremors do not necessarily occur in all mining situations, their maximum size is different in different areas, and their depth is usually close to that of mining excavations.

The first point is well illustrated in Fig. 1.1, where the level of seismicity in the coal mines of the Upper Silesia coal basin in Poland is shown for the period 1977–1979 (Ostrihansky and Gerlach, 1982). The shaded areas mark mines with high and low seismicity, whereas the unshaded areas indicate the mines where no seismicity is observed. The areas of high seismicity are believed to be those of tectonic compaction of rocks during past orogenies. In Fig. 1.2 such areas of tectonic compaction are shown for the central part of the Upper Silesia basin (Goszcz, 1986).

Figure 1.1 Level of seismicity, measured by the amount of seismic energy released per volume unit of mined-out coal, in the mines of the Upper Silesia basin, Poland, for the period 1977–1979. Heavy and light shading denotes mines with high and low seismicity, respectively, whereas unshaded areas indicate the mines where no seismicity is observed. [From Ostrihansky and Gerlach (1982, Fig. 1).]

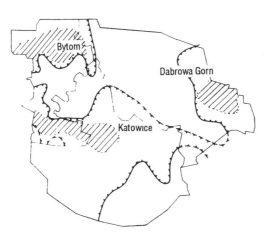

Figure 1.2 Area of tectonic compaction of rocks during the Asturian Phase of the Hercynian Orogeny, marked by thick indented lines, and the areas of seismicity induced by coal mining, denoted by shading, in the central part of the Upper Silesian basin, Poland. The boundaries of mining districts are marked by thin lines; the names of the principal towns in the region are also given. [From Goszcz (1986, Fig 1).]

It is not entirely clear whether the maximum size of seismic events in a given area is controlled mainly by mine geometry or by geologic and tectonic conditions, or by a combination of these factors. In some situations irregular mine configuration, including supporting pillars, can lead to large seismic events. In other situations large tremors can be connected with major geologic features such as faults and dikes. In different mining areas either one or a combination of these factors would control the overall seismic deformation for a given amount of mining.

The depth of mine tremors strongly depends on the type of rocks forming the roofs and floors of deposit seams. In the Polish coal mines, for example, mine tremors are usually located at the level of excavation. In the Polish copper mines, however, where the roof is formed by strong dolomite rocks, they appear above the ore seam in most cases, similar to the distribution pattern in some South African gold mines (e.g., Cook, 1976), although in the mines of the Orange Free State district a substantial number of seismic events are located more than 1 km above and below the mining operations (e.g., Lawrence, 1984).

The eastern Wasatch Plateau, Utah, area of underground mining is one of the very few areas where submine seismicity at a depth of several kilometers appears to be activated by mining (Smith *et al.*, 1974; Williams and Arabasz, 1989; Wong *et al.*, 1989). A relatively high level of natural seismicity in the area seems to be a dominant factor leading to the interaction at depth of mine-induced and tectonic stresses. A simplified theoretical model of crack propagation below the mining level, induced by mining extraction of a horizontal seam and leading to the generation of seismic events at depth, has been proposed by Gil and Litwiniszyn (1971), and a critical value of the extracted seam length, at which the propagation is initiated, has been estimated.

1.2   Seismicity Induced by Surface Mining

Seismicity induced by extensive quarrying operations is a rather unusual phenomenon in terms of the number of described cases. We are aware of such seismicity being observed only in four areas around the world.

The open-pit carbonate-pipe operations at Phalaborwa in the northeastern Transvaal, South Africa, may have generated measurable seismicity according to McGarr (unpublished comments on the nature of Witwatersrand mine tremors, 1987). In the earthquake catalog for South Africa of Fernandez and Guzman (1979), five events occurred in this area between 1967 and 1970 with magnitudes ranging from 2.3 to 2.7. This is a good example of possible induced seismicity not being recognized as such, because the seismic effects are usually associated with quarry blasts and no special studies are undertaken.

Another good example of such a situation is seismicity observed in the Cerro de Pasco in Peru, where microseismic events were recorded by chance during the study of blasting effects outside the quarry (Deza and Jaén, 1979). About 50 microtremors were recorded before and after a quarry blast in October 1973 by an accelerometer situated at a distance of less than 100 m from the shot point in the limestone quarry near the town of Cerro de Pasco. The recording was casual, as the accelerometer was installed for a short time only. The tremors were very small, with magnitudes ranging from -2 to 0. The quarry itself was 1200 m long, 750 m wide, and 60 m deep at the time.

One June 7, 1974 an earthquake of magnitude 3.3 occurred at Wappingers Falls, New York, followed by over 100 aftershocks during a 6-day period (Pomeroy et al., 1976). The aftershocks occurred within or beneath a block of dolomitic limestone on the Hudson River bank, which has been quarried extensively since the early 1900s. The composite fault-plane solution indicates the focal mechanism of thrust type in a region of high horizontal compressive stress. In 1974 the total depth of the quarry was of the order of 50 m. About 25 m of that depth had been quarried during the previous 22 years, since the last similar event in 1952 that released any accumulated nonlithostatic stresses. The unloading effect, or the stress corresponding to the weight of the rocks removed by quarry operations, was small compared to the failure strength of rocks or in relation to the regional state of stress. The area, therefore, must have been close to failure conditions, the unloading acting as a trigger to failure in prestressed rocks. Unloading leads toward the failure condition only when the vertical stress is the minimum principal stress (Pomeroy et al., 1976).

The most spectacular appearance of seismicity induced by surface mining is observed in the Belchatow area in Poland (Gibowicz et al., 1981, 1982). One of the largest deposits of brown coal in Poland is situated in the Belchatow trench, over 40 km long, 1.5–2 km wide, and 0.5 km deep. The trench is part of a deep tectonic fracture system running through the area. The mining started in 1976. The thickness of the overburden is from 100 to 200 m, and the rate of mining was about 300 m per year. By mid-1980 the pit was about 100 m deep, 1 km long, and 2 km wide. Groundwater extraction was started in October 1975, and there has been a gradual change in hydrological conditions in the mine area.

The first tremor in the open-pit Belchatow mine was recorded by the Polish seismic network in August 1979, and the first tremor was felt in February 1980. Further tremors with local magnitudes ranging from 2.8 to 3.6 were felt in March, April, and May 1980 (Gibowicz et al., 1981). On November 29, 1980 a magnitude 4.6 seismic event occurred in the area—so far the strongest manifestation of seismicity induced by surface mining

(Gibowicz *et al.*, 1982). The Belchatow earthquake was widely felt, even at a distance of more than 100 km at single sites. The most probable fault-plane solution corresponds to oblique-slip motion on a reversed fault, and the compressive stress is almost horizontal.

Surface mining in the Belchatow area could affect existing stress mainly through a decrease in vertical stress, caused by removal of the overburden (the unloading effect), and through an increase in the effective stress, caused by decreased pore pressure resulting from groundwater withdrawal (the pore–pressure effect). The value of the stress corresponding to the weight of the overburden removed from the Belchatow pit in the time of the occurrence of seismicity was about 2.5 MPa. This value is smaller even than the cumulative stress drop observed during larger events and much smaller than the failure strength of rocks. If an increase in effective stress caused by decreased pore pressure is adequately represented by the corresponding change in hydrostatic pressure, then this increase would be about twice as small as the estimated value of lithostatic pressure corresponding to the removal of overburden. Therefore, the stress changes corresponding to both effects, even if the two effects interact, seem to play the role of a triggering factor for inducing seismicity. Thus the observed focal mechanism of the oblique-slip thrust type can be explained by an interaction of the horizontal stress of tectonic origin with the vertical stress of mining origin (Gibowicz *et al.*, 1982).

The orientation of the principal stress in nature is different in different tectonic environments. Regions of thrust faulting are characterized by the minimum compressive stress being vertical, in regions of normal faulting the maximum compressive stress is vertical, and in regions of strike-slip faulting the intermediate stress is vertical (e.g., Simpson, 1986). Thus, a decrease in vertical stress, resulting from unloading effects in quarrying operations, will have the greatest effect in a thrust faulting environment.

1.3   The Scope of the Book

In this book methods, techniques, and results of studies related to mine-induced seismicity are described. Several other types of rock failure in deep mines, such as rockfalls (nonviolent falls or loose rock under the influence of gravity) and gas-related outbursts, result from processes fundamentally different from those generating earthquakes (e.g., Osterwald, 1970) and are not considered here.

Chapter 2 is a continuation, to some extent, of the introductory part of the book presented in Chapter 1. We describe here the principal types of mine tremors, universally observed in deep mines, corresponding to short- and long-term response of rockmass to induced stresses.

The knowledge of the generation and transmission of elastic waves in the rockmass is fundamental to seismology in general and to mining seismology in particular. Hence, a concise theory of seismic waves, as observed at short distances from the source, is described in Chapter 3.

In the next four chapters practical problems connected with travel times of seismic waves are considered. Chapter 4 deals with methods of location of seismic events in mines, recorded by three-dimensional networks. Here several approaches, such as least-squares iterative procedures, the Bayesian approach, the master event location technique, and the joint hypocenter determination method, are described in some detail. In Chapter 5 the theory of optimum configuration of underground seismic networks is given, with examples of its application in different environments. Some elements of seismic tomography and its possibilities and limitations in mines are discussed in Chapter 6. Seismic tomography is a very powerful technique for solving various problems encountered in mining practice. In Chapter 7 stress-induced anisotropy in a cracked medium is described. The shear-wave splitting effect observed in the rockmass with aligned cracks could possibly provide an important tool for prediction of large mine tremors and rockbursts.

The amplitudes, or in more technical terms the displacement field, of seismic waves and their observations, are at the core of the next several chapters. In Chapter 8 attenuation and scattering of seismic waves are described. Quality factor, coda waves, scattering models, measurement methods of attenuation and scattering effects, and other relevant topics are considered. Chapter 9 deals with the focal mechanism of mine tremors, based on a point source approach. Fault-plane solutions, double-couple (shear rock failure) versus non-double-couple mechanisms, and moment-tensor inversions are discussed in some detail. The moment-tensor inversion is of special interest for the studies of seismic events in mines, where nonshearing events are probably frequently generated. A more general description of seismic source modeling is given in Chapter 10. Dislocation and crack models, asperities and barriers, and faults and fractals are considered. Growing evidence shows that even mine tremors are often complex events and their source modeling in the time domain is of increasing importance. The most common technique, however, employed so far in studies of small seismic events, both natural and induced, is their source parameter estimation in the frequency domain. Seismic spectra and source parameters are described in Chapter 11. Methods of estimation of seismic moment and magnitude, source size, seismic energy, and stress release are given and scaling relations and characteristic tremors are discussed.

In the final chapters two major problems of practical importance are considered. Statistical assessment of seismic hazard in mines is presented in Chapter 12, and the other major problem, prediction and prevention of large seismic events in mines, is considered in Chapter 13. As in the case of natural earthquakes, the prediction problem of mine tremors is far from being solved. The significant difference in this respect, however, between natural and mine-induced seismicity is related to the manner of their observations. Various observations in mines are conducted either at short distances from the source or in the source area itself, thus forming a basis for the prediction of mine tremors more favorable than that for natural earthquakes. Unlike in the case of natural seismicity, the prevention of large mine tremors and rockbursts, or more correctly the mitigation of their severity, can be achieved by various methods and mining techniques employed in a number of mines in several countries. These methods are discussed only briefly at the end of Chapter 13, since they are clearly outside the scope of this book.

Chapter 2 | Principal Types of Mine Tremors

This chapter, slightly modified and extended, is taken from the review of seismicity induced by mining recently published in *Advances in Geophysics* (Gibowicz 1990b).*

From recent extensive studies of seismicity induced by mining it follows that two conclusions are now generally accepted. First, two broad types of mine tremors are observed almost universally—those directly connected with mining operations, that is, those associated with the formation of fractures at stope faces, and those associated with movement on major geologic discontinuities (Parysiewicz, 1966; Hurtig *et al.*, 1979; Joughin and Jager, 1983; Kijko *et al.*, 1987; Stankiewicz, 1989; Johnston and Einstein, 1990). Second, seismicity induced by mining is strongly affected by local geology and tectonics, that is, by medium inhomogeneities and discontinuities and interaction between mining, lithostatic, and residual tectonic stresses on a local and regional scale (e.g., Cook, 1976; Gibowicz *et al.*, 1979, 1989; Dempster *et al.*, 1983; Gay *et al.*, 1984; Gibowicz, 1984; Ortlepp, 1984; Potgieter and Roering, 1984; Al-Saigh and Kusznir, 1987; Kazimierczyk *et al.*, 1988; McGarr *et al.*, 1989; Lenhardt, 1990; Scott, 1990). Thus in general, seismicity in deep mines is affected by several factors such as depth, production, mining geometry, geologic structure, and geologic discontinuities. In situ stress measurements collected at the Underground Research Laboratory in Manitoba, Canada, indicate that major geologic discontinuities can act as boundaries for stress domains, and that the magnitude and direction of the stress field can change rapidly when these geologic features are traversed (Martin, 1990).

2.1 Seismic Events at Stope Faces

Mine tremors of the first type are of low to medium magnitude. Their number is generally a function of mining activity, measured by the excavation rate. In the Upper Silesian coal basin in Poland it was found that the energy of seismic events increases proportionally to the square of excavation rate,

---

*From S J. Gibowicz (1990). Seismicity induced by mining. *Adv. Geophys.* **32**, 1–74 with permission from Academic Press Inc.

and that the seismic energy release is much larger for longwall mining with caving than that with backfilling (Goszcz, 1988). The tremors occur within 100 m of the mining face or on some preexisting geologic discontinuities and zones of weakness near the face. Intact rocks can be ruptured when stresses induced by mining exceed the shear strength of the material (e.g., McGarr *et al.*, 1975; Johnston and Einstein, 1990).

The spatial distribution of seismicity is well illustrated in Fig. 2.1, reproduced from McGarr (1984), where the configuration of the ERPM gold mine in South Africa and the tremors located during a 100-day period in 1972 are

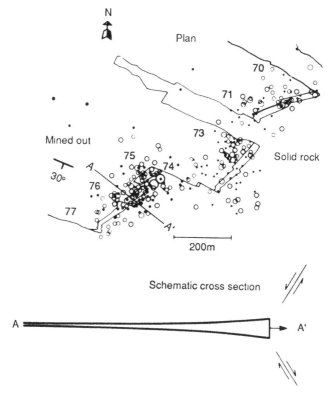

Figure 2.1   Plan and schematic cross-section view of the gold mine configuration in the lower East Hercules region of East Rand Propriety Mines, South Africa. The broad, tabular stopes are about 1.2 m thick at the face and extend more than 1 km The open circles represent tremors located during a 100-day period in 1972, the circle size corresponds to magnitude. The stop face positions at the beginning and end of the same 100-day period are indicated. [From McGarr (1984, Fig 2), copyright by The South African Institute of Mining and Metallurgy]

shown. The broad, tabular stopes are about 1.2 m thick at the face and extend more than 1 km.

Another illustration of the distribution of seismic events against the mining face is shown in Fig. 2.2, reproduced from Syrek and Kijko (1988). In this figure the distribution of the number of tremors and their seismic energy, summed up within a 10-m moving window, against the distance from a longwall face, is shown for one of the Polish coal mines in Upper Silesia. The curves are based on 843 well-located tremors that occurred between February 1979 and April 1981 in six longwalls mined with backfilling in a simple situation where no edges, previous workings, or geologic discontinuities were present. The number and energy of events attain the highest values in the direct vicinity of the face. The maximum number of tremors is exactly at the face, while the maximum energy release appears at a distance of some 12 m ahead of the face. For longwall mining with caving, the curve representing the number of tremors is much flatter and that of energy becomes more irregular, with a maximum at a distance of about 20 m behind the face (Syrek and Kijko, 1988).

Such curves show some degree of variability depending on the mining and geologic situations in different mining districts, but in principle they are of similar shape. Their depth distribution, although seldom as well constrained

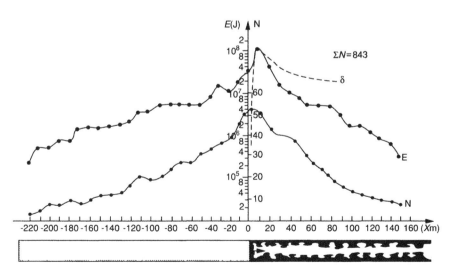

Figure 2.2   Distribution of the number N of tremors and their seismic energy E against the distance from the longwall face at the Wujek coal mine, Poland, for the period February, 1979–April, 1981. Theoretical values of the vertical stresses are shown by a dashed curve [From Syrek and Kijko (1988, Fig. 3.]

as horizontal distributions, strongly depends on the type of rocks forming the roofs and floors of deposit seams.

Daily and weekly distributions of mine tremors of the first type show excellent correlation with mining operations; good examples of such distributions can be found in Cook's (1976) review of seismicity associated with mining.

The division of seismic events in mines into two types is rather general, and several subsets of these events can be conceived. Six models of induced seismicity in underground mines in Canada have been proposed by Horner and Hasegawa (1978) and are sufficiently general to be considered for analyzing mine seismicity in other countries as well. The models are shown in Fig. 2.3, reproduced from Hasegawa *et al.* (1989). Three models of mine tremors correspond to a non-double-couple focal mechanism, whereas the others are of the shear–slip type.

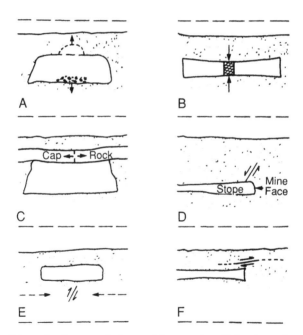

Figure 2.3 Schematic diagram of six possible ways in which mine-induced tremors can occur: (A) cavity collapse; (B) pillar burst, (C) tensional fault; (D) normal fault, (E) thrust fault; and (F) shallow (near-horizontal) thrust faulting. Solid arrows indicate mine-induced force direction on host rock during induced events Dashed arrows denote ambient tectonic stress. [From Hasegawa *et al* (1989, Fig. 3).]

2.2 Geologic Discontinuities and Seismicity

The second type of mine seismicity is not as well specified as the first type. The events are usually larger than those of the first type; they often occur at some distance from the mine faces; they are generally connected with major geologic discontinuities; their time distribution is erratic; and they are of a regional, more global, character in the sense that they seem to respond to stress changes on the scale of a whole mine and cannot always be pinpointed to any specific area of mining. They occur, however, within a mining district and are triggered by mining operations. The horizontal distribution of seismic events of both types in the Klerksdorp gold mining district in South Africa, observed between 1971 and 1981, shown in Fig. 2.4 (Gay et al., 1984), is a good example of the overall seismicity pattern in a complex and extensive mining situation.

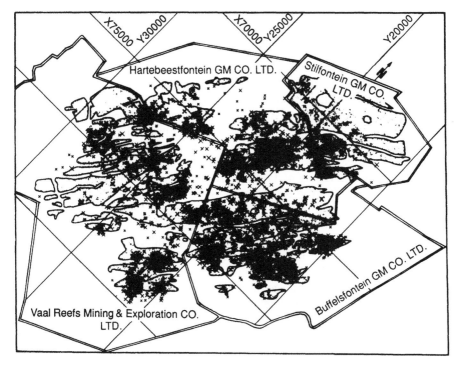

Figure 2.4 Horizontal distribution of seismic events in the Klerksdorp gold mining area, South Africa, observed between 1971 and 1981. Shading denotes mined-out areas. [From Gay et al. (1984, Fig. 4), copyright by The South African Institute of Mining and Metallurgy.]

In the Klerksdorp district the strata are offset by major normal faults, and the largest tremors, with magnitudes of up to 5.2, tend to be associated with the slip on these preexisting faults (Gay *et al.*, 1984; Potgieter and Roering, 1984; Syratt, 1990). Some large seismic events occur during the final stage of remnant extraction and are associated with geologically controlled remnants and not necessarily with the failure of the mined remnants themselves. Furthermore, the seismic source and ground-motion parameters are rather similar to those of natural earthquakes (McGarr *et al.*, 1989). Not all faults and dikes, however, are seismically activated. The behavior of a fault depends on its orientation to the reef horizon, extent and uniformity, and frictional and cohesive properties (Syratt, 1990). In general, however, seismicity is widely distributed throughout individual mines (Fig. 2.4).

Five major seismic events, with magnitude greater than 4.5, which occurred in the Welkom gold field between 1972 and 1989, have several characteristics in common (van Aswegen, 1990). These major events occurred along the major faults; they caused mainly shakedown-type damage (as opposed to classical "rockburst"-type damage) because of their "slow" nature; their source regions coincided with the area on the fault close to its intersection with mined out stopes.

In the Carletonville, South Africa, district, faults and dikes are also present, and the majority of large tremors occur in their vicinity. But the throws of faults there are several times smaller than those of major faults in the Klerksdorp district, and there appears to be an upper-magnitude limit of about 4. The associated stress drops and ground-motion parameters are higher than those normally observed from natural, shallow earthquakes (McGarr *et al.*, 1989). It is difficult, however, to assess whether these regional differences in stress drops result from different tectonic environments alone, as advocated by McGarr *et al.* (1989), or are associated with different mining techniques employed in the two districts as well. In the Carletonville area the longwall mining method is widely used at great depth, and pillars have been found to be subject to foundation failures, generating large seismic events (Lenhardt and Hagan, 1990). Shear stresses along pillars seem to rise to their shear strength, while the longwall is advancing in the area before a foundation failure occurs. The failure occurs in the area where major geologic features are not present and generates seismic events with very high stress drops (Leach and Lenhardt, 1990; Lenhardt and Hagan, 1990). In any case, these regional differences in mine-induced seismicity are of practical importance in the management of mining operations (Dempster *et al.*, 1983).

The Western Deep Levels gold mine in the Carletonville district, where mining operations are conducted at depths between 1600 and 3500 m, experiences an increase of seismicity with depth (Lenhardt, 1990). The seismic network records at present more than 400 events, with magnitude greater than zero, per month, which are located in the mine. Large seismic events with magnitude greater than 3 have been observed only on geologic discontinuities, abutments, and planned and isolated pillars. About 82 percent of all large seismic events occur close to dikes and faults, which were either approached or left by mining.

A statistical analysis of the relationship of geologic features to seismic events was carried out for Lucky Friday mine, Idaho (Scott, 1990). A set of 746 tremors that occurred from 1982 to 1986 at depths from 1600 to 1800 m was combined with geologic information about each event. Preliminary results indicate that 29 percent of the seismic events occurred on strike-slip or bedding plane faults, whereas 71 percent of the events were not associated with identified faults.

The effect of faults on seismicity has been studied for seismicity associated with coal extraction in North Staffordshire, Great Britain (Al-Saigh and Kusznir, 1987). It was found that seismic events occur if the mining operations cause the redistribution of strata pressure in such a way that sliding movements take place along major faults. In particular, movements occur when the active longwall face is driven into the footwall of the faults and parallel to the fault plane.

The horizontal distribution of large seismic events observed in the Lubin copper mining district in Poland, where mining was started at the end of the 1960s and seismicity appeared early in the 1970s, shows that all the seismic events are located within the areas of mine workings, but their distribution is different on the two sides of a major fault running through the area. On one side of the fault the tremors are the largest but infrequent, whereas on the other side they are smaller but more frequent. Only a few seismic events have occurred close to the fault (Kazimierczyk et al., 1988). Another point of interest is the distribution of large mine tremors with time. There is a steady increase in the number of large tremors with time as a result of the ever-increasing extent of mining excavations at depth. The number of smaller events with magnitudes between 2.5 and 2.9, however, tends to increase much faster than the number of large tremors with magnitudes between 3.0 and 4.5.

When considering the distribution of seismic events, those of the first type against a mining face or those of the second type against geologic discontinuities, the accuracy of their location becomes an important problem. Since the first underground seismic network was operated in South Africa (Cook, 1963), a vast improvement has been achieved in seismic equipment (e.g., Brink, 1990; Green, 1990; Mendecki, 1990; Mendecki et al., 1990) and in the methods of location of mine tremors (e.g., Kijko, 1975, 1978; Niewiadomski, 1989; Mendecki et al., 1990). Two types of three-dimensional seismic monitoring systems are employed in mines. The microseismic system is a dense seismic network used to monitor microseismic events, with local magnitudes down to −4, around the active mining face (e.g., Brink, 1990). The accuracy of location in this case is of the order of 10–20 m or less. The system is used in selected areas of interest in a given mine and only during restricted time intervals. In contrast, a seismic system covering a whole mine or a mining district is usually a permanent seismic network of considerable dimensions of up to several kilometers, with central facilities for recording and data processing (e.g., Mendecki *et al.*, 1990). The accuracy of location of seismic events in this case is of the order of 20–50 m or 50–100 m, depending on the number of sensors and the network size and geometry. The methods of location of mine tremors are described in detail in Chapter 4, and the methods of planning of the optimum network configurations in underground mines are presented in Chapter 5.

2.3 Bimodal Distribution of Mine Tremors: An Introduction

The bimodal distribution of seismic events and its theory and application are described in some detail in Chapter 12. Here only brief introduction to this approach seems to be relevant; the approach providing more precise evidence on the principal types of mine tremors than their qualitative description. In general, it is not clear how to distinguish between mine tremors of the first and second types. A probabilistic approach shows that the two types of seismic events behave differently.

During the study of the recurrence of large seismic events in Polish mines, it was found that the pattern of empirical distributions of the largest seismic events is more complex than might be expected from the most general theoretical considerations, such as the Gumbel distributions (Kijko *et al.*, 1982). A closer look at the observed distributions suggests that they are of bimodal character. Figure 2.5, reproduced from Kijko *et al.* (1987), shows the probability distribution of the monthly occurrence of maximum magnitude tremors at the Lubin copper mine, Poland, for the period 1972–1980. Similar relations have been found for the coal mines in the Upper Silesia basin in Poland.

The bimodal distribution results from the mixing of random variables generated by two different phenomena. The first is responsible for the low-energy component and the second for the high-energy component of the distribution. The horizontal lines in Fig. 2.5 indicate the maximum expected seismic events of both types, and the thick curves correspond to two mathematical models describing the observed bimodal distribution of seismic energy, proposed by Kijko *et al.* (1987). It is easier to construct mathematical models to approximate observed distributions than to explain their nature, but the phenomenon seems to be real in the sense that there is no way of approximating the observed distribution by a single probability curve. Two hypotheses as to the nature of the two sets of events can be considered (Kijko

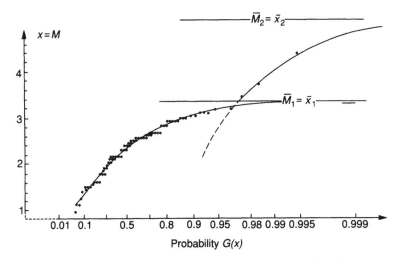

Figure 2.5 Probability distribution of the monthly occurrence of maximum magnitude tremors at the Lubin copper mine, Poland, for the period 1972–1980. [From Kijko *et al.* (1987, Fig. 3).]

et al., 1987): (1) the bimodal distribution is a result of the inhomogeneous and discontinuous structure of rockmass; and (2) the low-energy component of the distribution is a result of mine-induced stress release, and the high-energy component is a result of interaction between mine-induced and residual tectonic stresses in the area.

Whatever the explanation—and it can be supposed that the first hypothesis would be acceptable in some mining districts and the second hypothesis in others—geologic factors play by far the most important role in the generation of seismic events in mines.

The bimodal distribution of natural seismic events has also been found in a few areas. The seismicity associated with the eruption of Mount St. Helens, Washington, in May 1980 was bimodal, confirmed by the distribution of characteristic periods in the maximum amplitude signals and by the frequency–magnitude relations (Main, 1987). The bimodal character of seismicity there most probably results from the separation of source mechanisms into short-period tectonic earthquakes and long-period volcanic tremor associated with harmonic fluid oscillations in the magma chamber. Another striking example of a bimodal seismicity distribution has been found in the New Madrid, Missouri, zone, interpreted as a result of the superposition of two distinct seismogenic source types observed in the area (Main and Burton, 1984).

Chapter 3 | Seismic Waves at Short Distances

There are a number of excellent, more or less advanced, textbooks available describing elastic wave propagation in some detail or in more general fashion, such as the works of Ewing *et al.* (1957), Aki and Richards (1980), Ben-Menahem and Singh (1981), Lee and Stewart (1981), Kennett (1983), White (1983), Hanyga (1984), and Bullen and Bolt (1985), to mention only a few. Here we introduce only basic concepts on the propagation of seismic waves at short distances, necessary in the following chapters. For that a concise description of the theory of elasticity is given so far as it is necessary for the theory of seismic waves. A more detailed treatment of this subject can be found in relevant textbooks, such as the classic work of Jaeger and Cook (1976) for those who prefer a more rigorous approach, or the book by Means (1979) for a more general description.

3.1 Displacement and Strain

The theory of elasticity is concerned with the strain experienced by deformable material subjected to the application of a system of forces. It is assumed that the matter is continuous, and the classic theory is considered within framework of continuum mechanics. The presence of cracks, joints, bedding planes, and other disturbances in rocks could raise doubts on the continuity of a rockmass. Where such disturbances are small in comparison with the dimensions of a given body in a rock, they would alter the mechanical properties of the rockmass, but the body may still be considered as a continuum. Experience shows that continuum mechanics, even when applied to discontinuous materials, often provides correct results, and it is mathematically much simpler than the theory of discontinua.

The fundamental concept of continuum mechanics is that of the displacement of all particles of the material. The forces applied to the body cause its displacement from the initial position to a final position. Within the body, any point P with rectangular coordinates (x, y, z), fixed in the space, is then displaced to a new position, and its displacement components are u, v, w, respectively. For a neighboring point Q with coordinates

24

$(x + \delta x, y + \delta y, z + \delta z)$, the displacement components can be given by a Taylor expansion. And for the small strains associated with elastic waves, within the infinitesimal strain theory, higher-order terms can be neglected. Introducing the notations

$$\omega_z = \frac{1}{2}\left(\frac{\partial v}{\partial x} - \frac{\partial u}{\partial y}\right) \quad \text{and} \quad e_{xy} = \frac{1}{2}\left(\frac{\partial v}{\partial x} + \frac{\partial u}{\partial y}\right) \tag{3.1}$$

and similar others obtained by the cyclic change of x, y, z and u, v, w, respectively, the displacement components of the point Q can be written in the form

$$u + (\omega_y \delta z - \omega_z \delta y) + (e_{xx} \delta x + e_{xy} \delta y + e_{xz} \delta z),$$
$$v + (\omega_z \delta x - \omega_x \delta z) + (e_{yx} \delta x + e_{yy} \delta y + e_{yz} \delta z), \tag{3.2}$$
$$w + (\omega_x \delta y - \omega_y \delta x) + (e_{zx} \delta x + e_{zy} \delta y + e_{zz} \delta_z).$$

The first terms in expressions (3.2) are the components of displacement of point P, corresponding to a pure translation without rotation or deformation. It can be shown that the terms in the first parentheses correspond to a pure rotation without translation or deformation of a volume element and that the terms in the second parentheses are associated with deformation of this element.

The nine elements

$$\begin{matrix} e_{xx} & e_{xy} & e_{xz} \\ e_{yx} & e_{yy} & e_{yz} \\ e_{zx} & e_{zy} & e_{zz} \end{matrix} \tag{3.3}$$

are all associated with internal deformation or strain of the material, when the displacement of given particles results from changes in the dimension or in the volume, but not by rotation. They represent a symmetrical tensor (since $e_{xy} = e_{yx}$, $e_{xz} = e_{zx}$, and $e_{yz} = e_{zy}$) of the second order, which is called the *Cauchy strain tensor* at P. The three components $e_{xx} = \partial u/\partial x$, $e_{yy} = \partial v/\partial y$, and $e_{zz} = \partial w/\partial z$ represent simple, mutually perpendicular, extensions parallel to the x, y, and z axes, respectively. The other three elements, e_{xy}, e_{yz}, and e_{zx}, are called the *shear components of strain*. It can be shown that they are equal to half the angular changes in the xy, yz, and zx planes of an originally orthogonal volume element, preserving the lengths of its sides. This corresponds to an elementary case when an initially rectangular prism is distorted without change of side lengths into a prism whose section is a parallelogram. A deformation characterized by these properties is a shear, and angles $2e_{xy}$, $2e_{yz}$, and $2e_{zx}$ are called the *angles of shear*.

In the theory of elasticity it is shown that at any given instant there is one particular set of orthogonal axes through point P for which the shear components of strain at P vanish. These axes are known as the *principal axes of strain* at P, and the corresponding values of e_{xx}, e_{yy}, and e_{zz} are called the *principal extensions*. They completely determine the deformation at P. It can also be shown that the sum $e_{xx} + e_{yy} + e_{zz}$ is independent of the choice of the orthogonal coordinate system.

The cubical dilatation at point P, considering a small portion of matter enclosing P and undergoing deformation, is defined as the limit approached by the ratio of increase in volume to the initial volume of this matter when the area of the boundary surface approaches zero. If e_{xx}, e_{yy}, and e_{zz} are the principal extensions at P associated with the undergoing deformation, then the dilatation θ at P is equal to

$$\theta = e_{xx} + e_{yy} + e_{zz} = \frac{\partial u}{\partial x} + \frac{\partial v}{\partial y} + \frac{\partial w}{\partial z}, \qquad (3.4)$$

neglecting higher-order terms. This relation is valid even if e_{xx}, e_{yy}, and e_{zz} are other than the principal extensions. The result is true for any Cartesian coordinate system because of the invariance of the sum. A cubical compression is a negative dilatation.

3.2   Stress

The deformation of any body is caused by forces acting on this material. Two principal types of forces are considered. A body (volume) force is a force referred to a unit mass. It depends on the amount of material affected, and it can work at a distance. The most common body forces are the gravitational forces, which are always present. The other forces of importance are surface forces (per unit area) because they operate across surfaces of contact between adjacent parts of material. If the force is directed along the normal to the plane, it is called a *normal force*, and if it is directed perpendicular to the normal, it is called a *shearing force* or a *shear force*. In general, when a surface force is directed neither parallel nor perpendicular to a given plane, it can be resolved into normal and shearing components. The normal component of a force is either compressive or tensile according to the produced distortion.

The forces acting at a point P in the interior of a body may be described as follows. At a given instant we draw through P a unit vector $\mathbf{n}$ in any direction (Fig. 3.1). We consider the forces across a small plane element δS, containing P and normal to the unit vector, separating two small portions of a body. These forces are statically equivalent to a single resultant force acting at P and a couple. This single force is called the *traction* across the element δS. As δS is indefinitely diminished, the limit of the ratio of traction on δS

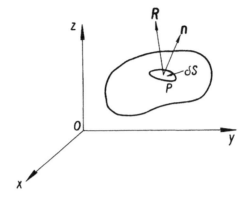

Figure 3.1   Traction **R** acting on an element of area δS.

to the area δS is finite and is called the *stress* at P corresponding to the direction of a unit vector. The ratio of the couple to δS is usually neglected, since the area δS is supposed to be very small. The stress appears as a surface force.

The analysis of stress is based on pure statics, and the appropriate relations assume nothing about the properties of the material, which may be elastic, plastic, viscous, or of any other type, as long as it is continuous. Similarly, if the strains are infinitesimal, the theory runs parallel to that for stress within the classical theory of linear elasticity.

In general, the stress at P varies with the direction of the normal to the small area δS. For a complete description of the stress at P, it is necessary to consider the traction acting on all planes passing through the point. All these tractions can be reduced to component tractions across planes parallel to the coordinate planes. Across each of these planes the tractions can be resolved into three components parallel to the axes, and then a set of nine elements of stress is obtained

$$
\begin{matrix}
p_{xx} & p_{xy} & p_{xz} \\
p_{yx} & p_{yy} & p_{yz} \\
p_{zx} & p_{zy} & p_{zz},
\end{matrix}
\tag{3.5}
$$

where the first suffixes represent a coordinate axis normal to a given plane and the second suffixes represent the axis to which the traction is parallel. The set of nine components (3.5) completely determines the stress across any small plane area containing the point and constitutes the *stress tensor* at point P, which is a tensor of the second order. The components with identical suffixes, p_{xx}, p_{yy}, and p_{zz}, are normal components of stress because they act along the normal of the given plane, and the components with different suffixes are shear components of stress because they act parallel to the plane in question.

Introducing the coordinate axes x_1, x_2, and x_3 instead of x, y, and z, and using letter suffixes such as i and j in place of the numerals 1, 2, and 3, which is the usual practice in tensor analysis, the set of nine stress components (3.5) can be simply written as p_{ij}; a quantity followed by two suffixes represents nine quantities, the suffixes taking the values 1, 2, and 3 in turn in all possible combinations. It is also convenient to introduce here the summation convention, where if any suffix occurs twice in a single term, it is to be put equal to 1, 2, and 3 in turn and the results added.

The components of any second-order tensor in one coordinate system can be transformed into the components in another coordinate system. The relations between the stress components p_{ij} in the old coordinate system and p_{kl} in the new coordinate system may be written as

$$p_{kl} = a_{ik} a_{jl} p_{ij}, \tag{3.6}$$

where a_{ik} and a_{jl} are the direction cosines relating the new coordinate axes to the old ones. The stress tensor is symmetric, that is $p_{ij} = p_{ji}$, and only six of its nine components are independent.

As was the case for the shear components of strain, at any given point three mutually perpendicular axes can always be found with respect to which the shear components of stress vanish. Then the three remaining components p_{11}, p_{22}, and p_{33} are the principal stresses, which are either compressions or tensions, corresponding to these axes called the *principal stress directions*. Thus the stress at a given point is completely specified by the three principal stresses. In general, however, the directions of the principal axes are different for different points and for describing stress in the whole-body six stress components are still needed.

In rock mechanics the three principal stresses are designated as $\vec{\sigma}_1$, $\vec{\sigma}_2$, and $\vec{\sigma}_3$ with magnitudes $\sigma_1 \geq \sigma_2 \geq \sigma_3$. Several terms are used to describe the stress at a point. *Triaxal stress* is a state in which all three principal stresses are nonzero, a state in which only two principal stresses are nonzero is called *biaxial stress*, and only one nonzero principal stress corresponds to *uniaxial stress*. Other terms are referred to the relative values of the principal stresses. *Polyaxial stress* has $\sigma_1 \neq \sigma_2 \neq \sigma_3$, *axial stress* has either $\sigma_1 > \sigma_2 = \sigma_3$ (axial compression) or $\sigma_1 = \sigma_2 > \sigma_3$ (axial extension), and *hydrostatic stress* has $\sigma_1 = \sigma_2 = \sigma_3$. Hydrostatic stress is also called *hydrostatic pressure*, and it is the only kind of stress that can exist in a fluid at rest. The spatial array of states of stress at every point in any body at an instant is called a *stress field*. The simplest stress field is a homogeneous one, where all components of the stress tensor are identical at every point.

3.3 Stress–Strain Relations

The occurrence of stress in deformable matter is accompanied by the occurrence of strain. The relation between the components of stress and the components of strain at a point at one instant of time depends on the material constitution of the body, and equations relating stress to strain are called *constitutive equations*. "Perfect elasticity" is when the components of strain at any point are linear functions of the components of stress at the point

$$p_{kl} = A_{ijkl} e_{ij}, \tag{3.7}$$

where A_{ijkl} are proportionality constants or elastic constants, in rock mechanics called the *stiffnesses*, constituting a set of 81 coefficients. Of these coefficients no more than 36 could be independent, since each of the tensors p_{ij} and e_{ij} is symmetric and has only six independent components. Thus it follows that $A_{ijkl} = A_{jikl} = A_{ijlk}$. Equations (3.7) describe a generalization of Hooke's law.

If the elastic behavior is entirely independent of any particular direction, then the material is called *isotropic*. For a perfectly elastic isotropic material the following relations may be written

$$p_{11} = Ae_{11} + B(e_{22} + e_{33}) + Ce_{23} + D(e_{31} + e_{12}), \tag{3.8}$$

$$p_{23} = Ee_{11} + F(e_{22} + e_{33}) + Ge_{23} + H(e_{31} + e_{12}), \tag{3.9}$$

where the coefficients A, B, C, D, E, F, G, and H depend only on the particular material and on the thermodynamic conditions. The coefficients are independent of the direction of particular axes.

It can be shown that if the axes are the principal axes of strain at point P, then $p_{12} = p_{13} = p_{23} = 0$, which means that the principal axes of stress at P coincide with the principal axes of strain at P. Similarly, it can also be shown that $C = D = E = F = H = 0$ and $A - B = G$. Thus it follows that for a perfectly elastic isotropic substance the eight coefficients can be expressed in terms of two elastic parameters. The Lamé elastic parameters λ and μ are usually taken as these two parameters, where $\lambda = B$ and $\mu = \frac{1}{2}(A - B)$. Now equations (3.8) and (3.9) can be replaced by

$$p_{11} = \lambda\theta + 2\mu e_{11}, \tag{3.10}$$

and

$$p_{23} = 2\mu e_{23}, \tag{3.11}$$

where the dilatation θ is given by formula (3.4), and the full stress–strain relations are

$$p_{ij} = \lambda\theta\delta_{ij} + 2\mu e_{ij}, \tag{3.12}$$

where δ_{ij} is a special second-order tensor, the Kronecker delta, whose components are equal to 1 for $i = j$ and equal to 0 for $i \neq j$.

Of the two Lamé constants, μ has a relatively simple physical meaning. It measures the resistance of the elastic body to shearing deformation and is called the *rigidity modulus*. The constant λ is not simply related to experimentally observed quantities, and its value is usually calculated from those of μ and of one of the other experimentally determined coefficients. If the material in the vicinity of a particle P of a perfectly elastic isotropic body is subjected to an additional stress in the form of hydrostatic pressure, then the ratio of this pressure to the produced compression is called the *bulk modulus* or *incompressibility k* at P. The bulk modulus k can be simply expressed in terms of λ and μ as

$$k = \lambda + \tfrac{2}{3}\mu. \tag{3.13}$$

Another pair of important elastic coefficients are Young's modulus E and Poisson's ratio σ. If point P is inside a small cylinder, the stress at each end of the cylinder is normal, and there is no traction across the lateral surface of the cylinder, then Young's modulus at P is the ratio of the stress at one of these ends to the longitudinal extension of the cylinder; and Poisson's ratio is the ratio of the lateral contraction to the longitudinal extension. The elastic parameters E and σ are readily expressible in terms of λ and μ

$$E = \frac{p_{11}}{e_{11}} = 2\mu + \frac{\lambda\mu}{\lambda + \mu} = 2\sigma(1 + \sigma) \tag{3.14}$$

and

$$\sigma = -\frac{e_{22}}{e_{11}} = \frac{\lambda}{2(\lambda + \mu)}. \tag{3.15}$$

Similarly, the elastic coefficients λ, μ, and k can be readily expressed in terms of E and σ.

Relations (3.12) describe stress p_{ij} in terms of strain e_{ij}, using the elastic parameters λ and μ. For the description of e_{ij} in terms of p_{ij}, the parameters E and σ are the most useful. For many solids, particularly for most rocks, the two elastic parameters λ and μ are nearly equal, and the Poisson relation $\lambda = \mu$ is occasionally used as a simplification. This corresponds to $k = \tfrac{5}{3}\mu$ and $\sigma = \tfrac{1}{4}$.

The strain energy or the potential energy stored in a strained body is of fundamental importance. In elementary physics, the notion of elastic energy is associated with a compressed or stretched spring. When the ends of the spring are released, the potential energy is converted into kinetic energy and does the work of motion. The notion of elastic energy accumulated in an elastic body under deformation is introduced in a similar manner. Considering the total work done by the forces in a small volume of material and that in this case the work is expressed as the potential energy of elastic strain accumulated in a unit volume, it can be shown that the total potential energy per unit volume W is

$$W = \tfrac{1}{2} p_{ij} e_{ij}. \tag{3.16}$$

The quantity W is also called the *strain–energy density* or the *strain–energy function*. It is sometimes called the *elastic potential* as well, since any component of the strain or stress can be obtained from partial derivatives of the function W.

From relations (3.4), (3.12), and (3.16) it follows that for the elastic behavior of an isotropic material the total elastic energy is given by

$$W = \tfrac{1}{2} \left(\lambda \theta^2 + 2\mu e_{ij}^2 \right), \tag{3.17}$$

where by e_{ij}^2 we understand $e_{ij} e_{ij}$ according to the summation convention.

3.4 Equations of Motion

If forces are not in equilibrium, the elastic body undergoes deformation and the equations of motion are needed to describe the motion of all points of the body. To derive the equations of motion, the tractions across the surfaces of a volume element corresponding to the stress components (3.5) and the body forces X_1, X_2, X_3 proportional to the mass in the volume element with the dimensions $\delta x_1, \delta x_2, \delta x_3$ are considered.

Introducing the components of acceleration $d^2 u_i / dt^2$ of point P, where u_i $(i = 1, 2, 3)$ are the components of displacement and t is time, and the density ρ of the medium, the forces of inertia can be expressed as $\rho (d^2 u_i / dt^2) \delta x_i$. The equations of motion are obtained by adding all the forces and the inertia terms for each component:

$$\rho \frac{d^2 u_i}{dt^2} = \rho X_i + \frac{\partial p_{ij}}{\partial x_i}. \tag{3.18}$$

This system of equations, called the *Cauchy equations*, relates the components of stress in an elastic medium to the acceleration. In many problems, however, direct relations with the displacement are needed. From equations

(3.12) and (3.18) it follows that

$$\rho \frac{du_i}{dt^2} = \frac{\partial}{\partial x_j}(\lambda\theta\delta_{ij} + 2\mu e_{ij}) + \rho X_i.$$ (3.19)

If the material is isotropic, in which the parameters λ and μ are constants, then the equations of motion in terms of displacement are as follows:

$$\rho \frac{\partial^2 u_i}{\partial t^2} = (\lambda + \mu)\frac{\partial\theta}{\partial x_i} + \mu\nabla^2 u_i + \rho X_i,$$ (3.20)

where ∇^2 is Laplace's operator $\partial^2/\partial x_i^2$, which is a sum of the second partial derivatives in Cartesian coordinates. In relation (3.20) d^2/dt^2 is replaced by $\partial^2/\partial t^2$ since the displacement u_i is very small and the second powers of components $\partial u_i/\partial x_j$ are already neglected.

Assuming that the body forces may be neglected, the equations of motion of the disturbance are

$$\rho \frac{\partial^2 u_i}{\partial t^2} = (\lambda + \mu)\frac{\partial\theta}{\partial x_i} + \mu\nabla^2 u_i.$$ (3.21)

Differentiating both sides of equations (3.21) with respect to x_i (which involves adding the results of separate differentiations for $i = 1, 2, 3$), applying the operation curl to the both sides, and recalling relation (3.4), the following equations are obtained:

$$\rho \frac{\partial^2\theta}{\partial t^2} = (\lambda + 2\mu)\nabla^2\theta$$ (3.22)

and

$$\rho \frac{\partial}{\partial t^2}\text{curl}(u_i) = \mu\nabla^2\text{curl}(u_i).$$ (3.23)

These equations are scalar and vector equations, respectively. They indicate that two types of disturbances may be propagated through an elastic solid. From equation (3.22) it follows that a dilatational (or irrotational) disturbance θ may be transmitted through the body with velocity α equal to

$$\alpha = \sqrt{\frac{\lambda + 2\mu}{\rho}},$$ (3.24)

and from equations (3.23) it follows that a rotational (or equivoluminal) disturbance may be transmitted with velocity β

$$\beta = \sqrt{\frac{\mu}{\rho}} \, . \tag{3.25}$$

From these two relations it follows that $\alpha > \beta$. In seismology, the two types of waves are called the *primary* (P) *waves* and the *secondary* (S) *waves*, respectively. If the rigidity modulus μ is zero, then the velocity β is also zero. Thus rotational waves are not transmitted through a fluid. If the Poisson relation $\lambda = \mu$ is acceptable, then $\alpha/\beta = \sqrt{3}$. This case is close to the real conditions for much of the Earth.

Numerous equations are associated with wave propagation. To introduce the classic wave equation, it is convenient to define for displacements in a solid body a scalar potential φ and a vector potential ψ_i as follows:

$$u_1 = \frac{\partial \varphi}{\partial x_2} + \frac{\partial \psi_3}{\partial x_3} - \frac{\partial \psi_2}{\partial x_1},$$

$$u_2 = \frac{\partial \varphi}{\partial x_2} + \frac{\partial \psi_1}{\partial x_3} - \frac{\partial \psi_3}{\partial x_1}, \tag{3.26}$$

$$u_3 = \frac{\partial \varphi}{\partial x_2} + \frac{\partial \psi_2}{\partial x_3} - \frac{\partial \psi_1}{\partial x_1},$$

or in vector form

$$u_i = \operatorname{grad} \varphi + \operatorname{curl} \psi_i. \tag{3.27}$$

From the definition of θ, as given by relation (3.4), it follows that $\theta = \nabla^2 \varphi$. It can be shown that the equations of motion (3.21) are satisfied if the functions φ and ψ_i are solutions of the equations

$$\frac{\partial^2 \varphi}{\partial t^2} = \alpha^2 \nabla^2 \varphi \tag{3.28}$$

and

$$\frac{\partial^2 \psi_i}{\partial t^2} = \beta^2 \nabla^2 \psi_i. \tag{3.29}$$

These are the wave equations describing the two types of disturbances in Cartesian coordinates, similarly as equations (3.22) and (3.23). By introduction of the potentials φ and ψ_i our problem has been reduced to that of solving the wave equations in a classic form. Two wave equations found for a solid represent the propagation of compressional and distortional waves with velocities α and β, respectively. The distortional waves, also known as *shear* or *transverse waves*, are described in general by three functions ψ_i, which are the solutions of equations (3.29). It is possible to include additional effects such as those of body forces or finite strains, but they are not considered here.

3.5 Plane Seismic Waves

The wave equations (3.28) and (3.29) are linear partial differential equations of the second order. Their general solution, which would be valid for all cases of wave propagation in a homogeneous and isotropic medium, is rather complex, and simpler cases should be considered first. If the waves are sufficiently distant from the source of a confined disturbance, they may be regarded as plane. Such an approximation is relevant to many seismological problems, and in this case the displacements associated with P and S waves, called the *far-field displacements*, are in effect longitudinal and transverse, respectively.

The simplest possible case is the one-dimensional wave equation. If the motion takes place along the x_1 axis, which is a particular one of the three rectangular coordinates x_i, then the displacement components u_i are not dependent on the coordinates x_2 and x_3, and we have

$$\theta = \frac{\partial u_1}{\partial x_1}, \quad \nabla^2 u_1 = \frac{\partial^2 u_1}{\partial x_1^2}, \quad \text{and} \quad \frac{\partial u_i}{\partial x_2} = \frac{\partial u_i}{\partial x_3} = 0. \qquad (3.30)$$

The system of equations (3.21) now takes the form

$$\frac{\partial^2 u_i}{\partial t^2} = c^2 \frac{\partial^2 u_i}{\partial x_1^2} \qquad (i = 1, 2, 3), \qquad (3.31)$$

where c is either α or β defined by relations (3.24) and (3.25). The general solution of equations (3.31) is either of the form

$$u_i = f_i(x_1 - ct) + F_i(x_1 + ct) \qquad (3.32)$$

or the form

$$u_i = f_i(t - x_1/c) + F_i(t + x_1/c), \qquad (3.33)$$

where f_i and F_i are arbitrary twice-differentiable functions, and corresponds to the superposition of plane waves traveling with velocity c in the positive and negative directions of the x_1 axis, respectively. If instead the direction of propagation has direction cosines ν_j ($j = 1, 2, 3$), then the corresponding equations are

$$u_i = f_i(v_j x_j - ct) + F_i(v_j x_j + ct). \tag{3.34}$$

Another approach to the solution of equations (3.21) is based on a trial substitution of the form

$$u_i = A_i \exp\left[i\frac{2\pi}{l}(v_j x_j - ct)\right], \tag{3.35}$$

where $v_j^2 = 1$, representing a simple harmonic (or sinusoidal) advancing plane wave with the wavelength l or the wavenumber $\kappa = 2\pi/l$ and with the period $T = l/c$. The quantities v_j can be considered as the direction cosines of a line L. Thus equations (3.35) represent a system of plane waves advancing along L with velocity c. The substitution of relations (3.35) into equations (3.21) leads to the three homogeneous linear equations in A_i

$$-\rho c^2 A_i + (\lambda + \mu)v_i(A_j v_j) + \mu A_i = 0. \tag{3.36}$$

The coefficients A_i determine components of the displacement vector, which may be considered as the resultant of the mutually perpendicular vectors **B**, **C**, and **D**, where **B** is in the direction of L. Taking the x_1 axis in the direction of L and the x_2 and x_3 axes in the direction of **C** and **D**, respectively, leads to the conditions $v_1 = 1$ and $v_2 = v_3 = 0$. If we put for the component **B** in relations (3.35) $A_1 = B$ and $A_2 = A_3 = 0$, for the component **C**, $A_2 = C$ and $A_1 = A_3 = 0$, and for the component **D**, $A_3 = D$ and $A_1 = A_2 = 0$, then equations (3.36) take the form

$$-\rho c^2 B + (\lambda + 2\mu)B = 0,$$
$$-\rho c^2 C + \mu C = 0,$$
$$-\rho c^2 D + \mu D = 0. \tag{3.37}$$

Thus the velocity $c = \alpha = [(\lambda + 2\mu)/\rho]^{1/2}$ is associated with the component of displacement parallel to the direction of propagation and the velocity $c = \beta = [\mu/\rho]^{1/2}$ is associated with the components in two mutually perpendicular directions normal to the propagation.

The two types of P and S waves are independent of each other, and the S waves may be plane-polarized. When an S wave is polarized in such a way that all particles of the medium move horizontally during its passage, it is denoted by SH, and when the particles move in vertical planes containing the direction of propagation, the wave is denoted by SV. Thus the system of plane waves traveling, for instance, along L consists of three independent parts that correspond to the P (compressional), SH (horizontally polarized shear), and SV (vertically polarized shear) seismic waves.

To describe the energy associated with elastic motion, two approaches can be used. At any instant, the kinetic and potential energies, E_k and E_p, of the vibrating medium between the planes $x = x_a$ and $x = x_b$, resulting from the displacements associated with the simplest one-dimensional wave with a displacement component u, can be expressed in the form

$$E_k = \frac{c_1}{2} \int_{x_a}^{x_b} \left(\frac{\partial u}{\partial t} \right)^2 dx, \qquad E_p = \frac{c_2}{2} \int_{x_a}^{x_b} \left(\frac{\partial u}{\partial x} \right)^2 dx, \qquad (3.38)$$

respectively, where c_1 and c_2 are constants. The corresponding wave equation is

$$\frac{\partial^2 u}{\partial t^2} = \frac{c_1}{c_2} \frac{\partial^2 u}{\partial x^2}, \qquad (3.39)$$

with the wave velocity $c = \sqrt{c_1/c_2}$. If the solution $f(x - ct)$, following relations (3.32), is substituted for u in formulas (3.38), then it can be seen that the results are equal. Thus, the energy at any instant in a propagating plane wave is half kinetic and half potential.

In some problems, it is convenient to use the energy density, which is the energy per unit volume in the medium. The concept of an elastic strain–energy density can be applied to describe the energy in plane-wave motion. The strain energy of a medium is its capacity to do work by virtue of its configuration, and relation (3.16) shows that the strain–energy density is equal to $\frac{1}{2} p_{ij} e_{ij}$. For a plane wave it is easy to show, using the stress–strain relations for an isotropic medium (3.12), that in the case of either a P wave or an S wave the strain–energy density equals the kinetic–energy density

$$\frac{\partial u_1}{2} p_{ij} e_{ij} = \frac{1}{2} \rho \left(\frac{\partial u_i}{\partial t} \right)^2. \qquad (3.40)$$

Energy densities depend on t and x_i only and the velocity of energy propagation is not different from the velocity of either P or S waves. It also follows that the flux rate of energy transmission in a plane wave, which is the amount of energy transmitted per unit time across unit area normal to the direction of propagation, is $\rho\alpha(\partial u_i/\partial t)^2$ for P waves and $\rho\beta(\partial u_i/\partial t)^2$ for S waves. This result is valid only for plane waves in homogeneous media, and it depends on material properties and on the planar nature of the wave only at the point at which the flux rate is evaluated.

3.6 Spherical Waves

In a spherical coordinate system with spherical symmetry, that is, putting $R^2 = x_1^2 + x_2^2 + x_3^2$ and assuming that $\varphi = \varphi(R, t)$, the equation of motion (3.28) takes the form

$$\frac{\partial^2 \varphi}{\partial t^2} = \alpha^2 \left(\frac{\partial^2 \varphi}{\partial R^2} + \frac{2}{R} \frac{\partial \varphi}{\partial R} \right). \tag{3.41}$$

Since R and t are independent, it can be deduced that the wave equation for a spherical wave with spherical symmetry may be written as

$$\frac{\partial^2 (\varphi R)}{\partial t^2} = \alpha^2 \frac{\partial^2 (\varphi R)}{\partial R^2}. \tag{3.42}$$

This equation for the product φR is identical with that for a plane wave, and its general solution has the form

$$\varphi = \frac{1}{R} [f(R - \alpha t) + F(r + \alpha t)], \tag{3.43}$$

where f and F are arbitrary functions. Each term in relation (3.43) has a constant value on a sphere $R = \mathrm{const}$ (constant) at a given time t. If f and F are periodic functions, then formula (3.43) represents infinite trains of spherical waves propagating toward and away from the common center of the spheres $R = \mathrm{const}$.

A similar treatment of equations (3.29) leads to the expression

$$\psi_t = \frac{1}{R} [g_t(R - \beta t) + G_t(R + \beta t)], \tag{3.44}$$

where g_t and G_t are again arbitrary functions. Functions f and g_t represent radiations from a point source at the origin, whereas functions F and G_t correspond to disturbances traveling toward the origin and are usually zero.

Special solutions of the type (3.43) (or (3.44)) are the functions

$$\varphi = \frac{A}{R} \exp[i(\kappa_\alpha R - \omega t)], \tag{3.45}$$

where A is a constant, $\kappa_\alpha = \omega/\alpha$ is the wavenumber, and $\omega = 2\pi/T$ is the angular frequency. Solutions in this form are used for an infinite medium or for a certain time interval in a finite domain until the effect of boundaries has to be considered. There are two approaches to such solutions. It can be shown that spherical waves are a superposition of either plane or cylindrical waves.

The Weyl integral uses plane waves as a basis, summing them to give the solution for a point source. The first exponential factor in formula (3.45) depending only on distance can be written in the form of a double integral

$$\frac{\exp(-i\kappa_\alpha R)}{-i\kappa_\alpha R} = \iint \exp(i\kappa_\alpha \nu_i x_i) \, ds, \tag{3.46}$$

where ds is a surface element. Here the integration is performed over the half sphere $\nu_3 > 0$ of the unit sphere $\nu_i^2 = 1$.

The Sommerfeld integral is the analogous result for cylindrical waves. The first exponential factor in relation (3.45) can be written in the form

$$\frac{\exp(i\kappa_\alpha R)}{R} = \int_0^\infty J_0(\kappa r) \exp(-\nu|z|) \frac{\kappa}{\nu} \, d\kappa, \tag{3.47}$$

where J_0 is the zero-order Bessel function, z and r are cylindrical coordinates, and $\nu^2 = \kappa^2 - \kappa_\alpha^2$, where κ is a parameter. The integrand here is a new kind of fundamental wave, the cylindrical wave with symmetry about a vertical axis, in which the dependence on r and z appears in separate factors.

3.7 Plane Seismic Waves in a Layered Medium

The previous considerations of the wave equations were related to elastic waves propagating in a medium of infinite extent in all three dimensions. The next step is to consider elastic waves in a homogeneous and isotropic half-space with a free plane boundary, which is the simplest case in a bounded medium.

The free boundary is a horizontal plane $x_1 x_2$ ($x_3 = 0$), and a train of plane P waves propagates in a direction AO in the $x_1 x_3$ plane (Fig. 3.2), under an angle e to the boundary or $i = \pi/2 - e$ with the normal to the boundary. Similarly, a train of plane SV waves propagates in a direction BO (Fig. 3.3), making an angle f with the boundary. The plane waves are independent of x_2, and the displacements corresponding to the P and SV waves [Equations (3.26)]

$$u_1 = \frac{\partial \varphi}{\partial x_1} - \frac{\partial \psi}{\partial x_3} \quad \text{and} \quad u_3 = \frac{\partial \varphi}{\partial x_3} + \frac{\partial \psi}{\partial x_1}, \tag{3.48}$$

are considered together, whereas the SH wave can be discussed separately. Functions φ and ψ satisfy the wave equations (3.28) and (3.29).

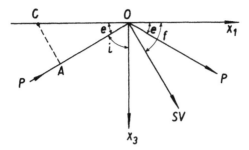

Figure 3.2 Reflection of P waves at the free surface of elastic solid.

The boundary condition is that there is no stress over the surface $x_1 x_2$; then at $z = 0$

$$p_{31} = \mu\left(\frac{\partial u_3}{\partial x_1} + \frac{\partial u_1}{\partial x_3}\right) = \mu\left(2\frac{\partial^2 \varphi}{\partial x_1 \partial x_3} + \frac{\partial^2 \psi}{\partial x_1^2} - \frac{\partial^2 \psi}{\partial x_3^2}\right) = 0, \quad (3.49)$$

$$p_{32} = \frac{\partial u_2}{\partial x_3} = 0, \quad (3.50)$$

$$p_{33} = \lambda\theta + 2\mu\frac{\partial u_3}{\partial x_3} = \lambda^2 \nabla\varphi + 2\mu\left(\frac{\partial^2 \varphi}{\partial x_3} + \frac{\partial^2 \psi}{\partial x_1 \partial x_3}\right). \quad (3.51)$$

Equations (3.49) and (3.51) contain potentials φ and ψ only and describe motions parallel to the plane $x_2 = 0$, whereas equation (3.50) contains only u_2 and gives motion perpendicular to the plane $x_2 = 0$. Thus the (φ, ψ) motion, corresponding to P and SV waves, and the u_2 motion, corresponding to SH wave, are independent.

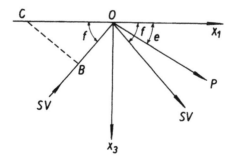

Figure 3.3 Reflection of SV waves at the free surface of elastic solid.

For an incident P wave, there may be reflected waves of both P and S types, and a solution of equations (3.49) and (3.51) has the form

$$\varphi = A_1 \exp[i\kappa(x_1 - x_3 \tan e - ct)] + A_2 \exp[i\kappa(x_1 + x_3 \tan e - ct)], \quad (3.52)$$

$$\psi = B_2 \exp[i\kappa(x_1 + x_3 \tan f - ct)]. \quad (3.53)$$

The first term on the right-hand side of (3.52) corresponds to an incident wave; and the second term, to a reflected wave. If the coefficients of x_1 and t were not taken the same in all waves, then it would not be possible to satisfy the conditions at $z = 0$ for all values of x_1 and t. It should be noted that the velocity c in the present problem is not the actual wave velocity, but the velocity of advance of the line in which a plane-wave front cuts the plane boundary surface; it is an apparent velocity along the surface. A similar interpretation is also relevant for the wavenumber κ. From Figs. 3.2 and 3.3, taking into account the distances traveled by P or S waves, it follows that $\tan e = (c^2/\alpha^2 - 1)^{1/2}$, $\tan f = (c^2/\beta^2 - 1)^{1/2}$, and $c = \alpha \sec e = \beta \sec f$ or

$$\frac{\cos e}{\alpha} = \frac{\cos f}{\beta}, \quad (3.54)$$

which is similar to Snell's law in geometric optics. The angle e is known as the *angle of emergence*, the complement of the angle of incidence ι, and $\pi/2 - f$ is the angle of reflection for S waves. If for simplicity the Poisson's relation $\lambda = \mu, \alpha/\beta = \sqrt{3}$, is accepted, then the boundary conditions take the form

$$2(A_1 - A_2)\tan e + B_2(\tan^2 f - 1) = 0,$$

$$(A_1 + A_2)(1 + 3\tan^2 e) + 2B_2 \tan f = 0. \quad (3.55)$$

From equations (3.55) the ratios A_2/A_1 and B_2/A_1 can be expressed by the angles e and f:

$$\frac{A_2}{A_1} = \frac{4\tan e \tan f - (1 + 3\tan^2 e)^2}{4\tan e \tan f + (1 + 3\tan^2 e)^2}, \quad (3.56)$$

$$\frac{B_2}{A_1} = \frac{-4\tan e(1 + 3\tan^2 e)}{4\tan e \tan f + (1 + 3\tan^2 e)^2}, \quad (3.57)$$

and f can be readily expressed by e from relation (3.54).

From equation (3.57) it follows that B_2 vanishes for normal incidence $e = \pi/2$ and for grazing incidence $e = 0$. In both cases the reflection consists of a P wave only. Thus a reflected disturbance of SV type exists for all angles of incidence except zero and $\pi/2$. The coefficient A_2 vanishes if

$(1 + 3 \tan^2 e)^2 = 4 \tan e \tan f$. This equation has two roots, $e = 12.8°$ and $e = 30°$, and for these angles of emergence no reflected P wave exists, whereas between them the ratio A_2/A_1 is very small. At least half of the energy goes into the reflected SV wave for angles of incidence between 12° and 63°.

For an incident SV wave (Fig. 3.3), the boundary conditions are satisfied if the incident wave gives rise to a reflected transverse wave and a reflected longitudinal wave

$$\varphi = A_2 \exp[i\kappa(x_1 + x_3 \tan e - ct)], \tag{3.58}$$

$$\psi = B_1 \exp[i\kappa(x_1 - x_3 \tan f - ct)] + B_2 \exp[i\kappa(x_1 + x_3 \tan f - ct)]. \tag{3.59}$$

Assuming again that Poisson's relation can be accepted, the following relations are obtained for the reflection coefficients:

$$\frac{A_2}{B_1} = \frac{4 \tan f (1 + 3 \tan^2 e)}{4 \tan e \tan f + (1 + 3 \tan^2 e)^2}, \tag{3.60}$$

$$\frac{B_2}{B_1} = \frac{4 \tan e \tan f - (1 + 3 \tan^2 e)^2}{4 \tan e \tan f + (1 + 3 \tan^2 e)^2}. \tag{3.61}$$

If $\tan e = 0$ or ∞, then complete reflection of SV occurs. The amplitude B_2 of the reflected SV wave vanishes for the angle $f = 55.7°$ and $f = 60°$, and for a range of angles between 55° and 75° more than half the energy goes into the reflected P wave. If $\cos f > \beta/\alpha$, then from relation (3.54) it follows that $\cos e > 1$ and e becomes imaginary. In such a case, for sufficiently small f, the reflected SV is equal in amplitude to the incident one, but differs in phase, and the reflected P wave is not a harmonic wave but a motion confined to the neighborhood of the free surface.

For a incident SH wave we take a solution of equation (3.50) in the form

$$u_2 = C_1 \exp[i\kappa(x_1 - x_3 \tan f - ct)] + C_2 \exp[i\kappa(x_1 + x_3 \tan f - ct)]. \tag{3.62}$$

From the boundary condition (3.50) it follows that $C_2 = C_1$. Thus all the energy is reflected as SH. The reflected wave is equal in amplitude to the incident wave, and the horizontal displacement of the free surface is twice that of the incident wave.

A more general case is the reflection and refraction of seismic waves at the interface between two perfectly elastic and isotropic solid media in welded contact, separated by a plane horizontal boundary and extending to indefinitely great distances from the boundary. Any incident wave at the interface will, in general, generate compressional and distortional waves in

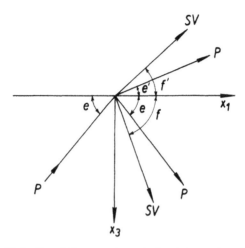

Figure 3.4 Reflection of P waves at an interface between two elastic solids.

both media. The boundary conditions are that the displacement and the stress across the interface be continuous at all times and places.

For incident P and SV waves four boundary conditions must be satisfied, requiring continuity of the two components of displacement u_1 and u_3 and the two stresses p_{33} and p_{31} across the boundary. As before, the interface is the plane $x_3 = 0$ and the waves approach from positive x_3. Indicating by the subscripts 1 and 2 quantities related to incident and reflected waves, respectively, and by accents quantities referred to refracted waves, we have the following solutions:

$$\varphi = A_1 \exp[i\kappa(x_1 - ax_3 - ct)] + A_2 \exp[i\kappa(x_1 + ax_3 - ct)], \quad (3.63)$$

$$\psi = B_1 \exp[i\kappa(x_1 - bx_3 - ct)] + B_2 \exp[i\kappa(x_1 + bx_3 - ct)], \quad (3.64)$$

$$\varphi' = A' \exp[i\kappa(x_1 - a'x_3 - ct)], \quad (3.65)$$

$$\psi' = B' \exp[i\kappa(x_1 - b'x_3 - ct)], \quad (3.66)$$

where the angles e, f, e', and f' are as defined in Figs. 3.4 and 3.5, and $a = \tan e$, $b = \tan f$, $a' = \tan e'$, and $b' = \tan f'$. The coefficients c and κ are the same in all four solutions, since the boundary conditions at the interface are assumed to be independent of x and t, and then the elementary laws of reflection and refraction are immediately deduced:

$$c = \frac{\alpha}{\cos e} = \frac{\beta}{\cos f} = \frac{\alpha'}{\cos e'} = \frac{\beta'}{\cos f'}. \quad (3.67)$$

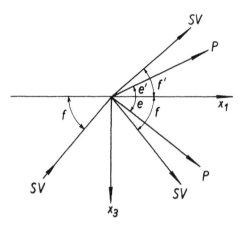

Figure 3.5 Reflection of *SV* waves at an interface between two elastic solids.

These conditions imply that for real angles e, f, e', and f' the velocity c must be greater than the velocities α, β, α', and β'. When any of the coefficients a, b, a', and b' is imaginary, complex reflection coefficients will occur, indicating phase changes.

The four boundary conditions $u_1 = u'_1$, $u_3 = u'_3$, $p_{33} = p'_{33}$, and $p_{31} = p'_{31}$ at $x_3 = 0$ give

$$A_1 + A_2 - b(B_1 - B_2) = A' - b'B', \tag{3.68}$$

$$a(A_1 - A_2) + (B_1 + B_2) = a'A' + B', \tag{3.69}$$

$$\rho\beta^2\{(b^2 - 1)(A_1 + A_2) + 2b(B_1 - B_2)\} = \rho'\beta'^2\{(b'^2 - 1)A' + 2b'B'\}, \tag{3.70}$$

$$\rho\beta^2\{-2a(A_1 - A_2) + (b^2 - 1)(B_1 + B_2)\} = \rho'\beta'^2\{-2a'A' + (b'^2 - 1)B'\}. \tag{3.71}$$

These equations have to be solved numerically. An incident wave of a single type usually occurs, and then either $A_1 = 0$ or $B_1 = 0$, and the four amplitude coefficients may be expressed in terms of the amplitude of the incident wave.

Although numerous cases have been worked out in detail, the behavior of incident waves on an interface is highly varied, defying general description. It is convenient to describe a *P* wave derived from an *SV* wave, or an *SV* wave from a *P* wave, as a transformed wave. The nearest approach to a general rule is that the transmitted transformed wave is usually small as long as the velocity ratios of *P* and *S* waves in two media are not greatly different. If one of the media is a liquid, only *P* waves are transmitted. In this case a large *P* wave can be derived from an SV wave incident on a surface between a solid and a liquid.

For an incident SH wave at the interface of two solid media, we take

$$u_2 = \begin{cases} C_1 \exp[i\kappa(x_1 - x_3 \tan f - ct)] \\ + C_2 \exp[i\kappa(x_1 + x_3 \tan f - ct)], & \text{for} \quad x_3 > 0, \quad (3.72) \\ C' \exp[i\kappa(x_1 - x_3 \tan f' x_3 - ct)], & \text{for} \quad x_3 < 0. \quad (3.73) \end{cases}$$

The continuity conditions are

$$C_1 + C_2 = C' \tag{3.74}$$

and [noting that $\mu = \rho\beta^2$ from relation (3.25)]

$$\mu \tan f(C_1 - C_2) = \mu' \tan f' C'. \tag{3.75}$$

If f' is real, then C' is also real; and if f' is imaginary, then there is total reflection with a change of phase. Complete transmission is also possible $(C_2 = 0)$ when

$$\mu \tan f = \mu' \tan f'. \tag{3.76}$$

This condition is usually satisfied if $\beta < \beta'$ for some intermediate values of the angle f'.

In addition to the direct, reflected, and refracted waves observed during the propagation of plane waves in two semiinfinite media separated by a plane interface, an important case occurs when a plane wave is incident on a plane boundary at the critical angle e_c (for a P wave or f_c for an S wave), such that $e_c = \cos^{-1}(\alpha/\alpha')$, where $\alpha < \alpha'$. After critical reflection at the interface, the wave travels parallel to the interface in the medium with the higher velocity. When an impulsive source and a receiver are located in a lower-velocity medium at a distance sufficiently greater than either of their distances from the plane of contact with the higher-velocity medium, the first arrivals correspond to propagation along the path shown in Fig. 3.6. This is

Figure 3.6   Propagation of P head waves when the wave velocity $\alpha < \alpha'$.

the well-known refraction arrival, first observed by Mohorovičić in 1909 from the records of near earthquakes. It is also the basis of the seismic–refraction method of exploration. No energy, however, would be expected for such a path within the geometric optics approach. The description of these waves traveling along the boundary, called *diffraction waves, head waves,* or *conical waves,* requires appropriate full-wave diffraction theory which can be found, for example, in the book by Kennett (1983). Surface motions from head waves are often small because their amplitude strongly decreases with distance along the interface.

In addition to the P and S waves, a solid medium with a free surface can transmit two types of surface waves, which are confined to the vicinity of the surface and give little movement at greater depths. The surface waves are especially important at great distances, because their amplitude varies inversely to the square root of distance from the source, whereas the amplitude of body waves decreases proportionally to the distance. Thus the body waves are more prominent at short distances and the surface waves may be the larger at great distances. The theory for surface waves on the free surface of a semiinfinite elastic solid was given by Rayleigh in 1887, and these waves are named after him. Their velocity is smaller than β, and when $\alpha/\beta = \sqrt{3}$ (Poisson's relation $\lambda = \mu$ is acceptable), the velocity of Rayleigh waves is 0.92β. The waves are polarized, and the particles of the medium move in vertical planes parallel to the direction of the wave motion, describing an ellipse with the maximum displacement in the vertical direction. The Rayleigh waves are the result of the superposition of P waves approaching the surface and wholly reflected as S waves and of S waves wholly reflected as P waves.

Rayleigh waves arise not only in a homogeneous half-space but also when a layer or a layered structure is present. Then they acquire a new property, dependence of the wave velocity on the wavenumber (and thus on the wavelength and period), called *dispersion*, and the wave velocity c becomes the phase velocity. In a dispersive system the original disturbance is continually sorting itself out into groups of simple waves, where each group is associated with a particular wavelength, and propagating with the group velocity C. The group velocity for a given period is the velocity at which an envelope of a wave packet with similar periods is propagated. The peaks and troughs of the wave packet travel with the phase velocity c, which in general is different from the group velocity and $C = c + \kappa \, dc/d\kappa$, where κ is the wavenumber. The wave energy in the dispersed waves is also transported with the group velocity.

When solids in welded contact have the same α and β but different densities, a wave of Rayleigh type in both media can travel along the interface. The existence theorem for such waves was first proved by Stoneley in 1924, and the waves are called *Stoneley waves.*

Surface waves of the *SH* type are observed on the Earth's surface. An explanation of these waves was provided by Love in 1911, who considered a surface layer of finite thickness overlying a deep layer, with no slipping over the interface. He showed that the waves, called *Love waves*, consist of horizontally polarized shear waves trapped in a superficial layer and propagated by multiple total reflections. Thus Love waves can be regarded as *SH* waves continually reflected between the outer surface and the interface, and therefore they can travel horizontally without continual loss of energy downward. This kind of wave exists only if the velocity of *S* is greater in the lower layer, and the wave velocities of the Love waves lie between the two velocities of the *S* body waves in the upper and lower layers. The wave velocity is not constant and is dependent on the wavenumber; dispersion of a general waveform will appear. The existence of Love waves is not limited to two uniform layers only; they are generated when the velocity of *S* waves increases with depth.

The equations of motion in homogeneous perfectly elastic and isotropic media are given by relations (3.21). In a heterogeneous elastic and isotropic medium, the equations of motion of elastic waves are more complex and, when body forces are neglected, they are

$$\rho \frac{\partial^2 u_i}{\partial t^2} = \frac{\partial}{\partial x_i}(\lambda \theta) + \frac{\partial}{\partial x_j}\left\{\mu\left(\frac{\partial u_j}{\partial x_i} + \frac{\partial u_i}{\partial x_j}\right)\right\}, \qquad (3.77)$$

where u_i is the displacement vector. The compressional wave motion is no longer purely longitudinal, and the shear wave motion is no longer purely transverse. In vector notation, equation (3.77) becomes

$$\rho \frac{\partial^2 \mathbf{u}}{\partial t^2} = (\lambda + \mu)\nabla(\nabla \cdot \mathbf{u}) + \mu\nabla^2 \mathbf{u} + \nabla\lambda(\nabla \cdot \mathbf{u})$$
$$+ \nabla\mu \times (\nabla \times \mathbf{u}) + 2(\nabla\mu \cdot \nabla)\mathbf{u}. \qquad (3.78)$$

The solution of the elastic wave equations with the appropriate initial and boundary conditions is very important for various seismological studies. Explicit and unique solutions are rather rare. In a heterogeneous medium, where the velocity is a function of the spatial coordinates, the concept of rays is especially useful. One approach is to transform the wave equation to the eikonal equation and to find solutions in terms of wavefronts and rays that are valid at high frequencies. The waves propagate along rays as in geometric optics, each ray proceeding normally outward from the wavefront at any instant. Even when the properties of the medium vary from point to point, as in the case of the Earth, the concept of rays may still be used.

The equation of a wavefront, for example, in the form

$$t = \tau(x_i), \tag{3.79}$$

defines the position of the wavefront at a given time or the curve for a ray as travel time varies. A series solution of equations (3.78), called the *ray series*, is usually sought

$$\mathbf{u}(x_i, t) = \sum_{k=0}^{\infty} \mathbf{u}_k(x_i) F_k(t - \tau), \tag{3.80}$$

where F_k are usually complex functions and τ is the phase. For high-frequency monochromatic waves, a convenient form is

$$\mathbf{u}(x_i, t) = \exp\left[-i\omega(t - \tau(x_i))\right] \sum_{k=0}^{\infty} (-i\omega)^{-k-\gamma} \mathbf{u}_k(x_i), \tag{3.81}$$

where γ is a constant parameter. Equations for τ and $\mathbf{u}_k$ may be found by substitution of formula (3.81) into equations (3.78) and equating coefficients. For $k = 0$, three equations are obtained:

$$\rho \mathbf{u}_0 = (\lambda + \mu) \nabla\tau(\mathbf{u}_0 \cdot \nabla\tau) + \mu(\nabla\tau)^2 \mathbf{u}_0. \tag{3.82}$$

This system of equations has a nontrivial solution if its determinant vanishes; then

$$(\nabla\tau)^2 = \frac{\partial\tau}{\partial x_i} \frac{\partial\tau}{\partial x_i} = c^{-2}, \tag{3.83}$$

where c is the velocity of P and S waves. This first-order partial-differential equation is called the phase or eikonal equation, and is often used as a starting point for ray theory, providing one way of numerically computing the coordinates of points along a specified ray $x_i = x_i(\tau)$.

It is convenient to use slowness instead of velocity of seismic waves. The slowness vector $s_i = \partial\tau/\partial x_i$ is normal to the wavefront and tangent to the ray, and $s_i s_i = c^{-2}$. An advantage of using slowness rather than velocity to describe the speed and direction of propagation of a wave is that slownesses may be added vectorially, but not necessarily velocities. In Cartesian coordinates the slowness of a given wave is the vectorial sum of its components s_1, s_2, and s_3 along each coordinate direction, and the slowness in direction $\mathbf{n}$ is simply $\mathbf{s} \cdot \mathbf{n}$.

Details of the problem of ray tracing in seismology may be found in a number of books, as in the monograph by Červeny *et al.* (1977) or in the book by Lee and Stewart (1981).

Chapter 4 | Location of Seismic Events in Mines

Location of seismic events is the first step in studies of seismicity in mines. Accuracy is one of the most important factors in such studies, and the requirements in this respect are demanding. In mining practice, the expected accuracy for hypocenter locations are a few tens of meters, or even a few meters in some cases. The location error might be considered to consist of two components: random location scatter and systematic bias (e.g., Jordan and Sverdrup, 1981; Chang *et al.*, 1983; Pavlis, 1986). The first type is caused by errors in arrival-time measurements, and the second is generated by the differences between the rockmass structure at the source and receiver and the velocity model used in the location procedure. Under mining conditions, the error of the second type is time-dependent as a result of the stress migration.

Our discussion of various approaches to the location of seismic events in mines is based on one criterion only: the treatment of time residuals. For a given velocity model a time residual is defined as the difference between the observed and calculated arrival time of a seismic wave. According to such a criterion, we discuss two conceptually different approaches.

In the first approach, the time residuals are considered regardless of their causes. No attempt is made to distinguish between the time residuals caused by time reading errors and velocity model inaccuracy. Typical representatives of such an approach are the classic least-squares procedures attributed to Geiger (1912), which are described in Section 4.1, and the Bayesian procedure, formulated and described in Section 4.2.

In the second approach, the time residuals are split into two components: random ones related to the reading errors of arrival times, and travel-time residuals caused by insufficient knowledge of a velocity model. Several location procedures are in use, depending on the assumptions made with regard to the nature of velocity model uncertainty.

The first such procedure is Fedorov's (1974) extension of classic least-squares procedure to that with controllable variables (in our case, the velocity model parameters) subject to random errors. The physical assumption behind such an approach is that the velocity model is so complex that it behaves like a structure with random inhomogeneities. The idea is close to Chernov's description of wave propagation in random media (Chernov, 1960). This approach is possible if the fluctuations of velocity model parameters, from an average model, can be expressed in statistical terms. In mining practice, a knowledge of the rockmass seismic velocities and their variances is usually adequate for this technique. The method deserves special attention because of its physical clarity and efficiency. Another procedure employs the concept of relative location of seismic events and is based on arrival-time differences between a reference (master) event and nearby events. The travel-time residuals caused by a limited knowledge of the velocity model are not treated as random values; they have systematic character and are removed implicitly. The method is commonly used in mines where large production blasts are fired during mining excavations.

Finally, the last and the most difficult location technique is the procedure of simultaneous inversion of seismic velocity structure and location of a set of seismic events. In this case, from a conceptual point of view, the travel time residuals are treated in the same way as in the relative location procedure. A clear distinction between simultaneous inversion and velocity inversion studies should be made: in the latter case the hypocentral coordinates are not considered as unknown. This approach is called seismic active tomography and is discussed in Chapter 6.

4.1 Classic Approach and Its Computational Aspects

The classic Geiger (1912) location procedure modified for local seismic events can be formulated as follows (Buland, 1976; Lee and Steward, 1981). Given n observations of arrival times $t_1, \ldots, t_n$, find the origin time t_0 and the hypocenter, in Cartesian coordinates (x_0, y_0, z_0), such that the sum of squared time residuals r_i

$$\Phi(t_0, x_0, y_0, z_0) = \sum_{i=1}^{n} r_i^2 \qquad (4.1)$$

becomes minimum, where r_i is equal to

$$t_i - t_0 - T_i(x_0, y_0, z_0), \qquad (4.2)$$

defined as the difference between the observed t_i and the calculated $t_0 + T_i(x_0, y_0, z_0)$ arrival times. $T_i(x_0, y_0, z_0)$ is the calculated travel time from the hypocenter (x_0, y_0, z_0) to the ith station.

Introducing vector notation

$$\mathbf{t} = \begin{pmatrix} t_1 \\ \vdots \\ t_n \end{pmatrix}, \mathbf{T}(\mathbf{h}) = \begin{pmatrix} T_1(\mathbf{h}) \\ \vdots \\ T_n(\mathbf{h}) \end{pmatrix}, \mathbf{h} = \begin{pmatrix} x_0 \\ y_0 \\ z_0 \end{pmatrix}, \boldsymbol{\theta} = \begin{pmatrix} t_0 \\ x_0 \\ y_0 \\ z_0 \end{pmatrix}, \tag{4.3}$$

we can write the sum of squared time residuals (4.1) as

$$\Phi(\boldsymbol{\theta}) = \mathbf{r}^T \mathbf{r}, \tag{4.4}$$

where the matrix operator T stands for the transposition and $\mathbf{r}$ is the n-dimensional column vector of the time residuals equal to

$$\mathbf{t} - t_0 \mathbf{1} - \mathbf{T}(\mathbf{h}), \tag{4.5}$$

in which $\mathbf{1}$ denotes the n-dimensional column vector of ones. In general, the problem is overdetermined and $n \geq 4$. The Geiger location approach is an application of the Gauss–Newton method of minimization of the misfit function $\Phi(\boldsymbol{\theta})$.

Assuming that approximate values of the sought focal parameters are known and denoted by

$$\boldsymbol{\theta}^* = \begin{pmatrix} t_0^* \\ x_0^* \\ y_0^* \\ z_0^* \end{pmatrix}, \tag{4.6}$$

and applying the first-order Taylor expansion of travel times T_i, the observed arrival time t_i can be approximated by

$$t_i = t_0^* + \delta t_0 + T_i(\mathbf{h}^*) + \frac{\partial T_i}{\partial x_0}\delta x_0 + \frac{\partial T_i}{\partial y_0}\delta y_0 + \frac{\partial T_i}{\partial z_0}\delta z_0 \tag{4.7}$$

and rewritten as

$$r_i = \delta t_0 + \frac{\partial T_i}{\partial x_0}\delta x_0 + \frac{\partial T_i}{\partial y_0}\delta y_0 + \frac{\partial T_i}{\partial z_0}\delta z_0, \tag{4.8}$$

where $r_i = t_i - t_0^* - T_i(\mathbf{h}^*)$ and $i = 1, \ldots, n$. After adopting the matrix notation, the set of equations (4.8) can be expressed as

$$\mathbf{A}\delta\boldsymbol{\theta} = \mathbf{r}. \tag{4.9}$$

The minimization of the sum of squared time residuals (4.1) is equivalent to the minimization of $(\mathbf{A}\delta\boldsymbol{\theta} - \mathbf{r})^T(\mathbf{A}\delta\boldsymbol{\theta} - \mathbf{r})$, where the adjacent vector $\delta\boldsymbol{\theta}$ is equal to $(\delta t_0, \delta x_0, \delta y_0, \delta z_0)^T$, $\mathbf{r}$ is the column vector of time residuals r_i, and $\mathbf{A}$ is the $(n \times 4)$ matrix of partial derivatives equal to

$$\mathbf{A} = \begin{vmatrix} 1, & \partial T_1/\partial x_0, & \partial T_1/\partial y_0, & \partial T_1/\partial z_0 \\ \vdots & \vdots & \vdots & \vdots \\ 1, & \partial T_n/\partial x_0, & \partial T_n/\partial y_0, & \partial T_n/\partial z_0 \end{vmatrix}, \qquad (4.10)$$

in which the partial derivatives are evaluated at a starting point $\mathbf{h}^*$.

Our purpose is to adjust the trial vector $\boldsymbol{\theta}^*$ by $\delta\boldsymbol{\theta}$ in such a way that the sum of the squares of residuals (4.1) becomes minimal. It is known that the adjustment vector $\delta\boldsymbol{\theta}$, to be added to the trial vector $\boldsymbol{\theta}^*$, is the solution of the set of linear equations (e.g., Draper and Smith, 1981)

$$\mathbf{B}\delta\boldsymbol{\theta} = \mathbf{b} \qquad (4.11)$$

and is given by

$$\delta\boldsymbol{\theta} = \mathbf{B}^{-1}\mathbf{b}, \qquad (4.12)$$

where $\mathbf{B} = \mathbf{A}^T\mathbf{A}$ and $\mathbf{b} = \mathbf{A}^T\mathbf{r}$. Hence, the classic location problem involves the following steps, described for example by Lee and Stewart (1981):

(1) Guess (assume) a trial origin time t_0^* and a trial hypocenter $\mathbf{h}^* = (x_0^*, y_0^*, z_0^*)^T$.
(2) Compute time residuals r_i $(i = 1, \ldots, n)$ and derivatives (4.10) at point $\boldsymbol{\theta}^* = (t_0^*, x_0^*, y_0^*, z_0^*)^T$.
(3) Solve the system of four linear equations as given by (4.11) for the hypocenter parameters adjustments δt_0, δx_0, δy_0, and δz_0.
(4) Apply an adjustment vector to the origin time and the hypocenter using $\delta t_0 + t_0^*$, $\delta x_0 + x_0^*$, $\delta y_0 + y_0^*$, and $\delta z_0 + z_0^*$. These values become now the new trial of the origin time and hypocenter parameters.
(5) Repeat steps 2–4 until some termination criteria are met. At that point we set $\hat{t}_0 = t_0^*$, $\hat{x}_0 = x_0^*$, $\hat{y}_0 = y_0^*$, and $\hat{z}_0 = z_0^*$ as the solution for the origin time and the event hypocenter.

These calculations are performed under an assumption that the reliability of all observed arrival times t_i $(i = 1, \ldots, n)$ is the same. If some arrival times are considered more reliable than the others, the procedure must be modified.

An obvious measure of reliability is the standard deviation. Let σ_i be the standard deviation of time residuals at station i $(i = 1, \ldots, n)$. Assuming statistical independence between the residuals at different stations, the covariance matrix of time residuals is

$$\mathbf{C}_r = \operatorname{diag}\left(\sigma_{r_1}^2, \ldots, \sigma_{r_n}^2\right) \tag{4.13}$$

where diag stands for diagonal matrix. The misfit function $\Phi(\boldsymbol{\theta})$ is equal to the weighted sum of the squared time residuals and takes the form

$$\Phi(\boldsymbol{\theta}) = \mathbf{r}^{\mathrm{T}}\mathbf{C}_r^{-1}\mathbf{r}. \tag{4.14}$$

It is easy to show that the weighted system of normal equations (4.11) becomes

$$\left(\mathbf{A}^{\mathrm{T}}\mathbf{C}_r^{-1}\mathbf{A}\right)\delta\boldsymbol{\theta} = \mathbf{A}^{\mathrm{T}}\mathbf{C}_r^{-1}\mathbf{b}. \tag{4.15}$$

The set of equations (4.15) can be written as a least squares problem without weighting by letting

$$\tilde{\mathbf{A}} = \mathbf{C}_r^{-1/2}\mathbf{A}, \qquad \tilde{\mathbf{b}} = \mathbf{C}_r^{-1/2}\mathbf{b}, \tag{4.16}$$

where $\mathbf{C}^{1/2}$ is an $(n \times n)$ diagonal matrix with elements that are the square roots of the corresponding diagonal elements of $\mathbf{C}_r$. Then, set (4.15) becomes

$$\left(\tilde{\mathbf{A}}^{\mathrm{T}}\tilde{\mathbf{A}}\right)\delta\boldsymbol{\theta} = \tilde{\mathbf{A}}^{\mathrm{T}}\tilde{\mathbf{b}} \tag{4.17}$$

and has exactly the same form as the respective relations without weighting.

The described iterative procedure can occasionally be problematic for two common reasons: a poor choice of the initial guess hypocenter parameters $\boldsymbol{\theta}^*$ and singularity or near-singularity of the matrix $\mathbf{A}^{\mathrm{T}}\mathbf{A}$.

Since the travel times are usually nonlinear in relation to the hypocenter parameters $\mathbf{h}$, more than one minimum of the misfit function $\phi(\boldsymbol{\theta})$ may exist. Consequently, if the initial guess is poor, the search may not terminate at the global (the lowest) minimum of Φ but at some other minimum. Figure 4.1 shows, in two dimensions, what might happen with a poor choice of the starting focal parameters vector $\boldsymbol{\theta}^*$. In the mining practice, the approximate values of the hypocenter parameters $(x_0^* y_0^* z_0^*)$ are often known from the observations of macroseismic effects. Some computer algorithms routinely used in mines calculate the guess hypocenter parameters using linear location procedures (discussed in Section 4.6), which require neither initial guess of $\boldsymbol{\theta}$ nor iterations. An alternative approach is to try several starting points $\boldsymbol{\theta}^*$ in the feasible range and to ascertain whether they all provide the same value for the minimum of Φ (Himmelblau, 1972).

The least-squares solution of equation (4.9) can always be found provided that the matrix $\mathbf{A}^{\mathrm{T}}\mathbf{A}$ is nonsingular. As $\mathbf{A}^{\mathrm{T}}\mathbf{A}$ becomes near-singular, the corrections $\delta\boldsymbol{\theta}$ oscillate and become very large, leading to instability and divergence of the iteration procedure. The source of difficulty is easily seen when a fundamental decomposition theorem (Lanczos, 1961) is applied to the

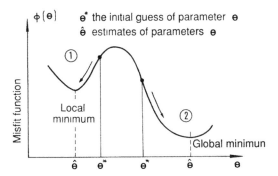

Figure 4.1 Influence of the initial guess value of θ^* on the final solution: (1) the initial value θ^* is too far from the global minimum and the iterative procedure may converge to a local minimum, (2) the procedure converges to the global minimum.

matrix **A**. In terms of generalized inverse, the $(n \times m)$ matrix **A** can be decomposed into

$$\mathbf{A} = \mathbf{U}\Lambda\mathbf{V}^\mathrm{T}, \tag{4.18}$$

where the $(n \times n)$ matrix **U** consists of orthogonalized eigenvectors of the product $\mathbf{AA}^\mathrm{T}$, the $(m \times m)$ matrix **V** consists of the orthonormalized eigenvectors of the product $\mathbf{A}^\mathrm{T}\mathbf{A}$, and Λ is a $(m \times m)$ diagonal matrix whose diagonal elements λ_i, known as singular values, are equal to non-negative square roots of the eigenvalues of the product $\mathbf{A}^\mathrm{T}\mathbf{A}$. For the problems considered here, the matrix **A** is of full rank and there are no zero singular values. In the discussed case, $m = 4$ and is equal to the number of unknown parameters. The factoring of matrix **A** into **V**, **U** and Λ is known as the *singular value decomposition* (SVD) technique (e.g., Aki and Richards, 1980).

In terms of generalized inverse, relation (4.12) takes the form

$$\delta\theta = \mathbf{V}\Lambda^{-1}\mathbf{U}^\mathrm{T}\mathbf{r}. \tag{4.19}$$

The decisive role in solving our problem belongs to the matrix Λ. From an examination of the eigenvalues λ_i, one can tell at a glance whether a reliable estimate of $\delta\theta$ can be obtained. Since

$$\Lambda^{-1} = \mathrm{diag}(1/\lambda_1, \ldots, 1/\lambda_m), \tag{4.20}$$

any very small values of λ (or very large values of $1/\lambda$) will have a dominant effect on the $\delta\theta$ evaluation. This could be of no consequence, unless it is admitted that a small λ is likely to be predominantly composed, for example, of roundoff errors (Hatton *et al.*, 1986). It can be demonstrated (e.g., Buland, 1976) that the condition number $\mathscr{K}(\mathbf{B})$ of the matrix **B** provides an estimate for the relative error in the solution. The higher the number, the lower the probability that a solution of the system (4.11) will lead to satisfactory results. Since in our case $\mathbf{B} = \mathbf{A}^{\mathrm{T}}\mathbf{A}$, the condition number of the matrix $\mathbf{B}$ is

$$\mathscr{K}(\mathbf{B}) = \frac{\lambda_{\max}^2}{\lambda_{\min}^2}, \qquad (4.21)$$

where $\lambda_{\max}$ and $\lambda_{\min}$ are the largest and smallest values of λ_i.

Several studies (Buland, 1976; Niewiadomski, 1989) indicate that even for a simple half-space model of constant velocity and for a dozen or so seismic stations, the condition number $\mathscr{K}(\mathbf{A}^{\mathrm{T}}\mathbf{A})$ is large near the array and increases rapidly away from the array, both laterally and vertically. Buland (1976) pointed out that the use of more than one seismic phase at some stations drastically reduces the condition number $\mathscr{K}(\mathbf{A}^{\mathrm{T}}\mathbf{A})$. It is also known that the condition number strongly depends on the space configuration of seismic stations. The more unsatisfactory the configuration, the higher the condition number. The problem of the optimum distribution of seismic stations will be discussed in Chapter 5. Here the most important procedures leading to the reduction of the condition number are considered.

4.1.1 Centering and Scaling

Centering and scaling techniques lead to the simplification of inverse problems, as well as to the improvement of numerical stability. The centering of the location procedure is considered here following Lienert *et al.* (1986).

From relation (4.7) it follows that the mean value of the arrival time is

$$\langle t \rangle = t_0^* + \delta t_0 + \langle T \rangle + \left\langle \frac{\partial T_i}{\partial x_0} \right\rangle \delta x_0 + \left\langle \frac{\partial T_i}{\partial y_0} \right\rangle \delta y_0 + \left\langle \frac{\partial T_i}{\partial z_0} \right\rangle \delta z_0, \quad (4.22)$$

where the brackets $\langle \cdot \rangle$ represent the mean value of all the observations. Substituting equation (4.22) into (4.7), we obtain the set of n centered linear equations

$$r_{ci} = \left(\frac{\partial T_i}{\partial x_0} \right)_c \delta x_0 + \left(\frac{\partial T_i}{\partial y_0} \right)_c \delta y_0 + \left(\frac{\partial T_i}{\partial z_0} \right)_c \delta z_0, \qquad (4.23)$$

where

$$r_{ci} = t_i - \langle t \rangle - (T_i - \langle T \rangle), \qquad (4.24)$$

and

$$\left(\frac{\partial T_l}{\partial x_0}\right)_c = \frac{\partial T_l}{\partial x_0} - \left\langle\frac{\partial T_l}{\partial x_0}\right\rangle,$$

$$\left(\frac{\partial T_l}{\partial y_0}\right)_c = \frac{\partial T_l}{\partial y_0} - \left\langle\frac{\partial T_l}{\partial y_0}\right\rangle, \qquad (4.25)$$

$$\left(\frac{\partial T_l}{\partial z_0}\right)_c = \frac{\partial T_l}{\partial z_0} - \left\langle\frac{\partial T_l}{\partial z_0}\right\rangle.$$

Introducing the centered ($n \times 3$) dimensional matrix $\mathbf{A}_c$ equal to

$$\mathbf{A}_c = \begin{vmatrix} (\partial T_1/\partial x_0)_c, & (\partial T_1/\partial y_0)_c, & (\partial T_1/\partial z_0)_c \\ \vdots & \vdots & \vdots \\ (\partial T_n/\partial x_0)_c, & (\partial T_n/\partial y_0)_c, & (\partial T_n/\partial z_0)_c \end{vmatrix}, \qquad (4.26)$$

we can write the set of equations (4.23) as

$$\mathbf{A}_c \delta\boldsymbol{\theta}_c = \mathbf{r}_c. \qquad (4.27)$$

The least-squares solution of (4.27) is

$$\delta\boldsymbol{\theta}_c = \left(\mathbf{A}_c^T \mathbf{A}_c\right)^{-1} \mathbf{A}_c^T \mathbf{r}_c, \qquad (4.28)$$

where $\mathbf{d}\boldsymbol{\theta}_c = \delta\mathbf{h} = (\delta x_0, \delta y_0, \delta z_0)^T$. It should be noted that the centering procedure reduces the location problem to a simultaneous search for the three hypocenter parameters only: x_0, y_0, and z_0. The origin time is calculated separately as

$$\hat{t}_0 = \langle t \rangle - \langle T \rangle. \qquad (4.29)$$

The centering of location procedure is used at Canadian potash mines (Prugger and Gendzwill, 1988) and at South African gold mines (Mendecki, 1990).

In addition to the centering, the reduction of the condition number $\mathcal{K}(\mathbf{A}^T\mathbf{A})$ can be obtained by the so-called scaling technique. The scaling, like centering, is a standard technique used in statistical applications (Draper and Smith, 1981). In general, the requirement for the scaling procedure appears when the sensitivity of the misfit function (4.1) to the value of one of the sought parameters is of an order of magnitude different from that to the other parameters. In the case of the location of mine tremors from practically planar underground networks with the hypocenter situated at the level of the network, the derivatives $\partial T/\partial z_0$ in comparison to $\partial T/\partial x_0$ and $\partial T/\partial y_0$ are very small. This means that the standard location procedure can be significantly improved by an implementation of the scaling. It is useful, therefore, to introduce the scaling matrix $\mathbf{S}$, where the new (scaled) parameters $\delta\boldsymbol{\theta}_\searrow$ are

$$\delta\boldsymbol{\theta}_\searrow = \mathbf{S}\delta\boldsymbol{\theta}, \qquad (4.30)$$

and, accordingly, the set of conditional equations (4.9) is given by

$$\mathbf{r} = \mathbf{A}\delta\boldsymbol{\theta} = \mathbf{A}\mathbf{S}^{-1}\delta\boldsymbol{\theta}_\searrow = \mathbf{A}_\searrow\delta\boldsymbol{\theta}_\searrow. \qquad (4.31)$$

The choice of the scaling matrix $\mathbf{S}$ is nonunique and can be performed in different ways. The calculations are improved by weighting the parameters so that they are all of a similar order. The $\mathbf{S}$ matrix therefore can be built, for example, as (Lienert *et al.*, 1986)

$$\mathbf{S} = \mathrm{diag}\left\{ n^{-1}, \left(\sum_{i=1}^{n} A_{i2}^2 \right)^{-1}, \left(\sum_{i=1}^{n} A_{i3}^2 \right)^{-1}, \left(\sum_{i=1}^{n} A_{i4}^2 \right)^{-1} \right\}. \qquad (4.32)$$

Niewiadomski (1989) proposed a more advanced approach in which the scaling matrix $\mathbf{S}$ depends on the certain scaling factor s. Then, the best scaling is equivalent to the choice of a value of s that would minimize the condition number $\mathscr{K}(\mathbf{A}^T\mathbf{A})$. Having the least-squares solution of equations (4.31) for $\delta\boldsymbol{\theta}_\searrow$, the value of $\delta\boldsymbol{\theta}$ is obtained from

$$\delta\boldsymbol{\theta} = \mathbf{S}^{-1}\delta\boldsymbol{\theta}_\searrow. \qquad (4.33)$$

The scaling technique was used by Lienert *et al.* (1986) and is applied in all the location programs in the ISS packet (Mendecki, 1990), used in South African gold mines. The application of these two techniques in location algorithms used at Polish coal mines shows that the centering reduces the condition number on average by a factor of 10, whereas the proper scaling improves the condition by a factor of 100.

4.1.2   Modification of Initial Equations

The essence of stabilizing the location procedure can be stated as follows (Crosson, 1976). We seek the solution procedure that results in suppression of those variable components that are associated with insufficient information, that is, with small or near-zero eigenvalues of the matrix $\mathbf{A}$.

In practice, the stabilization of the location procedure is realized by the following modification of normal equations (4.11):

$$(\mathbf{A}^T\mathbf{A} + \mathbf{E})\delta\boldsymbol{\theta} = \mathbf{A}^T\mathbf{r}. \qquad (4.34)$$

With regard to the introduced assumptions and available information, the elements of the matrix **E** can take three different forms.

(1) If the damping least-squares technique is applied (Marquardt, 1963), then

$$\mathbf{E} = \alpha^2 \mathbf{I}, \tag{4.35}$$

where α is a stabilizing (damping) coefficient to be adjusted to the requirements of a problem and **I** is a unit matrix. In terms of generalized inverse scheme, the solution of the modified normal equations (4.34) takes the form

$$\delta\boldsymbol{\theta} = \mathbf{V}(\boldsymbol{\Lambda}^2 + \alpha^2 \mathbf{I})^{-1}\boldsymbol{\Lambda}\mathbf{U}^{\mathrm{T}}\mathbf{r}. \tag{4.36}$$

Let us compare the condition numbers of the matrix $(\mathbf{A}^{\mathrm{T}}\mathbf{A} + \alpha^2 \mathbf{I})$ with those of the original matrix $\mathbf{A}^{\mathrm{T}}\mathbf{A}$. It is not difficult to show that the new condition number is

$$\mathcal{H}(\mathbf{A}^{\mathrm{T}}\mathbf{A} + \alpha^2 \mathbf{I}) = \frac{\lambda_{\max}^2 + \alpha^2}{\lambda_{\min}^2 + \alpha^2}. \tag{4.37}$$

The result is that for any stabilizing factor $\alpha \neq 0$, the condition number of modified equations is always less than the original one: $\mathcal{H}(\mathbf{A}^{\mathrm{T}}\mathbf{A} + \alpha^2 \mathbf{I}) < \mathcal{H}(\mathbf{A}^{\mathrm{T}}\mathbf{A})$. The inclusion of a damping factor ensures that no eigenvalue is less than α, and hence the existence of a least-squares solution is assured. The choice of the value of α is important in controlling the resolution and variance of unknown parameters. By using the damping least-squares procedure, we may obtain a numerically stable solution, but this does not necessarily mean that the solution has a physical meaning. These and other aspects of the damping least-squares procedure have been discussed, for example, by Crosson (1976) and Herrmann (1979). Aki and Lee (1976) introduced an adaptive damping technique in which the parameter α is analyzed and possibly modified after each iteration.

(2) If the second-order partial derivatives of travel times with respect to the focal parameters are taken into account, then the matrix **E** takes the form (Thurber, 1985)

$$\{\mathbf{E}\}_{ij} = \sum_{k=1}^{n} \frac{\partial^2 T_k}{\partial\theta_i \partial\theta_j} r_k, \tag{4.38}$$

where T_k and r_k are the travel time and the corresponding time residual at the kth station respectively, $\theta_1 = t_0$, $\theta_2 = x_0$, $\theta_3 = y_0$, and $\theta_4 = z_0$. It is clear that in formula (4.38), for all i or j equal to one, corresponding to the origin time t_0, the elements $\{\mathbf{E}\}_{ij}$ are equal to zero.

(3) If some information about the hypocenter coordinates is available from other sources than the arrival times t_i $(i = 1, \ldots, n)$, then from the Bayesian estimation technique (Tarantola, 1987) it follows that the matrix $\mathbf{E}$ can be expressed as

$$\mathbf{E} = \mathbf{C}_h^{-1}, \tag{4.39}$$

where the covariance matrix $\mathbf{C}_h$ characterizes the reliability of the a priori information on the hypocenter location. Often in practice

$$\mathbf{C}_h = \mathrm{diag}\left(\sigma_{x_0}^2, \sigma_{y_0}^2, \sigma_{z_0}^2\right), \tag{4.40}$$

where σ_{x_0}, σ_{y_0}, and σ_{z_0} are the standard deviations of the a priori information relative to each hypocenter coordinate. Since the Bayesian technique is routinely used for the location of seismic events in some mines, all aspects of such an approach are discussed in detail in Section 4.2.

4.1.3   Singular Value Decomposition of Matrix A

Significant reduction of the condition number can be achieved by the application of the singular value decomposition (SVD) technique. Instead of solving the set of normal equations (4.11) with the condition number equal to $\lambda_{max}^2/\lambda_{min}^2$, the application of SVD makes it possible to solve the set of equations (4.9) with the condition number

$$\mathscr{K}(\mathbf{A}) = \frac{\lambda_{max}}{\lambda_{min}}. \tag{4.41}$$

Such an approach can be applied because the SVD decompositions exhibit all the properties of least-squares estimation. The SVD technique requires a modest increase in storage and execution time. Numerical algorithm of SVD has been devised by Golub and Reinsch (1971), and now the SVD computer subroutine is available in most numerical libraries, such as Numerical Recipes (Press *et al.*, 1989).

4.2   Bayesian Approach

Even a few days practice at any seismic network in mines is sufficient for an observer to be able to find from a glance at seismic records that the tremor occurred in a particular area of the mine. It is also well known that seismic activity usually concentrates in the vicinity of mining works. The areas of the most probable location of seismic events in mines, therefore, are generally known. This a priori information is not used in classic algorithms: the only input data are the arrival times of seismic waves recorded by a given network. In this section an algorithm based on the Bayesian estimation theory is formulated for the location of seismic events in mines. The algorithm permits the combination of a priori information with that contained in the arrival times of seismic waves, and the application of the algorithm in mining practice is described.

Strict formulation and solution of the Bayesian location procedure was presented by Tarantola and Valette (1982). The most elegant solution of the problem was given by Matsu'ura (1984) and Jackson and Matsu'ura (1985). The problem that we have to solve is as follows. The sought parameters θ are evaluated from two types of information: the arrival times $\mathbf{t} = (t_1, \ldots, t_n)^T$ of P waves and a priori information about the location of event. Let us assume that the time residuals have random Gaussian character with the mean value of zero and the covariance matrix $\mathbf{C}_r$. Then, the likelihood function of the unknown parameters θ becomes (e.g., Menke, 1989)

$$L(\theta|\mathbf{t}) = \text{const} \exp\left[-\tfrac{1}{2}\Phi(\theta)\right], \qquad (4.42)$$

where, according to (4.14), the misfit function $\Phi(\theta)$ is $\mathbf{r}^T\mathbf{C}_r^{-1}\mathbf{r}$. If no additional information is available, such values of θ that maximize the likelihood (4.42), or alternatively minimize the sum of squared residuals $\Phi(\theta)$, are taken as an evaluation of the unknown parameters.

Let us assume additionally that we are able to indicate the area where the tremor occurred. The a priori information may be written in the following form (Matsu'ura, 1984)

$$\mathbf{h}^P = \mathbf{h} + \Delta\mathbf{h}, \qquad (4.43)$$

where $\mathbf{h}^P$ are the prior estimates of the hypocenter coordinates, which are subject to the unknown errors $\Delta\mathbf{h}$. Assuming Gaussian nature of the errors $\Delta\mathbf{h}$, with the zero mean value and the covariance matrix $\mathbf{C}_h$, the joint probability density function of the hypocenter coordinates prior to the observed arrival times is given by

$$p(\mathbf{h}) = \text{const} \exp\left[-\tfrac{1}{2}(\mathbf{h}^P - \mathbf{h})^T\mathbf{C}_h^{-1}(\mathbf{h}^P - \mathbf{h})\right]. \qquad (4.44)$$

Relations (4.42) and (4.44) present two different sources of available information about the location of the event. To consider both of them, we apply the Bayesian theorem (e.g., Tarantola, 1987). Then the joint probability density function of the event hypocenter parameters is proportional to the likelihood function $L(\mathbf{t}|\theta)$ multiplied by the a priori probability density function $p(\mathbf{h})$:

$$p(\theta|\mathbf{t}) = \text{const}\, L(\mathbf{t}|\theta)p(\mathbf{h}). \qquad (4.45)$$

The parameter const is the normalizing factor, which assures that the integral $p(\theta|t)$ over the whole domain of the vector θ is equal to one.

According to the Bayesian estimation theory, the best estimates of the parameters $\hat{\theta}$ are those that maximize relation (4.45). It is well known that for a seismic event with depth at the level of seismic network or/and for an epicenter outside the array it is difficult to determine independently the two parameters origin time t_0 and focal depth z_0. To eliminate such a problem, Tarantola and Valette (1982) and Matsu'ura (1984) propose an elimination of the origin time by the determination of the marginal distribution:

$$p(\mathbf{h}|\mathbf{t}) = \int_{-\infty}^{+\infty} p(\theta|\mathbf{t}) \, dt_0. \tag{4.46}$$

Matsu'ura (1984) showed that the maximization of (4.46) is equivalent to an iterative search of the adjustment vector $\delta\mathbf{h}$

$$\delta\mathbf{h} = \mathbf{B}^{-1}\mathbf{b}, \tag{4.47}$$

where $\mathbf{B} = (\mathbf{A}^T\mathbf{V}_r\mathbf{A} + \mathbf{V}_h)$, $\mathbf{b} = \mathbf{A}^T\mathbf{V}_r[\mathbf{t} - \mathbf{T}(\mathbf{h})] + \mathbf{V}_h(\mathbf{h}^p - \mathbf{h})$; $\mathbf{h}$ denotes the current location of the hypocenter coordinates, $\mathbf{A}$ is an $(n \times 3)$ matrix of the form

$$\mathbf{A} = \begin{vmatrix} \partial T_1/\partial x_0, & \partial T_1/\partial y_0, & \partial T_1/\partial z_0 \\ \vdots & \vdots & \vdots \\ \partial T_n/\partial x_0, & \partial T_n/\partial y_0, & \partial T_n/\partial z_0 \end{vmatrix}, \tag{4.48}$$

$\mathbf{V}_r = \mathbf{C}_r^{-1}(\mathbf{I} - \mathbf{1}\mathbf{1}^T\mathbf{C}_r^{-1}/a)$, $\mathbf{I}$ is the identity matrix of order n, $\mathbf{1}$ denotes an n-dimensional column vector of ones, $\mathbf{V}_h = \mathbf{C}_h^{-1}$, and the constant $a = \mathbf{1}^T\mathbf{C}_r^{-1}\mathbf{1}$. The origin time t_0, corresponding to the hypocenter location $\hat{\mathbf{h}}$, can be determined from the expression

$$\hat{t}_0 = \mathbf{1}^T\mathbf{C}_r^{-1}\left[\mathbf{t} - \mathbf{T}(\hat{\mathbf{h}})\right]/a, \tag{4.49}$$

and the covariance matrix of the vector $\hat{\mathbf{h}}$ becomes

$$\left(\mathbf{A}^T\mathbf{V}_r\mathbf{A} + \mathbf{V}_h\right)^{-1}. \tag{4.50}$$

The diagonal elements of matrix (4.50) are asymptotic evaluations of the variances of the vector $\hat{\mathbf{h}}$ components, that is, the coordinates of the hypocenter.

To ensure the maximum clarity, we did not show that the elements of the matrix $\mathbf{A}$, and in some cases the elements of $\mathbf{V}_r$, are functions of the current values of the sought hypocenter coordinates $\mathbf{h}$. The dependence of the matrix $\mathbf{V}_r$ on $\mathbf{h}$ depends on the manner by which the covariance matrix $\mathbf{C}_r$ is defined. If the elements of $\mathbf{C}_r$ are function of location of the hypocenter [as it is assumed, for example, by Uhrhammer (1982)], then the matrix V_r is also updated after each iteration. If the matrix C_r, on the other hand, is diagonal with elements depending on time residuals (as it is the case in most location programs used in mines), it can be readily shown that the complex form of the matrix V_r is reduced to (Kijko, 1988)

$$\{V_r\}_{ij} = \begin{cases} w_i(1 - w_i/a), & i = j, \\ -w_i w_j/a, & i \neq j, \end{cases} i, j = 1, \ldots, n, \tag{4.51}$$

and

$$\hat{t}_0 = \sum_{i=1}^{n} w_i \left[t_i - T_i(\hat{h}) \right] / a, \tag{4.52}$$

where $a = \sum_{i=1}^{n} w_i$ and each w_i is the reciprocal of its corresponding diagonal element from the matrix C_r. It is interesting to note that although the covariance matrix C_r is diagonal, its counterpart V_r^{-1} is not.

To demonstrate the effectiveness of the Bayesian approach and its superiority over the conventional location procedure, we show the results of its application in three different situations typical for mining conditions, numerically simulated. We assume that in a given mine with plan area 2×4 km there are eight evenly distributed seismic stations. However, their vertical distribution is highly limited. We also assume that event occurred practically in the center of the seismic network at the mining level. The initial position of the hypocenter h^* is placed at a distance of 500 m to the north and 500 m to the east from the true epicenter and at the depth only 20 m above the true depth of the event.

In the first experiment the event hypocenter was located without any a priori information. Figures 4.2A–C illustrate, respectively, the process of

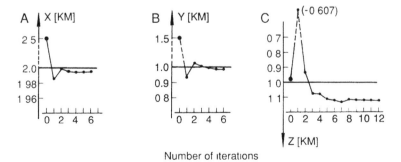

Number of iterations

Figure 4.2 Illustration of the hypocenter location without a priori information Correct hypocenter position ($x_0 = 2.0$ km, $y_0 = 1.0$ km, and $z_0 = 1.0$ km) is marked by thick lines. The epicentral coordinates (A, B) are determined with insignificant errors, whereas the depth (C) is found 120 m below the true position.

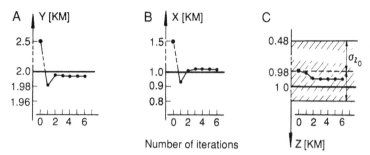

Figure 4.3 Illustration of the hypocenter location with a priori information about the focal depth. The true hypocenter position is marked by thick lines. The epicentral coordinates are determined correctly (A, B). The error in depth is reduced from 120 to 10 m (C), where the dashed area corresponds to the standard deviation of the prior information about the depth coordinate equal to $\sigma_{z_0} = 0.5$ km.

iterative estimation of the parameters x_0, y_0, and z_0. As a result of highly favorable situation (the epicenter inside the array), the event epicenter was located just after a few iterations. The focal depth, on the other hand, was estimated rather poorly and after 12 iterations. From this numerical experiment it follows that if the event occurs at the level of a planar seismic network, the conventional methods of location do not ensure precise depth determination, irrespective of the fact that the nearest station is relatively close to the hypocenter.

The second experiment was made to answer the question whether even approximate information on the focal depth can improve its determination. Similarly to the previous experiment, the iterations start from a point placed 20 m above the simulated event. Furthermore, it was accepted that the average depth ranges within a ± 500-m interval. We also assume that there is no a priori information about the event epicenter. Figure 4.3 illustrates the location procedure in such a case. After five iterations the solution became stable, and the focal depth was determined with an error of about 10 m. This experiment is a good illustration of the efficiency of the Bayesian algorithm. Even relatively uncertain information leads to a considerable improvement in the hypocenter location.

The purpose of the third experiment was to study the effect of wrong a priori information. Using additional information, we assume that the most probable position of the epicenter is 500 m to the north and 500 m to the east from the simulated true location. The standard deviations of the a priori epicenter coordinates are 450 m. We also accept the same a priori information about the focal depth as in the previous experiment. The results of the event location under such conditions are shown in Fig. 4.4. Wrong a priori

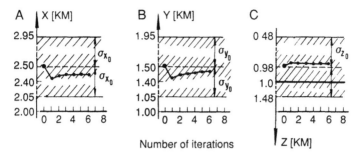

Figure 4.4 Illustration of the process of the event hypocenter determination with wrong a priori information about the epicenter. It was assumed that the standard deviations of the prior epicentral coordinates marked by dashed areas (A, B) do not exceed 450 m. The starting point of the iterative search for the epicenter, understood also as the prior location of the event epicenter, was set at 500 m to the east and 500 m to the north from the actual focus. From the calculations it follows that the overoptimistic a priori information (with errors $\sigma_{x_0} = \sigma_{y_0} = 450$ m) leads to a completely wrong location, although the error in depth is not dramatic (C).

information holds the epicenter close to the point to which too much importance was attributed.

The location procedure described above is used in about 30 coal and copper mines in Poland. Some experiments with Bayes-based procedures were also performed in gold mines in South Africa, and at the Men-Tou-Gou coal mine near Beijing in China.

4.3 Fedorov's Generalization of Least-Squares Procedure: Location with Approximate Velocity Models

In order to introduce the location procedure based on the Fedorov's generalization of classic least-squares algorithm, let us recall our original assumptions and then slightly change the notation. In general, the arrival times of seismic waves are described by the formula

$$t_i = t_0 + T(\mathbf{h}, \mathbf{m}_i) + \varepsilon_i, \qquad (4.53)$$

where $T(\mathbf{h}, \mathbf{m}_i)$ is the travel time from the hypocenter $\mathbf{h} = (x_0, y_0, z_0)^{\mathrm{T}}$ to the station i, $\mathbf{m}_i$ is column vector of velocity model parameters, ε_i is the measurement error of arrival times, $i = 1, \ldots, n$, and n is the number of seismic stations. The velocity model parameters $\mathbf{m}$ are never known exactly. We assume that the actual values of the velocity model parameters are random variables. The values of these parameters oscillate around the known mean values of the velocity model $\langle \mathbf{m} \rangle$ and can be expressed as

$$\mathbf{m}_i = \langle \mathbf{m}_i \rangle + \delta \mathbf{m}_i, \tag{4.54}$$

where $\delta \mathbf{m}_i$ are the errors reflecting the deviation of a specific velocity model from the average one. We assume that these errors are of random Gaussian character with the mean value equal to zero and a covariance matrix $\mathbf{C}_m$. Assuming further that the errors of the velocity model parameters are mutually independent, the matrix $\mathbf{C}_m$ becomes diagonal with the elements

$$\mathbf{C}_m = \mathrm{diag}(\sigma_{m1}^2, \ldots, \sigma_{mk}^2), \tag{4.55}$$

where σ_{mi} are the known standard deviations of the velocity model parameters m_i, $i = 1, \ldots, k$, and k is the number of parameters describing the velocity model. We accept that the time reading errors ε_i are Gaussian random values with their mean equal to zero and standard deviations σ_{ti} for $i = 1, \ldots, n$. Assuming again mutual independence of time reading errors at different stations, the covariance matrix $\mathbf{C}_t$ of time reading errors becomes diagonal with the elements σ_{ti}^2.

It can be shown that the application of Fedorov's theory (Fedorov, 1974) and the introduction of the errors of velocity model leads to a specific disturbance of the travel times T (Kijko, 1975; 1977b). The mean value of such an anomaly is

$$\Delta T \equiv \Delta T(\mathbf{h}, \langle \mathbf{m} \rangle) = \frac{1}{2} \mathrm{Sp} \left\{ \mathbf{C}_m \frac{\partial^2 T(\mathbf{h}, \mathbf{m})}{\partial \mathbf{m} \, \partial \mathbf{m}^T} \Big|_{\mathbf{m} = \langle \mathbf{m} \rangle} \right\} \tag{4.56}$$

and its variance is

$$\sigma_T^2 \equiv \sigma_T^2(\mathbf{h}, \langle \mathbf{m} \rangle)$$
$$= \mathrm{Sp} \left\{ \mathbf{C}_m \left[\frac{\partial T(\mathbf{h}, \mathbf{m})}{\partial \mathbf{m}} \frac{\partial T(\mathbf{h}, \mathbf{m})}{\partial \mathbf{m}^T} + \frac{\sigma_t^2(\mathbf{h}, \mathbf{m})}{\partial \mathbf{m} \, \partial \mathbf{m}^T} \right] \Big|_{\mathbf{m} = \langle \mathbf{m} \rangle} \right\}. \tag{4.57}$$

These relations are of a general nature. If the travel times are described by the formula $T(\mathbf{h}, \mathbf{m}) = d / V_P$, relations (4.56) and (4.57) take the simple form

$$\Delta T = \frac{1}{2} \sigma_v^2 \frac{\partial^2}{\partial^T V_P^2} \left(\frac{d}{V_P} \right) \Big|_{V_P = \langle V_P \rangle} = \frac{d}{\langle V_P \rangle} q^2 = Tq^2, \tag{4.58}$$

and

$$\sigma_T^2 = \sigma_v^2 \left[\frac{\partial}{\partial V_P} \left(\frac{d}{V_P} \right) \right]^2 \Big|_{V_P = \langle V_P \rangle} = \left(\frac{d}{\langle V_P \rangle} \right)^2 q^2 = T^2 q^2. \tag{4.59}$$

where $q = \sigma_v / \langle V_P \rangle$, in which d is the hypocentral distance and V_P is the P-wave velocity. From relations (4.56) and (4.57) it follows that the misfit function is of the form $\Phi(\theta) = \mathbf{r}^T \mathbf{C}_r^{-1} \mathbf{r}$, in which

$$\mathbf{r} = \mathbf{t} - t_0 \mathbf{1} - \mathbf{T}(\mathbf{h}, \langle \mathbf{m} \rangle) - \Delta \mathbf{T}(\mathbf{h}, \langle \mathbf{m} \rangle) \tag{4.60}$$

and

$$\mathbf{C}_r = \mathbf{C}_t + \mathbf{C}_T, \tag{4.61}$$

where $\mathbf{C}_t$ is the diagonal matrix with variances of the arrival-time reading errors $\sigma_{t_i}^2$, and $\mathbf{C}_T$ is the diagonal matrix with diagonal elements equal to the travel-time variances $\sigma_{T_i}^2$ defined by relation (4.57), and $\Delta \mathbf{T}$ is the n-dimensional column vector of mean values of travel times anomalies with the elements defined by relation (4.56).

The minimization of the misfit function $\Phi(\theta) = \mathbf{r}^T \mathbf{C}_r^{-1} \mathbf{r}$ is equivalent to a successive solution of the set of normal equations (4.11) in which $\mathbf{B} = \mathbf{A}^T \mathbf{C}_r^{-1} \mathbf{A}$ and $\mathbf{b} = \mathbf{A}^T \mathbf{C}_r^{-1} \mathbf{r}$. Equations (4.60) and (4.61) define the vector $\mathbf{r}$ of time residuals and its covariance matrix $\mathbf{C}_r$, respectively. The matrix $\mathbf{A}$ is

$$\mathbf{A} = \begin{vmatrix} 1, & \partial(T_1 + \Delta T_1)/\partial x_0, & \partial(T_1 + \Delta T_1)/\partial y_0, & \partial(T_1 + \Delta T_1)/\partial z_0 \\ \vdots & \vdots & \vdots & \vdots \\ 1, & \partial(T_n + \Delta T_n)/\partial x_0, & \partial(T_n + \Delta T_n)/\partial y_0 & \partial(T_n + \Delta T_n)/\partial z_0 \end{vmatrix}. \tag{4.62}$$

The meaning of these relations is simple. As a result of the uncertainty in the velocity model, the observed and calculated travel times differ by some random values. Knowing the principal statistical characteristics of velocity uncertainty and using relations (4.56) and (4.57), we are able to estimate the mean value and the variance of the random component of travel times. In addition, according to (4.61), the variances σ_r^2 of the time residuals can be divided into the variance σ_t^2 of arrival time reading errors and the variance σ_T^2 of travel-time fluctuations.

The described procedure has been routinely used since 1975 at several coal mines in Poland and is also used at a number of coal mines in the Beijing mining district in China.

4.4 Relative Location Technique

The relative location of seismic events, known also as a "master event" procedure or arrival-time difference (ATD) technique, is an old and highly popular method. One of the reasons for its wide use is that the travel time anomalies resulting from velocity model uncertainties are removed implicitly in this method, during the formation of equations (4.9). Excellent references to this subject can be found in the paper by Spence (1980). The ATD location procedure has been also used in studies of microearthquakes forming a single seismic sequence (e.g., Oncescu and Apolozan, 1984; Slunga *et al.*, 1984; Thorbjarnrdottir and Pechmann, 1987; Console and Di Giovambattista, 1987).

In the ATD procedure, a set of P-wave arrival-time differences $t_i - t_{Ri}$ is used. These differences are obtained from stations $i = 1, \ldots, n$, which recorded a reference event R with known focal parameters θ_R and a nearby event with unknown parameters θ. Let us assume that the arrival time at an ith station can be written as

$$t_i = t_0 + T(\mathbf{h}, \langle \mathbf{m}_i \rangle) + \Delta T(\mathbf{h}, \mathbf{m}_i) + \varepsilon_i, \qquad (4.63)$$

where $T(\mathbf{h}, \langle \mathbf{m}_i \rangle)$ is the theoretical travel time calculated from an average velocity model $\langle \mathbf{m}_i \rangle$, $\Delta T(\mathbf{h}, \mathbf{m}_i)$ is the unknown travel time anomaly caused by the differences between the true unknown velocity model $\mathbf{m}_i$ and the known average velocity model $\langle \mathbf{m}_i \rangle$ used during the location procedure, and ε_i is the unknown arrival-time reading error of Gaussian nature with mean equal to zero and the known variance σ_{ti}^2.

Accordingly, the arrival time of P wave at the ith station, generated by the reference event R, is

$$t_{Ri} = t_{R0} + T(\mathbf{h}_R, \langle \mathbf{m}_{Ri} \rangle) + \Delta T(\mathbf{h}_R, \mathbf{m}_{Ri}) + \varepsilon_{Ri}. \qquad (4.64)$$

Applying the first-order Taylor expansion, we can approximate the arrival times (4.63) as

$$t_i = t_{R0} + \delta t_0 + T(\mathbf{h}_R, \langle \mathbf{m}_i \rangle) + \frac{\partial T_i}{\partial x_0} \delta x_0 + \frac{\partial T_i}{\partial y_0} \delta y_0 + \frac{\partial T_i}{\partial z_0} \delta z_0$$

$$+ \Delta T(\mathbf{h}, \mathbf{m}_i) + \varepsilon_i, \qquad (4.65)$$

where δt_0 and $\delta x_0, \delta y_0, \delta z_0$ are the corrections to the origin time and the hypocenter coordinates of the reference event R and an obvious notation $T_i = T(\mathbf{h}_R, \langle \mathbf{m}_i \rangle)$ is introduced. We assumed that the ray paths from the reference hypocenter $\mathbf{h}_R$ and the located hypocenter $\mathbf{h}$ traverse the same structure. This means that the time anomalies $\Delta T(\mathbf{h}_R, \mathbf{m}_{Ri})$ and $\Delta T(\mathbf{h}, \mathbf{m}_i)$ are approximately the same. Subtracting equations (4.64) from equations (4.65), we obtain the following set of n linear equations

$$t_i - t_{Ri} = \delta t_0 + \frac{\partial T_i}{\partial x_0} \delta x_0 + \frac{\partial T_i}{\partial y_0} \delta y_0 + \frac{\partial T_i}{\partial z_0} \delta z_0, \qquad (4.66)$$

which in matrix form can be written as

$$\mathbf{A}\delta\boldsymbol{\theta} = \delta\mathbf{t}, \tag{4.67}$$

where the matrix $\mathbf{A}$ is defined by relation (4.10), its elements are computed at the reference hypocenter $\mathbf{h}_R$, and $\delta\mathbf{t}$ is an n-dimensional column vector with elements equal to the time differences $\delta t_i = t_i - t_{R_i}$.

Two elements contribute to the uncertainty of each time difference δt_i: the time reading error of the located event equal to ε_i and the time reading error of the reference event equal to ε_{R_i}. Assuming that the time reading errors at each station are mutually independent and that the respective variances of reading times are known $[E(\varepsilon_i^2) = \sigma_{t_i}^2$ and $E(\varepsilon_{R_i}^2) = \sigma_{Rt_i}^2$, where E stands for the expected value operator], the covariance matrix of the time differences δt_i is

$$\mathbf{C}_r = \mathbf{C}_t + \mathbf{C}_{Rt}, \tag{4.68}$$

where $\mathbf{C}_t$ and $\mathbf{C}_{Rt}$ are the $(n \times n)$ diagonal matrices with the elements $\sigma_{t_i}^2$ and $\sigma_{R_i}^2$, respectively. The covariance matrix (4.68) takes the simple form

$$\mathbf{C}_r = 2\mathbf{C}_t, \tag{4.69}$$

if the reliability of arrival time measurements at all stations are the same $(\sigma_{t_i} = \sigma_t, \sigma_{Rt_i} = \sigma_{Rt})$, and both the events are recorded with the same reliability $(\sigma_t = \sigma_{Rt})$.

Finally, taking into account that the elements of the matrix $\mathbf{A}$ are determined at the reference hypocenter $\mathbf{h}_R$, the least-squares solution of (4.67) does not require the iterative procedure and is equivalent to the solution of (4.11), where $\mathbf{B} = \mathbf{A}^T\mathbf{C}_r^{-1}\mathbf{A}$ and $\mathbf{b} = \mathbf{A}^T\mathbf{C}_r^{-1}\delta\mathbf{t}$. The diagonal elements of the matrix $(\mathbf{A}^T\mathbf{C}_r^{-1}\mathbf{A})^{-1}$ are the asymptotic evaluations of the variances of components of the vector $\boldsymbol{\theta}$.

It should be noted that theoretical travel times do not appear in the set of conditional equations (4.67). Only their partial derivatives with respect to the hypocentral coordinates are found in these equations. In mining practice, the velocity model $\mathbf{m}$ is very often described by one parameter only: the first arrival velocity V_P. In general, V_P can depend on the hypocentral distance, and the formula describing the travel time has a simple form such as $T_i = d_i/V_{P_i}$. The travel-time derivatives with respect to the hypocenter coordinates, taken at the reference event hypocenter $\mathbf{h}_R$, are then of the form

$$\begin{aligned}
\partial T_i/\partial x_0 &= (x_R - x_i)/(d_{R_i}V_{P_i}), \\
\partial T_i/\partial y_0 &= (y_R - y_i)/(d_{R_i}V_{P_i}), \\
\partial T_i/\partial z_0 &= (z_R - z_i)/(d_{R_i}V_{P_i}),
\end{aligned} \tag{4.70}$$

where (x_R, y_R, z_R) are the hypocenter coordinates of the reference event,

(x_i, y_i, z_i) are the coordinates of the ith station, d_{Ri} is the distance from the reference event to the ith station, and V_{Pi} is the average P-wave velocity between the reference event and the ith station. Taking into account that the velocity V_{Pi} can be approximated by an expression $d_{Ri}/(t_{Ri} - t_{R0})$, the matrix $\mathbf{A}$ in relation (4.67) takes the form

$$\mathbf{A} = \begin{vmatrix} 1, & c_1(x_R - x_1), & c_1(y_R - y_1), & c_1(z_R - z_1) \\ \vdots & \vdots & \vdots & \vdots \\ 1, & c_n(x_R - x_n), & c_n(y_R - y_n), & c_n(z_R - z_n) \end{vmatrix}, \qquad (4.71)$$

where coefficients c_i are equal to $d_{Ri}^2/(t_{Ri} - t_{R0})$.

The application of the ATD procedure in everyday mining practice has several advantages. The method, in general, is more accurate than the conventional approach. The improvement follows from the fact that the ATD approach implicitly removes travel-time anomalies, which are the main source of the focal parameter errors. The procedure is extremely fast, as it does not require iterations and most of the intermediate results can be calculated only once, stored, and then recalled during the location procedure. The procedure can be used for the location of seismic events regardless of the information available on the absolute location of the reference event. If the reference event location is poorly known, then only a shape of the active area, related to the reference event, can be found.

The described ATD technique has been routinely used at potash mines in Germany and at coal and copper mines in Poland (Kijko et al., 1986).

The results of comprehensive tests performed at a copper mine in the Lubin district in Poland (Król and Kijko, 1991) are briefly described to demonstrate the efficiency of the ATD location procedure. More than 350 blasts from eight mining sectors were analyzed. Every blast was located twice, by the ATD and conventional method. The average errors of epicenter locations are shown in Fig. 4.5. From these data it follows that the reduction of errors was 30 percent by the ATD method.

The purpose of the second test was to answer the question of whether the ATD procedure can be used when no origin time of the reference event is known. The ATD procedure requires the knowledge of four parameters from the reference event: the three hypocenter coordinates and the origin time. Although the hypocenter coordinates of blasts are usually known, the firing time is never measured. To use, therefore, the blasts as reference events, their firing time must be estimated. Since the mean V_P velocity in the whole Lubin mining district is known (approximately 5.5 km/s), the origin time of reference event is obtained simply by subtracting from the P-wave recorded time at the nearest station the value of d_{min}/V_P, where d_{min} is the distance between the nearest station and the blast hypocenter. The adoption of an

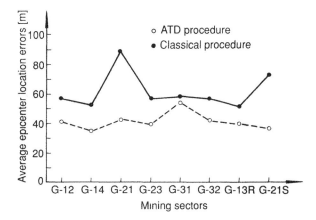

Figure 4.5 Comparison of the average errors of epicenter locations of 354 dynamite blasts fired in the Polkowice and Rudna copper mine in Poland, determined by the ATD and classic procedures. On the average, the ATD approach provides location errors 30 percent smaller than those provided by the conventional method.

erroneous velocity leads to a wrong estimate of firing time of the reference blast, which in turn is an additional source of error in the location of real seismic events. To examine the effect of an adopted P-wave velocity on the errors of hypocenter coordinates, 10 blasts were selected and each of them was located three times with three different velocity values: V_P equal to 5.0, 5.5, and 6.0 km/s. The differences between the epicenter coordinates were less than 5 m. Little influence of a given velocity V_P on the origin time estimate is another argument in favor of the ATD location method in mining practice.

4.5 Simultaneous Hypocenter and Velocity Determination

Simultaneous location of a group of seismic events and determination of a velocity model is founded on the same assumptions as the ATD approach: the trave-time anomalies from a set of close seismic events tend to be strongly correlated and have nearly constant values at the same stations. In contrast to the ATD approach, in which the hypocenter of only one event is located, in this procedure a set of hypocenters and velocity model parameters are jointly determined. The procedure is known as the *simultaneous structure and hypocenter* (SSH) determination method and was first formulated for local events by Crosson (1976). Crosson's approach originally formulated for a one-dimensional structure was extended for a three-dimensional velocity inversion by Aki and Lee (1976). Further references on this subject can be found, for example, in the work of Koch (1985a, 1985b).

The SSH procedure under mine conditions was applied by Mendecki (1981). Arrival times were used to evaluate the focal parameters and velocity anisotropy of the rockmass. In recent applications of the SSH technique in mines, Jech (1989) used the time difference between P- and S-wave arrivals to relocate seismic events and to improve the velocity model in the Ostrava-Karvina mining district in Czechoslovakia.

The SSH procedure can be formulated as follows. Let us consider a situation where a network of n_s stations is situated in an area in which n_e seismic events occurred. For simplicity, only the first P-wave arrivals are taken into account, and each event is recorded by all n_s stations. Additionally, the times t_{ij} ($i = 1, \ldots, n_e$; $j = 1, \ldots, n_s$) are functionally related to the velocity parameters in a known manner. For each seismic event i, the arrival times may be written in a matrix form as

$$\mathbf{t}_i = t_{0i}\mathbf{1} + \mathbf{T}(\mathbf{h}_i, \mathbf{m}) + \boldsymbol{\epsilon}_i, \qquad (4.72)$$

where $\mathbf{t}_i$, $\mathbf{T}(\mathbf{h}_i, \mathbf{m})$ and $\boldsymbol{\epsilon}_i$ are n_s-dimensional column vectors; $\mathbf{t}_i = (t_{i1}, \ldots, t_{in_s})^T$ are the first arrivals; t_{0i} is the unknown origin time; $\mathbf{T}(\mathbf{h}_i, \mathbf{m}) = [T_1(\mathbf{h}_i, \mathbf{m}), \ldots, T_{n_s}(\mathbf{h}_i, \mathbf{m})]^T$ are the travel times; $\boldsymbol{\epsilon}_i = (\varepsilon_{i1}, \ldots, \varepsilon_{in_s})^T$ is a random vector involved in the determination of arrival times $\mathbf{t}_i$; $\mathbf{h}_i = (x_{0i}, y_{0i}, z_{0i})^T$ is a three-dimensional column vector of unknown hypocentral coordinates of the ith event; and $\mathbf{m}$ is an unknown n_m-dimensional column vector of possible velocity model parameters describing the travel times.

Let us assume that the reading errors $\boldsymbol{\epsilon}_i$ show a normal distribution with a mean value equal to zero and the known covariance matrix $\mathbf{C}_{ti}$. It is also common to assume that there is no correlation between time reading errors at different stations. Then the matrices $\mathbf{C}_{ti}$ become diagonal

$$\{\mathbf{C}_{ti}\}_{jk} = \begin{cases} \sigma_{tij}^2, & j = k, \\ 0, & j \neq k, \end{cases} \qquad (4.73)$$

where $i = 1, \ldots, n_e$, $j, k = 1, \ldots, n_s$, σ_{tjk} is the standard deviation of the first arrival time determination from the ith earthquake recorded at the jth station. For convenience, notation $\mathbf{C}_t = \mathrm{diag}(\mathbf{C}_{t1}, \ldots, \mathbf{C}_{tn_e})$ is introduced. As the estimation is carried out by a least-squares procedure, such values as $\hat{\mathbf{t}}_0$, $\hat{\mathbf{h}}$, and $\hat{\mathbf{m}}$ that minimize the sum of the squared time residuals

$$\Phi(\boldsymbol{\theta}, \mathbf{m}) = [\mathbf{t} - \mathbf{t}_0 - \mathbf{T}(\mathbf{h}, \mathbf{m})]^T \mathbf{C}_t^{-1}[\mathbf{t} - \mathbf{t}_0 - \mathbf{T}(\mathbf{h}, \mathbf{m})] \qquad (4.74)$$

are taken as an evaluation of the unknown parameters, where $\boldsymbol{\theta}$, $\mathbf{t}$, $\mathbf{t}_0$, and

$T(h, m)$ are multivectors equal to

$$
\boldsymbol{\theta} = \begin{pmatrix} \boldsymbol{\theta}_1 \\ \vdots \\ \boldsymbol{\theta}_{n_c} \end{pmatrix}, \boldsymbol{\theta}_i = \begin{pmatrix} t_{0i} \\ h_i \end{pmatrix},
$$

$$
\mathbf{t} = \begin{pmatrix} \mathbf{t}_1 \\ \vdots \\ \mathbf{t}_{n_c} \end{pmatrix}, \mathbf{t}_0 = \begin{pmatrix} t_{01}\mathbf{1} \\ t_{0n_c}\mathbf{1} \end{pmatrix}, T(h, m) = \begin{pmatrix} T(h_1, m) \\ \vdots \\ T(h_{n_c}, m) \end{pmatrix}. \tag{4.75}
$$

In most cases the travel-time functions $T(h, m)$ are not linear with respect to the hypocenter coordinates and the velocity model parameters. A minimization of the misfit function (4.74), therefore, can be achieved by iterations only. Following the extension of a single event approach to the perturbations of the focal parameters of n_e events and n_m parameters of the velocity model, relation (4.8) takes the form

$$
r_{ij} = \delta t_{0i} + \frac{\partial T_{ij}}{\partial x_{0i}} \delta x_{0i} + \frac{\partial T_{ij}}{\partial y_{0i}} \delta y_{0i} + \frac{\partial T_{ij}}{\partial z_{0i}} \delta z_{0i}
$$

$$
+ \sum_{k=1}^{n_m} \frac{\partial T_{ij}}{\partial m_k} \delta m_k, \tag{4.76}
$$

where $T_{ij} = T_j(h_i, m)$, and the time residuals $r_{ij} = t_{ij} - t_0^* - T_{ij}^*$ and the partial derivatives of travel times are calculated for the first approximation of the origin times t_{0i}^*, hypocenter coordinates h_i^*, and velocity model parameters m^*. The system of equations (4.76) can be written in a matrix form as follows

$$
A_1 \delta\boldsymbol{\theta} + A_2 \delta m = r, \tag{4.77}
$$

where $r = t - t_0^* - T(h_i^*, m^*)$, $A_1 = \text{diag}(A_{11}, \ldots, A_{1n_c})$, and

$$
A_2 = \begin{pmatrix} A_{21} \\ \vdots \\ A_{2n_c} \end{pmatrix}. \tag{4.78}
$$

The minimization of the sum of time residuals is equivalent to the minimization of $(r - A_1\delta\boldsymbol{\theta} - A_2\delta m)^T C_t^{-1}(r - A_1\delta\boldsymbol{\theta} - A_2\delta m)$. Each A_{1i} is an $(n_s \times 4)$ matrix of the ith hypocenter partial derivatives equal to (4.10).

Each A_{2i} is an $(n_s \times n_m)$ matrix of the partial derivatives of velocity model parameters

$$\{A_{1i}\}_{kl} = \frac{\partial T_{ik}}{\partial m_l}, \tag{4.79}$$

where $i = 1, \ldots, n_e$, $k = 1, \ldots, n_s$, and $l = 1, \ldots, n_m$.

The least-squares estimation of the focal parameters θ and velocity model m is equivalent to the solution of the system of linear equations

$$\begin{cases} B_1 \, \delta\theta + E \, \delta m = F_1, \\ B_2 \, \delta\theta + E^T \, \delta m = F_2, \end{cases} \tag{4.80}$$

where $B_1 = A_1^T C_t^{-1} A_1$, $B_2 = A_2^T C_t^{-1} A_2$, $E = A_1^T C_t^{-1} A_2$, $F_1 = A_1^T C_t^{-1} r$, and $F_2 = A_2^T C_t^{-1} r$. Because of the nonlinearity of the problem, the matrices B_1, B_2, E, F_1, and F_2 are functions of the hypocenter coordinates and velocity model, and the problem must be solved iteratively. A solution of the system of equations (4.80) leads to a highly unstable iterative search, particularly in the case of the absence of a priori information. This difficulty can be overcome by introducing one of the stabilization techniques described in Section 4.1. Additionally, the straightforward solution uses inefficient numerical algorithms and is time consuming.

Spencer and Gubbins (1980) introduced a technique that resolves such problems. A simple rearrangement of (4.80) leads to

$$\delta m = \left(B_2 - E^T B_1^{-1} E \right)^{-1} \left(F_2 - E^T B_1^{-1} F_1 \right),$$
$$\delta\theta = C_1^{-1} \left(F_1 - E \, \delta m \right). \tag{4.81}$$

Solution of (4.81) requires the inversion of two matrices: the $(4n_e \times 4n_e)$ matrix B_1 and the $(n_m \times n_m)$ matrix $(B_2 - E^T B_1^{-1} E)$. The matrices E and F_1 have a block structure

$$E = \begin{vmatrix} E_1 \\ \vdots \\ E_{n_e} \end{vmatrix}, F_1 = \begin{vmatrix} F_{11} \\ \vdots \\ F_{1n_e} \end{vmatrix}, \tag{4.82}$$

and B_1 is a block diagonal

$$B_1 = \begin{vmatrix} B_{11} & & & 0 \\ & B_{12} & & \\ & & \ddots & \\ 0 & & & B_{1n_e} \end{vmatrix}, \tag{4.83}$$

where for each event i, $E_i = A_{1i}^T C_{ti}^{-1} A_2$, $F_{1i} = A_1^T C_{ti} r_i$, and $B_{1i} = A_{1i}^T C_{t1}^{-1} A_1$. The system of equations (4.81) therefore can be rearranged as

$$\delta m = \left[\sum_{i=1}^{n_c} \left(B_{2i} - E_i^T B_{1i}^{-1} E_i \right) \right]^{-1} \left[\sum_{i=1}^{n_c} \left(F_{2i} - E_i^T B_{1i}^{-1} F_{1i} \right) \right], \quad (4.84)$$

$$\delta \theta_i = B_{1i}^{-1} (F_{1i} - E_i \delta m), \quad i = 1, \ldots, n_e,$$

where $B_{2i} = A_{2i}^T D_{ti} A_{2i}$, $F_{2i} = A_{2i}^T D_{ti} r_i$, and $\delta \theta_i$ and δm are the correction vectors to the first guess of the ith focal parameters θ_i^* and velocity model parameters m^*, respectively. The procedure (4.84) is repeated until some cutoff criteria are met. The solution of our problem requires n_e inversions of the (4×4) matrices B_{1i} and one inversion of the $(n_m \times n_m)$ matrix $\Sigma(B_{2i} - E_i^T B_{1i}^{-1} E_i)$, where the summation is from $i = 1$ to n_e.

The advantage of using the SSH procedure is obvious. The procedure does not require velocity calibration blasts, is fast, and can be performed even by small computers. The application of SSH procedure in a region of high seismicity for a specified period of time also makes it possible to detect spatial and temporal variations of the velocity model parameters caused by stress migration.

As an example of the application of the presented procedure, we describe the results of the SSH inversion obtained by Mendecki (1987) from a very small seismic network in a South African gold mine in the Klerksdorp district. In the pillar, situated 2160 m below the surface, a seismic network composed of eight geophones was installed, extending over the area with a radius of 15 m. In order to obtain a reliable pattern of seismic activity, several calibration blasts were fired and the average values of P-wave velocities to individual geophones were determined. For the velocity model inversion, seven events were selected. All of them formed a relatively tightly spaced group. The inverted P-wave velocities were compared with the measured velocities from 7 blasts located in the same area as the seismic events. According to our notation, the travel times from the ith event to the jth station is equal to $T_{ij} = d_{ij}/V_j$, where $i = 1, \ldots, n_e$, $j = 1, \ldots, n_s$, $n_e = 7$, $n_s = 8$, and d_{ij} is the hypocentral distance between the ith event and the jth station. The velocity model m is described by eight parameters ($n_m = 8$), equal to the average velocities along the distances from the selected area to each of eight geophones. The comparison of the velocities obtained from the inversion with the velocities determined from calibration blasts is shown in Table 4.1. The velocities determined by the SSH inversion are in good agreement with those determined by the calibration blasts. The wide range of velocity changes, from 4739 to 5650 m/s, indicates strong velocity anisotropy within the pillar.

Table 4.1

Comparison of the Average P-Wave Velocities Obtained from an Inversion
of Seismic Observations with those Determined from Calibration Blasts [1]

| | Average P-wave velocity | |
Geophone	SSH procedure	Calibration blasts
1	5109	5070
2	4814	4830
4	5149	5270
5	4739	4840
7	5589	5710
8	5650	5780
9	5245	5440
11	5280	5230
Mean:	5197	5271

[1] From Mendecki, 1987.

4.6 Other Location Methods

4.6.1 Linear Methods

In some cases, the procedure of focal hypocenter determination can be significantly simplified. This can be achieved if a single-velocity model is accepted. For example, if seismic travel times $T(\mathbf{h})$ are calculated using the straight-line slant distance from the unknown hypocenter $\mathbf{h} = (x_0, y_0, z_0)^T$ to the station (x_i, y_i, z_i), $(i = 1, \ldots, n)$, then

$$T(\mathbf{h}) = \left[(x_i - x_0)^2 + (y_i - y_0)^2 + (z_i - z_0)^2 \right]^{1/2} / V, \qquad (4.85)$$

where V is the constant velocity for the whole area. After simple algebraic transformations, the focal parameters $\boldsymbol{\theta} = (t_0, x_0, y_0, z_0)^T$ are least-squares solutions of the following set of $n - 1$ linear equations

$$\mathbf{A\theta} = \mathbf{r}, \qquad (4.86)$$

where

$$\{\mathbf{A}\}_{ij} = \begin{cases} 2(t_{i+1} - t_i)V^2 & \text{for } j = 1, \\ 2(x_{i+1} - x_i) & \text{for } j = 2, \\ 2(y_{i+1} - y_i) & \text{for } j = 3, \\ 2(z_{i+1} - z_i) & \text{for } j = 4, \end{cases} \qquad (4.87)$$

$\{\mathbf{r}\}_i = (x_{i+1}^2 - x_i^2) + (y_{i+1}^2 - y_i^2) + (z_{i+1}^2 - z_i^2) + (t_{i+1}^2 - t_i^2)V^2$, and i denotes the station number. In such a case, the arrival times from at least five ($n = 5$) stations are required. Blake *et al.* (1974) and recently Eccles and Ryder (1984) give a comparative evaluation of these methods.

Linear methods used for hypocenter locations are very attractive. They are fast and free from all the problems characteristic for iterative procedures. An obvious limitation to these methods follows from the fact that they ignore complex velocity structures, and their application to real mining conditions may often lead to unreasonable solutions. At present, the linear methods are used for determination of a starting point for further iterative computations.

4.6.2 Large Time Residuals and L_1 Norm

The use of least-squares procedures is equivalent to the assumption that the arrival time residuals are of Gaussian nature. Jeffreys (1932), who was developing the global travel-time tables, first recognized that the time residuals do not follow the Gaussian distribution and occasional large residuals strongly affect the location of earthquakes. The problem of large time residuals is also important for seismic events in mines. An introduction of automatic arrival-time detectors, characteristic for modern mining recording systems, increases probability of picking up noise spikes, or confusing P and S arrivals on a particular channel.

To solve this problem, instead of least-squares misfit function (4.1), the sum of the absolute values of time residuals, known as the L_1 norm, is introduced. The corresponding misfit function is

$$\Phi(\boldsymbol{\theta}) = \sum_{i=1}^{n} |r_i|, \qquad (4.88)$$

where $|\cdot|$ denotes the absolute values. The misfit criterion (4.88) is less sensitive than the L_2 (least-squares) norm and tends to decrease the effects of a few large time residuals. A comprehensive comparison of the L_1 and L_2 norms and other misfit functions is given by Anderson (1982).

For the location of seismic events in mines, the L_1 norm was introduced by Prugger and Gendzwill (1988) and used at Saskatchewan potash mines, Canada, and it is also implemented in the ISS software packet used in South Africa (Mendecki, 1990). Unfortunately, the location procedures based on the L_1 norm are rather seldom applied in practice, because of the difficulty involved in the problem formulation in terms of matrix presentation and inversion. The L_1-norm-based location programs, therefore, are often accomplished by direct minimization procedure of the misfit functions as the Nelder–Mead simplex subroutine.

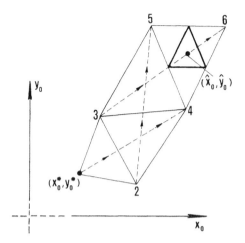

Figure 4.6 Simple illustration of the simplex location algorithm. Starting from a point (x_0^*, y_0^*), after several successive reflections, the simplex $(4, 5, 6)$ starts to "collapse" toward the minimum at a point $(\hat{x}_0, \hat{y}_0)$. [From Rabinowitz and Kulhánek (1988, Fig. 2).]

4.6.3   The Nelder–Mead Simplex Procedure of Misfit Function Minimization

In general, the Gauss–Newton procedure of minimization of the sum of squared residuals, described in Section 4.1, is efficient and fast. Occasionally, however, the Gauss–Newton approach becomes inefficient and unstable. These situations include ill-conditional inverse problem, strong mutual correlation between the unknown focal parameters θ, and numerical evaluation of travel-time derivatives. In such cases and for L_1-norm-based location programs, the simplex optimization procedure, which does not require the calculation of derivatives, is more appropriate. The detailed description of the method can be found in Himmelblau (1972) and in more recent textbooks on numerical algorithms such as that of Press *et al.* (1989).

In our brief description of the simplex algorithm, we follow the work of Rabinowitz and Kulhánek (1988). Let us assume that our problem of location contains an evaluation of two parameters only: the epicenter coordinates x_0 and y_0. Suppose that the minimum of the misfit function $\Phi(x_0, y_0)$ is at the point $(\hat{x}_0, \hat{y}_0)$ and the trial of epicenter is chosen to be at the point (x_0^*, y_0^*) (Fig. 4.6). Starting from the point (x_0^*, y_0^*), the algorithm constructs the triangle, having (x_0^*, y_0^*) as one of its vertices. Such a triangle in a two-dimensional space is called "simplex," and in general, a simplex is a figure having one more vertex than the number of dimensions for which it is defined. The misfit function $\Phi(x_0, y_0)$ is evaluated for each of the three vertices, the vertex for which the misfit function attends the highest value is then reflected through the gravity center of the triangle. If the value of $\Phi(x_0, y_0)$ at the reflected point (say 4 in Fig. 4.6) decreases with respect to the initial vertex, then a new simplex is formed by adding vertex 4 and deleting vertex 1. The process is repeated until the reflection fails to produce a vertex with a lower value of Φ. Then the simplex can contract itself and changes the search direction or "collapses" toward the minima according to the termination criterion.

In practice, the simplex iteration stops when the simplex size reaches a predetermined minimum value. Although the simplex-based optimization procedure has been known since the early 1960s (Spendly and Hest, 1962), only universal use of computers made it an attractive tool. This is especially evident for the problems in which the misfit function exhibits local minima and/or analytical calculation of derivatives is laborious. The modification of the algorithm proposed by Nelder and Mead (1965), the version used in all known seismological applications, has substantially improved the performance capability of the original algorithm. In the Nelder–Mead simplex procedure, the minimization (in general optimization) process is adaptive, causing the simplex to be revised to conform in the best way to the nature of the misfit function. In many cases, the Nelder–Mead simplex optimization procedure is more effective than the Gauss–Newton one: it is derivative-free, and it does not require matrix inversions in problems containing ill-conditioned matrices. The procedure can be used for any velocity model and any misfit function.

The Nelder–Mead simplex procedure was first introduced for the location of mining events by Prugger and Gendzwill (1988). The same technique is also used for the location of local (Rabinowitz, 1988) and distant earthquakes (Rabinowitz and Kulhánek, 1988). The comparison test of the simplex procedure with the Gauss–Newton procedure and the Thurber (1985) approach [known as a *Newton optimization procedure* (equation (4.38)], was carried out by Prugger and Gendzwill (1988). They assumed that the network is composed of eight surface seismic stations; a half-space velocity model was accepted, and only the first P arrival times were used for the location. Two seismic events were analyzed: the first had the hypocenter located in the center of the network, and the second had the coordinates outside of the array. The Nelder–Mead simplex procedure with a L_1 norm misfit function provided reasonable locations and in most cases was found to be more accurate than either the Newton [applied by Thurber (1985)] or the Gauss–Newton procedure.

Despite the fact that direct search optimization methods (such as simplex procedure) are slower than those that operate with derivatives of travel times, the location programs that use the Nelder–Mead algorithm are fast enough and can be used routinely. In general, an application of misfit function based on the L_1 norm, together with the Nelder–Mead simplex algorithm, is as good or even better than most of the other approaches.

Chapter 5 | Optimal Planning of Seismic Networks in Mines

The degree of accuracy attainable for mine tremor locations depends on several factors, such as spacial configuration of seismic networks, accuracy of readings of seismic-wave arrivals, and adequacy of the assumed velocity model. It was shown in Chapter 4 that the event location errors consist of systematic and random components. The influence of systematic errors can be eliminated by a detailed analysis of travel-time anomalies, or by the simultaneous location of a group of seismic events and velocity model determination, described in Section 4.5. The values of random errors of focal parameters can therefore be used as a quality criterion of the spacial distribution of seismic stations. For a given velocity model, random errors depend on the accuracy of the readings of seismic wave arrivals and the geometry between the hypocenter and seismic stations. Consequently, the problem of optimum event location is equivalent to an analysis of the spatial distribution of seismic stations, ensuring the minimum values of random errors in the source location procedure.

Although seismic networks are widely used in mines, it is remarkable how little attention has been paid to the optimum design of such networks. The performance of a network is usually evaluated after its installation and its operation during some interval of time. New stations are then added to the network, or occasionally the number of stations is kept fixed, while some of them are moved to different locations. In most cases the rearranging is based on personal judgment. This process is expensive in financial terms and in the loss of irretrievable observations as well. In a few cases only, the optimal spacial configuration of seismic stations is predesigned and theoretically tested.

The earliest attempts to optimize the spatial distribution of seismic stations were undertaken in Japan by modeling the process of seismic wave recording by the Monte Carlo method (Sato and Skoko, 1965; Skoko and Sato, 1966). In these early works, the approximate character of equations describing the propagation of seismic waves was not taken into account. These difficulties were overcome by and Peters and Crosson (1972), who described the method of estimating the accuracy of regional earthquake locations by the so-called prediction analysis, which is a generalization of the least-squares method for the case when equations describing physical process are of an approximate (random) character (Fedorov, 1974).

The optimal distribution of seismic stations has been also studied by various alternative approaches. The nonstatistical analysis of seismic network was suggested, for example, by Burmin (1986). An important conclusion of Burmin's work is that the optimal configuration of seismic stations for certain location procedures ceases to be optimal if the procedure changes. Some aspects of optimal configuration of seismic stations in mines have been also discussed by Salamon and Wiebols (1972), Båth (1984), and Eccles and Ryder (1984).

A general approach to the distribution of seismic stations, based on statistical theory of optimal experiments, was proposed by Kijko (1977a, 1977b, 1978). His approach was used for the optimum design of seismic networks in several countries (Savarenskiy *et al.*, 1979; Trifu, 1983; Ghalib *et al.*, 1985; Garcia-Fernandez *et al.*, 1988). An interesting extension of Kijko's ideas to the case of the correlation of time residuals between two different stations was described by Rabinowitz and Steinberg (1990).

Several general rules concerning the distribution of seismic stations in relation to a given hypocenter are well known. Stations should be evenly distributed around the hypocenter. It is advantageous to have stations distributed over a range of distances from the earthquake epicenter. In order to estimate the focal depth, several P phases should be recorded at epicentral distances smaller than the focal depth. The inclusion of S-wave arrivals leads to the hypocenter estimates with smaller standard errors than those determined with P-wave arrivals only (Buland, 1976; Uhrhammer, 1982; Gomberg *et al.*, 1990). These "rules" can be applied in practice only when it is possible to define a relatively small area of high seismic activity. If, however, the seismic area is large or several highly seismic areas are to be covered, it is necessary to examine the properties of the system of arrival-time equations to find more specific rules governing the optimum geometry of seismic stations.

The remaining part of this chapter is divided in two sections. In Section 5.1 some elements of the theory of optimal distribution of seismic stations are described, while in Section 5.2 several special cases and applications are discussed.

5.1 Theoretical Background

In general, the choice of one or another station configuration should depend on the value of some index related to a given configuration. Then, the best network would be defined by the value of the index, which could be called the "quality" when its maximum value defines the best configuration. The value of this index, denoted by Q, should depend on a covariance matrix of the sought parameters θ of the seismic focus. Thus in the most general case, the choice of the optimum network can be expressed in the form

$$\text{extremum} \left\{ Q = L\left[\mathbf{C_\theta(x)}\right] \right\}, \qquad (5.1)$$
$$\mathbf{x} \in \mathbf{\Omega}_x$$

where $\mathbf{C_\theta(x)}$ is the covariance matrix of the sought parameters $\mathbf{\theta} = (t_0, x_0, y_0, z_0)^T$, $\mathbf{x} = (\mathbf{x}_1, \ldots, \mathbf{x}_n)$ are the seismic station coordinates, each $\mathbf{x}_i = (x_i, y_i, z_i)^T$ is the Cartesian coordinate of the ith seismic station $\mathbf{\Omega}_x$ is the space domain of possible station locations, $i = 1, \ldots, n$, and n is the number of seismic stations.

The choice of the functional $L[\cdot]$ depends on the specific character of the considered problem. The most common is the so-called D-optimum planning, which minimizes the determinant of the matrix $\mathbf{C_\theta(x)}$. The D criterion is based on the following argument (e.g., John and Draper, 1975). An approximate confidence ellipsoid for the $\mathbf{\theta}$ parameters is of the form

$$(\mathbf{\theta} - \hat{\mathbf{\theta}})^T \mathbf{C}_\theta^{-1}(\mathbf{x})(\mathbf{\theta} - \hat{\mathbf{\theta}}) \leq \text{const}, \qquad (5.2)$$

where $\hat{\mathbf{\theta}}$ is the least-squares estimate of $\mathbf{\theta}$ and constant is an appropriate quantity from the κ_{n-4}^2 distribution. The content of this ellipsoid is proportional to $\sqrt{\det\left[\mathbf{C_\theta(x)}\right]}$. An obvious optimality criterion therefore is to make this ellipsoid as small as possible by minimizing $\det[\mathbf{C_\theta(x)}]$. A configuration $\mathbf{x}$ that minimizes $\det[\mathbf{C_\theta(x)}]$ is called D-optimal. This criterion has several attractive features, including low variances and low correlations between the nondiagonal elements of $\mathbf{C_\theta(x)}$. Also, if properties of determinants are used, it is not necessary to calculate the covariance matrix. Assuming that the focal parameters $\mathbf{\theta}$ are estimated by a least-squares method, the covariance matrix of these parameters is $\mathbf{C_\theta(x)} = [\mathbf{A}^T\mathbf{C}_r^{-1}\mathbf{A}]^{-1}$, where $\mathbf{A}$ is the partial-derivative matrix of computed arrival times with respect to the focal parameters $\mathbf{\theta}$ and $\mathbf{C}_r$ is the covariance matrix of time residuals. Since, by definition, $\det\mathbf{C_\theta}(x) = [\det \mathbf{C}_\theta^{-1}(\mathbf{x})]^{-1}$, minimizing $\det[\mathbf{C_\theta(x)}]$ will maximize $\det[\mathbf{A}^T\mathbf{C}_r^{-1}\mathbf{A}]$, satisfying the D-optimality criterion.

The D optimization of the observational points $\mathbf{x}$ is much more complex when the process (travel times of seismic waves in our case) is described by nonlinear equations, and the vector $\mathbf{\theta}$ is of a random variable character. In this case the elements of the matrix $\mathbf{A}$ and, equivalently, the covariance matrix $\mathbf{C_\theta(x)}$ are functions of true values of the unknown parameters $\mathbf{\theta}_{\text{TRUE}}$. Since $\mathbf{\theta}_{\text{TRUE}}$ is not known, the matrix $\mathbf{C_\theta}$ cannot be calculated and the whole procedure is not applicable. The technique frequently used in such a case is the calculation of the partial-derivative matrix $\mathbf{A}$ at a point $\langle\mathbf{\theta}\rangle$, which is an initial guess of $\mathbf{\theta}_{\text{TRUE}}$ (Box and Lucas, 1959).

The location of seismic events, based on the travel times of seismic waves recorded at several stations, is of such nature. The described approach to planning of a seismic network can be used in practice only when it is possible to define a relatively small area of high seismic activity. If, however, a seismic area is large and divided into several subregions, the described procedure of the optimum planning must be modified. Let us assume that from our knowledge of the previous seismicity and/or from the location of future mining areas we are able to determine the function $p(\mathbf{h})$, which is the space distribution of the event hypocenter coordinates $\mathbf{h} = (x_0, y_0, z_0)^{\mathsf{T}}$ such that $\int_{\Omega_h} p(\mathbf{h}) \, d\mathbf{h} = 1$, where Ω_h is the space domain of the occurrence of seismic events. Since the hypocenter coordinates have a random character, the value of $L[\mathbf{C}_\theta(\mathbf{x})]$ can be replaced by its mean value in the whole domain Ω_h

$$\langle L[\mathbf{C}_\theta(\mathbf{x})] \rangle = \int_{\Omega_h} p(\mathbf{h}) L[\mathbf{C}_\theta(\mathbf{x}, \mathbf{h})] \, d\mathbf{h}. \tag{5.3}$$

This expression averages the value of the Q parameter over all possible events in a given area of seismicity. Thus, the best estimation of the focal parameters θ, taking into account different probability of event occurrence at different points of the domain Ω_h, is secured by the distribution of observational points $\mathbf{x}$ fulfilling the following condition

$$\int_{\Omega_h} p(\mathbf{h}) L[\mathbf{C}_\theta(\mathbf{x}, \mathbf{h})] \, d\mathbf{h} = \text{extremum}. \tag{5.4}$$
$$\mathbf{x} \in \Omega_x$$

Criterion (5.4) is a natural generalization of the optimum planning methods described so far. For the case of linear initial equations, when $L[\mathbf{C}_\theta(\mathbf{x})]$ is independent of the focal parameters θ, condition (5.4) takes a form of the classic planning criterion

$$L[\mathbf{C}_\theta(\mathbf{x})] = \text{extremum}. \tag{5.5}$$
$$\mathbf{x} \in \Omega_x$$

Similarly, in the case of concentrated seismicity, when the space hypocenter distribution $p(\mathbf{h})$ can be approximated by the Dirac δ function $p(\mathbf{h}) = \text{const}$ $\delta(\mathbf{h} - \langle \mathbf{h} \rangle)$, where $\langle \mathbf{h} \rangle$ is the most probable event hypocenter location, condition (5.4) describes the so-called locally optimum criterion

$$L[\mathbf{C}_\theta(\mathbf{x}, \langle \mathbf{h} \rangle)] = \text{extremum}, \tag{5.6}$$
$$\mathbf{x} \in \Omega_x$$

introduced by Chernoff (1953). Condition (5.4) can be also interpreted as a maximization of a posterior probability of seismic event occurrence when the prior probability is described by $p(\mathbf{h})$.

Relation (5.4), which extends the optimal criterion to the design of seismic networks to monitor several "hot spots" of seismicity, is far from unique. For example, Chaloner and Larnitz (1989) suggest the maximization of posterior entropy of the sought parameters $\boldsymbol{\theta}$, which is equivalent to the replacement of the functional $L[\cdot]$ in relation (5.4) by its logarithm.

5.2 Special Cases and Applications

5.2.1 Three-Dimensional Seismic Networks around Concentrated Seismicity

Let us assume that in a given mine it is possible to distinguish a relatively small area with particularly high seismic activity. The problem consists of surrounding this area by seismic stations in such a configuration that the error in focal parameters, for a given number of stations, is the smallest. Obviously, the design of seismic networks in mines always involves technical restrictions, such as access to the selected sites, data transmission, and power supply. The problem of network configuration without constraints, however, is not only of academic interest. Theoretical studies of an "ideal" seismic network provide useful insight into the geometry of station distributions and can be considered as a "guide" to the design of seismic networks under real conditions. In addition, theoretical investigations provide simple formulas for quantitative evaluation of the expected location errors from ideal networks. Such an evaluation for a given number of seismic stations can be considered as a lower bound of the location errors for any real network.

The problem of optimal configuration of seismic networks in mines will be discussed in two steps. Firstly, we analyze the optimal configuration in isotropic and homogeneous media. In the second step, the solution is generalized for the case of isotropic but randomly inhomogeneous media. It is also assumed that the events are located by least-squares technique, using P-wave arrival times only.

Seismic Networks in Isotropic Homogeneous Media   The P-wave arrival times are described by the formula

$$t_i = t_0 + T(\mathbf{h}, \mathbf{m}, \mathbf{x}_i) + \varepsilon_i, \tag{5.7}$$

where for an homogeneous and isotropic velocity model

$$T(\mathbf{h}, \mathbf{m}, \mathbf{x}_i) = \left[(x_0 - x_i)^2 + (y_0 - y_i)^2 + (z_0 - z_i)^2 \right]^{1/2} / V_P \tag{5.8}$$

is the travel time from the hypocenter $\mathbf{h} = (x_0, y_0, z_0)^T$ to the ith station with coordinates $\mathbf{x}_i = (x_i, y_i, z_i)^T$, t_0 is the origin time, t_i is the onset time of a P arrival at ith station bearing a reading error ε_i, the vector $\mathbf{m}$ of velocity model parameters is described by one parameter only equal to the P-wave velocity V_P, $i = 1, \ldots, n$, and n is the number of seismic stations.

Following formula (5.6) and taking into account that the travel times $T(\mathbf{h}, \mathbf{m}, \mathbf{x})$ are nonlinear with respect to the hypocenter coordinates $\mathbf{h}$, the D-optimal planning criterion takes the form

$$\det[\mathbf{C}_{\theta}(\mathbf{x}, \langle \mathbf{h} \rangle)] = \min, \tag{5.9}$$

where $\langle \mathbf{h} \rangle$ is the known most probable hypocenter position equal to $\langle \mathbf{h} \rangle = (\langle x_0 \rangle, \langle y_0 \rangle, \langle z_0 \rangle)^T$. The planning criterion so obtained is usually called *locally D-optimal* (John and Draper, 1975), to underline its dependence on actual local values of the sought parameters $\mathbf{h}$. Assuming that the "true" hypocenter coordinates $\mathbf{h}$ are equal to $\langle \mathbf{h} \rangle + \delta \mathbf{h}$, expanding the travel times (5.8) into Taylor series at the point $\langle \mathbf{h} \rangle$ up to linear components of the expansion, and placing the origin of the system of coordinates X, Y, Z at that point, we obtain

$$r_i = t_0 + X_{i1} \frac{\delta x_0}{V_P} + X_{i2} \frac{\delta y_0}{V_P} + X_{i3} \frac{\delta z_0}{V_P}, \tag{5.10}$$

where $r_i = t_i - \langle t_0 \rangle - T(\langle \mathbf{h} \rangle, \mathbf{m}, \mathbf{x}_i)$.

$$\begin{cases} X_{i1} = x_i/d_i, \\ X_{i2} = y_i/d_i, \\ X_{i3} = z_i/d_i, \end{cases} \tag{5.11}$$

$d_i = x_i^2 + y_i^2 + z_i^2$, and $i = 1, \ldots, n$. Since $\sum_{i=1}^{3} X_{ij}^2 = 1$, the expansion of travel times into the Taylor series leads to the transformation of station coordinates $\mathbf{x}$ into new variables $\mathbf{X}$, where the domain over which the new variables range is a sphere with the unit radius. Thus the problem is reduced to finding the D-optimum design of the first order, where the controlled variables $\mathbf{X}$ range over a unit sphere. As a consequence, all considerations will be limited to that space.

First, we analyze an optimal design for which the number of stations n is the same as that of unknown parameters m and in discussed case equal to 4. The problem of optimal designs when the number of observations is equal to the number of unknowns has been considered by Box (1957) and Gorsky and Brodsky (1965). Their considerations lead to the notion of simplexial designs. For the sake of clarity, we assume that the time reading errors ε_i at different stations are mutually independent and have the same variances equal to $\sigma_{ti}^2 = \sigma_t^2$. Then, the covariance matrix $\mathbf{C}_{\theta}(\mathbf{x})$ of the focal parameters $\boldsymbol{\theta}$ is

$$\mathbf{C}_{\theta}(\mathbf{x}) = \sigma_t^2 \left([\mathbf{1} : \mathbf{X}]^T [\mathbf{1} : \mathbf{X}]\right)^{-1}, \tag{5.12}$$

where $[1:X]$ is the matrix obtained from the matrix X by adding the column of units on the left side. Thus, the minimization of the determinant (5.9) is equivalent to the maximizing of $\det([1:X]^T[1:X])$. It has been shown by Box (1957) that the design X that maximizes such a determinant should be sought among the designs with orthogonal rows. One example of such matrix designs, which range over the unit sphere for $n = m = 4$ variables, is a regular simplex with its center in the origin of the coordinate system.

Thus, it is easy to show that in the case of optimum distribution of four seismic stations, the diagonal elements of the covariance matrix C_θ are

$$\{C_\theta\}_{11} \equiv \sigma_{t_0}^2 = \sigma_t^2/4,$$

$$\{C_\theta\}_{22} \equiv \sigma_{x_0}^2 = 3V_P^2\sigma_t^2/4,$$

$$\{C_\theta\}_{33} \equiv \sigma_{y_0}^2 = 3V_P^2\sigma_t^2/4,$$

$$\{C_\theta\}_{44} \equiv \sigma_{z_0}^2 = 3V_P^2\sigma_t^2/4,$$

(5.13)

and the remaining elements of that matrix are zeros. In addition, it is possible to determine the coordinates for optimal distribution of four stations. From relation (5.11) and the known optimum designs on the sphere it follows that four stations should be distributed along the rays originating at the most probable hypocenter position $\langle h \rangle$ and passing through the vertices of a regular tetrahedron with a center at $\langle h \rangle$ (Fig. 5.1). Within the assumed

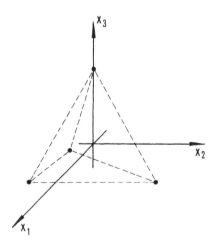

Figure 5.1 The rows of the matrix X are the vertices of a regular simplex, with a center in the origin of the coordinates system.

model, there are no restrictions on hypocentral distances to particular stations and these distances can be arbitrary.

The obtained solutions can be generalized for a larger number of stations. For $n > m$ recording points, the design matrix $\mathbf{X}$ has been described by Fedorov (1972). He proved that the recording points $\mathbf{X}$ should be allocated in the vertices of an arbitrary regular polyhedron inscribed into an $(m - 1)$ dimensional hyposphere. Thus all the formulas obtained for $n = 4$ can be generalized for the case when $n > 4$. For example, the minimal variances of the unknown focal parameters described by relations (5.13) now turn into

$$\{\mathbf{C}_\theta\}_{11} \equiv \sigma_{t_0}^2 = \sigma_t^2/n,$$

$$\{\mathbf{C}_\theta\}_{22} \equiv \sigma_{x_0}^2 = 3V_P^2\sigma_t^2/n,$$

$$\{\mathbf{C}_\theta\}_{33} \equiv \sigma_{y_0}^2 = 3V_P^2\sigma_t^2/n,$$

$$\{\mathbf{C}_\theta\}_{44} \equiv \sigma_{z_0}^2 = 3V_P^2\sigma_t^2/n,$$

$$(5.14)$$

where n is the number of stations.

Since the multiplication of the coordinates $\mathbf{x}_i$ of stations by an arbitrary constant is equivalent to the displacement of stations with respect to the design center without changing the radius of the sphere $\mathbf{X}$, the accuracy of the event location does not depend on the distance from the station to the most probable hypocenter position. In the considered case, the Fedorov design can be supplemented by adding additional points, for example, all centers of edges and all "centers" of faces of arbitrary regular polyhedrons.

Seismic Networks in Isotropic Randomly Inhomogeneous Media The isotropic randomly inhomogeneous medium is understood to be a medium in which the resultant P-wave velocities between the event hypocenter and seismic stations are different, but oscillate around the same value of the average P-wave velocity $\langle V_P \rangle$. In the first approximation, the variances $\sigma_{v_i}^2$ ($i = 1, \ldots, n$) of the resultant velocities are assumed to be constant ($\sigma_{v_i}^2 = \sigma_v^2$), that is, independent of the direction of seismic-wave propagation and the seismic ray path.

In Section 4.3 it was shown that such a definition of the medium inhomogeneity leads to the appearance of travel time perturbations with the mean value ΔT equal to $q^2 d/\langle V_P \rangle$ [equation (4.58)], and the variance σ_T^2 equal to $q^2 d^2/\langle V_P \rangle^2$ [equation (4.59)], where $q = \sigma_v/\langle V_P \rangle$, in which $\langle V_P \rangle$ is the mean value of P-wave velocity and σ_v^2 is its variance. Thus, the equation describing the mean value of the travel time in a randomly inhomogeneous medium is

$$T(\mathbf{h}, \mathbf{m}, \mathbf{x}_i) = d_i/\langle V_P \rangle + q^2 d_i/\langle V_P \rangle = d_i/V_P^*,$$

$$(5.15)$$

where $V_P^* = \langle V_P \rangle/(1 + q^2)$. This equation, has the same form as the equation (5.8) describing the travel time in homogeneous and isotropic media. The only difference between the two equations is related to the velocity. In equation (5.15) the seismic wave velocity was substituted by a fictitious velocity V_P^*.

Another effect of the medium inhomogeneity is the form of the arrival time variances. In our case, the covariance matrix of arrival times (4.61) is diagonal with elements equal to

$$\{\mathbf{C}_r\}_{ii} = \sigma_t^2 + q^2 \frac{d_i^2}{\langle V_P \rangle^2}, \tag{5.16}$$

where $i = 1, \ldots, n$, and n is the number of seismic stations. Thus, the covariance matrix of the focal parameter $\boldsymbol{\theta}$ is of the form

$$\mathbf{C}_{\boldsymbol{\theta}}(\mathbf{x}) = \left([1:\mathbf{X}]^T \mathbf{C}_r^{-1} [1:\mathbf{X}]\right)^{-1}. \tag{5.17}$$

The analysis of the determinant of such a matrix is difficult because of the presence of the matrix $\mathbf{C}_r$ with elements depending on the hypocentral distance d_i.

Since the matrix $\mathbf{C}_r$ is diagonal, the assumption that all its elements on the diagonal are the same and equal to $c_r = \{\mathbf{C}_r\}_{ii}$, $i = 1, \ldots, n$, simplifies formula (5.17) into

$$\mathbf{C}_{\boldsymbol{\theta}}(\mathbf{x}) = c_r \left([1:\mathbf{X}]^T [1:\mathbf{X}]\right)^{-1}, \tag{5.18}$$

where $c_r = \sigma_t^2 + q^2 d^2/\langle V_P \rangle^2 = $ constant.

The covariance matrix (5.18) differs only by a constant from a similar matrix for homogeneous media. The search for an optimal configuration of seismic stations for the case of isotropic randomly inhomogeneous media is, therefore, reduced to the known case of isotropic homogeneous media.

Thus, the optimal configuration of seismic networks requires an even distribution of stations over a sphere, placing them in vertices of a polyhedron that is symmetric with respect to each direction X, Y, Z. In general, each configuration of sensors located along the rays originating at the center of this sphere (equal to the most probable hypocenter position) and passing through the vertices of any regular polyhedrons inscribed into the sphere is D-optimal. The only difference between this configuration and the configuration for homogeneous and isotropic media is that in the case discussed above, in order to fulfill the condition $c_r = $ const, the stations cannot be placed in the centers of edges and faces of regular polyhedrons.

Seismic stations situated according to the described principles ensure the diagonality of the covariance matrix (5.18) with the elements equal to

$$\{\mathbf{C_\theta}\}_{11} \equiv \sigma_{t_0}^2 = c_r/n,$$

$$\{\mathbf{C_\theta}\}_{22} \equiv \sigma_{x_0}^2 = 3\langle V_P \rangle^2 c_r/n,$$

$$\{\mathbf{C_\theta}\}_{33} \equiv \sigma_{y_0}^2 = 3\langle V_P \rangle^2 c_r/n,$$

$$\{\mathbf{C_\theta}\}_{44} \equiv \sigma_{z_0}^2 = 3\langle V_P \rangle^2 c_r/n.$$

(5.19)

Hence, the errors in the origin times and coordinates of seismic events depend on d, the radius of a sphere over which the sensors are distributed.

From these considerations practical conclusions can be drawn. It can be readily shown that the parameter q is a velocity invariant of the medium and can serve as a measure of its inhomogeneity. By definition, the variance of P-wave velocity has the form

$$E\left[(V_P - \langle V_P \rangle)^2\right] = \sigma_{vP}^2,$$

(5.20)

where E stands for the expected value operator. For the velocity of S waves, which is related to the P-wave velocity through the equation $\langle V_S \rangle = \rho \langle V_P \rangle$, we have

$$E\left[(V_S - \langle V_S \rangle)^2\right] = \rho^2 \sigma_{vP}^2 = \sigma_{vS}^2.$$

(5.21)

From the definition of the parameter q and from relation (5.21) it follows that $q_S = q_P$.

According to our estimations for homogeneous media [equation (5.14)], the accuracy of the hypocenter location increases when smaller velocities are used, such as those of S waves. It follows from equations (5.19), however, that such a rule is not evident for the randomly inhomogeneous medium. The ratio of the variances of the hypocenter coordinates errors corresponding to P and S waves, based on equations (5.19), is

$$\frac{[\sigma_\alpha^2]_P}{[\sigma_\alpha^2]_S} = \frac{\langle V_P \rangle^2 \sigma_t^2 + d^2 q_P^2}{\langle V_S \rangle^2 \sigma_t^2 + d^2 q_S^2},$$

(5.22)

where $\alpha = x_0, y_0, z_0$. This ratio as a function of distance between the hypocenter and seismic stations, for typical mining conditions, is shown in Fig. 5.2. It was assumed that $\langle V_P \rangle = 4.0$ km/s, $\langle V_S \rangle = \langle V_P \rangle / \sqrt{3}$, $\sigma_{vP} = 0.15$ km/s, $\sigma_t = 0.01$ s, and that q_S is equal to q_P. From this figure it follows that

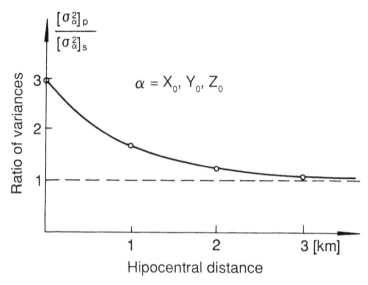

Figure 5.2 The ratio of the error variances of hypocenter coordinates estimated using P or S waves as a function of the distance between the most probable hypocenter and seismic sensors.

the classic approach, according to which the accuracy in seismic event location can be improved by using waves with lower velocity, is no longer true. Even at very short distances, the accuracy of location with the S waves is the same as that with the first arrivals. Furthermore, in equations (5.22) we accepted a highly optimistic assumption that the reading accuracy of arrival times of S and P waves is the same.

The obtained theoretical results can also be used to evaluate the quality of any existing seismic network. According to Flinn (1965), for any seismic event with the covariance matrix C_θ, the ratio $\min \det[C_\theta]/\det[C_\theta]$, which in the case discussed above is

$$\frac{27\sigma_t^8 V_P^6}{n^4 \det[C_\theta]}, \tag{5.23}$$

has the value between 0 (when the variances of focal parameters become infinite and no location is possible) and 1 (when the best possible distribution of stations around the hypocenter is achieved) and according to (5.14), $\min \det[C_\theta] = 27\sigma_t^8 V_P^6/n^4$. Relation (5.23) therefore can be used as a quantitative description of the network capability to detect and locate seismic events within a given area.

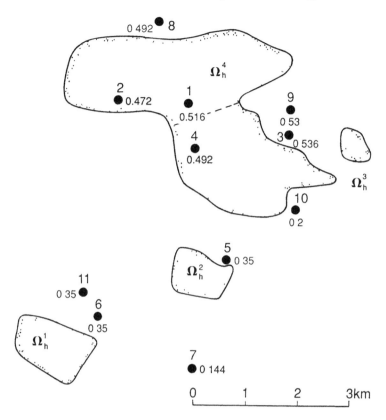

Figure 5.3 Regional seismic network in the Ostrava Coal Basin, Czechoslovakia. The areas of high seismicity Ω_h^i ($i = 1, \ldots, 4$) and 11 possible sites for seismic stations are marked. The depth of these sites below the surface is given in kilometers. [From Kijko (1978, Fig. 8).]

5.2.2 Optimum Configuration of a Regional Network in the Ostrava Coal Basin in Czechoslovakia

In the Ostrava Coal Basin, where six coal mines are situated, 11 points x_i ($i = 1, \ldots, 11$) were selected as possible sites for seismic sensors, shown in Fig. 5.3 (Kijko, 1978). Four regions Ω_h^i, $i = 1, \ldots, 4$, of the highest seismic activity are also known. The space distribution of event occurrence $p(\mathbf{h})$ at any point of the active area Ω_h was assumed to be independent of epicenter coordinates and changing with depth only. Seismic network was to be made by 7 stations placed at 7 points out of the 11 possible sites shown in Fig. 5.3. Two criteria were accepted for the selection of seismic network configuration, and, correspondingly, two configurations were obtained. The first criterion is denoted by *xyz*. The network is optimal with respect to the criterion *xyz* when it ensures maximum information on all three coordinates of the focus location. This criterion is used when no particular importance is attached to any of the sought coordinates. The network is optimal with respect to the second criterion, denoted by *xy*, when it ensures maximum information on the coordinates *x* and *y* of a given event.

To solve the problem of *D*-optimal planning for the best estimates of selected parameters, the results of theoretical study of Hill and Hunter (1974) were used. In the literature, the considered case of *D* planning is called the D_S *criterion*. According to our assumptions, the sought parameters θ can be divided into two sets. The first set θ_1 is formed by m_1 parameters that should be determined with the best possible accuracy, whereas the second set θ_2 contains m_2 other parameters. Thus $\theta = \begin{pmatrix} \theta_1 \\ \theta_2 \end{pmatrix}$ and the matrix **A** is

$$\mathbf{A} = [\mathbf{A}_1 : \mathbf{A}_2], \tag{5.24}$$

where $\mathbf{A}_1$ is an $(n \times m_1)$ dimension matrix whose columns are formed by partial derivatives of arrival times with respect to the m_1 parameters of the

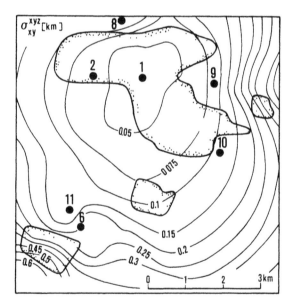

Figure 5.4 Expected standard errors of the epicenter location by a regional seismic network in the Ostrava Coal Basin, Czechoslovakia, optimized according to the criterion *xyz*. The areas of high seismicity and selected sites for seismic sensors are marked. [From Kijko (1978, Fig. 9).]

vector $\boldsymbol{\theta}_1$, and $\mathbf{A}_2$ is an $(n \times m_2)$ dimension matrix whose columns are formed by the derivatives of the other m_2 parameters of the vector $\boldsymbol{\theta}$. According to the results of Hill and Hunter (1974), the D-optimum estimation of the vector $\boldsymbol{\theta}_1$ leads to the distribution of the observation points $\mathbf{x}$ fulfilling the condition

$$L\left[\mathbf{C}_{\boldsymbol{\theta}}(\mathbf{x})\right] = \det\left[\mathbf{A}_1^{\mathrm{T}}(\mathbf{x})\mathbf{A}_3(\mathbf{x})\mathbf{A}_1(\mathbf{x})\right] = \max, \qquad (5.25)$$

$$\mathbf{x} \in \boldsymbol{\Omega}_x$$

where

$$\mathbf{A}_3(\mathbf{x}) = \mathbf{I} - \mathbf{A}_2(\mathbf{x})\left[\mathbf{A}_2^{\mathrm{T}}(\mathbf{x})\mathbf{A}_2(\mathbf{x})\right]^{-1}\mathbf{A}_2^{\mathrm{T}}(\mathbf{x}), \qquad (5.26)$$

and $\mathbf{I}$ is the diagonal unit matrix of dimension n. Thus, taking into account that the travel times are nonlinear with respect to the hypocenter coordinates

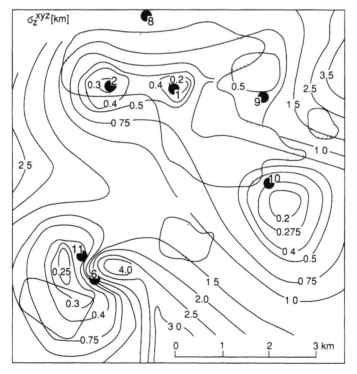

Figure 5.5 Expected standard errors of the focal depth determination by a regional network in the Ostrava Coal Basin, Czechoslovakia, optimized according to the criterion xyz. The areas of high seismicity and selected sites for seismic sensors are marked. [From Kijko (1978, Fig. 11).]

h, the optimum criterion (5.4) takes the final form

$$\int_{\Omega_h} p(\mathbf{h})\det\left[\mathbf{A}_1^{\mathsf{T}}(\mathbf{x},\mathbf{h})\mathbf{A}_3(\mathbf{x},\mathbf{h})\mathbf{A}_1(\mathbf{x},\mathbf{h})\right]d\mathbf{h} = \max. \qquad (5.27)$$

$$\mathbf{x} \in \Omega_x$$

It was assumed that in the first approximation an homogeneous and isotropic velocity model can be used. Thus, the P arrival times are described by formula (5.8) and the velocity model **m** is described by one parameter: the velocity V_p. The matrix $\mathbf{A}_1$ in formulas (5.25) and (5.26) is formed either by the last three columns of the matrix **A** in the case of the criterion xyz, or by two middle columns for the criterion xy. The domain Ω_x is formed by all combinations of possible station sites $(\mathbf{x}_{s_1},\ldots,\mathbf{x}_{s_7})$, $s_t = 1,\ldots,11$, whose total number is expressed by a Newton symbol $\binom{11}{7}$ equal to 330. Thus for each criterion 330 configurations were considered, selecting that one which fulfilled condition (5.27).

The best configuration in a sense of the criterion xyz was found for stations located at points 1, 2, 6, 8, 9, 10, and 11. Similarly, the best configuration in a sense of the criterion xy is for the stations located at points 1, 3, 6, 8, 9, 10, and 11. The quality of configurations can also be estimated from the maps of accuracy of epicenter and focal depth determinations. In Figs. 5.4 and 5.5 the maps of expected standard error of epicenter and depth determinations are shown for a network that is optimal in the sense of the criterion xyz. Similar maps for a network optimal in the sense of the criterion xy can also be drawn.

Chapter 6 | Selected Topics from Seismic Tomography in Mines

Seismic tomographic imaging can be used to map stress changes, faults, and the density variations in rocks. Periodic repetition of tomographic imaging in a given area would provide monitoring of rapidly changing stress conditions in mines. A combination of monitoring results and distress blasting techniques can be used to recognize, locate and control highly stressed rock masses (McGaughey *et al.*, 1987; Young *et al.*, 1989b). The seismic tomography technique can also be used before and after blasting to evaluate the efficiency of distressing operations (Blake, 1984; Young and Hill, 1985).

In general, information related to the state of the rock volume in a mine that has been traversed by seismic waves can be extracted from two kinds of seismic sources: controlled, artificially generated shots with known source positions and induced seismic events with unknown origin times and locations. The first approach is called *active tomography*, and the second is known as *passive tomography*. Some variants of passive tomography—known also under the name of *simultaneous inversion*—were discussed in Chapter 4. In this chapter we concentrate on the problems of active tomography.

The word tomography comes from the Greek word *tomos*, meaning section or slice. Thus tomography is based on the idea that an observed data set consists of integrals along lines or rays of some physical quantity (Bording *et al.*, 1987). The purpose of tomography is the production of an image of a medium such that the model's data agree approximately with measurements (McMechan, 1983). The aim of mining tomography is a reconstruction of a velocity model of the rockmass from the observed travel times and attenuation of seismic waves. In order to determine the basic parameters of rockmass and its physical state, access must be gained to at least two sides of a plane, such as between boreholes, mine walls, and levels or around pillars.

One of the first works related to seismic tomography was published by Bois *et al.* (1972), who used seismic wave travel times to detect a major structure between boreholes. Since then the technique has been successfully applied in a wide range of geophysical situations. In coal mining, for example, Mason (1981) applied travel-time tomography to a seam cross tunnel to reconstruct two-dimensional images of the stress distribution in a coal panel. One of the first successful applications of the explicit two-point ray tracing technique in mines was reported by Hermann *et al.* (1982). Cosma

(1983) has used crosshole seismic tomography at the Otomnaki ore mine, Finland. In his approach, the traced rays of P and S waves were restricted to travel in straight lines within blocks, changing direction according to Snell's law at blocks boundaries. Bodoky *et al.* (1985) used crosshole tomography to detect faults and dikes within coal seams in Hungary. Another application of travel time tomography in Hungarian coal mines, for monitoring of stress concentration areas, was described by Ivansson (1987).

Comprehensive tests of various algorithms in reconstructing P-wave velocity distributions was performed at the Restoff salt mine (U.S.A), known to contain a collapse structure (Peterson *et al.*, 1985). Jackson (1985) used this technique in the U.K. (British) coal industry to evaluate coal-seam conditions. By selecting an underdetermined system of equations, he estimated corrections corresponding to the variations in seam thickness. Gustavsson *et al.* (1986) have used crosshole tomography for mapping an ore body at the Kiruna Research Mine in Sweden. Travel-time seismic tomography has been also applied in Polish coal mines to detect seismic velocity anomalies and location of zones of rockburst hazard (Dubinski and Dworak, 1989). For this purpose, a technique of curved-ray tracing and inversion developed by Hermann *et al.* (1982) was used. Travel-time tomography was applied for mapping of coal seam structures in Australian mines (Rogers *et al.*, 1987). In Canada, Young *et al.* (1989b) have used P and S arrival times from small shallow blasts to monitor velocity anomalies and the integrity of rockmass in an underground hard-rock pillar. An attempt to utilize amplitude of seismic waves (amplitude tomography) in U.K. mines was described by New (1985). To locate small faults, Wong *et al.* (1985) have performed amplitude tomography at the Underground Research Laboratory in Manitoba.

The most comprehensive review of all aspects of seismic tomography in mines was prepared by McGaughey *et al.* (1987). Unfortunately, the work was published as a research report of the Department of Geological Sciences, Queen's University, Kingston, Ontario, and is not readily accessible. Other reviews of seismic tomography development and corresponding problems can be found in several publications, such as, for example, those by Ivansson (1987), Thurber and Aki (1987), Roberts *et al.* (1989), and Young *et al.* (1989b).

6.1   Mathematical Principles

6.1.1   Travel-Time Tomography

The essence of travel-time tomography is the fact that the travel time associated with a given seismic ray (i.e., the total transit time from the source to a receiver) is the integrated slowness along that ray. The travel time is the integral

$$T(\text{ray}) = \int_{\text{ray}} s(x, y, z)\, dl, \tag{6.1}$$

where $s(x, y, z)$ is the slowness (reciprocal of seismic wave velocity) along the ray path and dl is the differential distance along the ray.

A fundamental difficulty with the seismic tomography problem is that the ray path itself depends on the unknown slowness. Equation (6.1) is therefore nonlinear in respect to s (Bording *et al.*, 1987; Nolet, 1987). The approach traditionally used is to linearize equation (6.1) about some initial slowness model $s(x, y, z) \cong s_0(x, y, z) + \delta s(x, y, z)$ and instead of solving (6.1) for s, some approximation to (6.1) is solved for the perturbations in slowness δs. Equation (6.1) then becomes, to the first order,

$$\delta T = \int_{\text{ray}} \delta s(x, y, z)\, dl, \tag{6.2}$$

where δT and $\delta s(x, y, z)$ are the perturbations in travel time and slowness, respectively. If the medium is divided into blocks, equation (6.2) can be written as

$$\delta T_i = \sum_{j=1}^{m} l_{ij}\, \delta s_j, \tag{6.3}$$

where δT_i is the time delay associated with the ith ray, δs_j is the slowness perturbation of the jth block, and l_{ij} is the length of the ith ray segment in the jth block (Fig. 6.1).

In matrix form the discrete representation (6.3) can be written as

$$\delta \mathbf{T} = \mathbf{L}\, \delta \mathbf{s}, \tag{6.4}$$

where $\mathbf{L}$ is an $(n \times m)$ matrix with elements l_{ij} $(i = 1, \ldots, n; j = 1, \ldots, m)$, for n being the number of seismic rays and m being the number of model blocks; $\delta \mathbf{s}$ is an m-dimensional column vector with components δs_j (where $j = 1, \ldots, m$); and δT is an n-dimensional column vector with components δT_i (where $i = 1, \ldots, n$).

6.1.2 Attenuation Tomography

The formalism introduced by equations (6.1)–(6.4) is also applicable to amplitude tomography for determination of the attenuation factor Q (e.g., Evans and Zucca, 1988; Justice and Vassiliou, 1990). Let $Q(x, y, z)$ and $s(x, y, z)$ denote the Q and slowness functions, respectively. If f is the frequency and $A(f)$ is the amplitude spectrum recorded by a given receiver

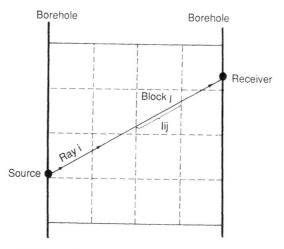

Figure 6.1 Linearized ray path between the source and the receiver and its discretization into blocks. The ith ray travels the distance l_{ij} in the jth block

and corrected for the source spectrum, geometric spreading, and instrumental response, we have (Ivansson, 1987)

$$A(f) = \exp\left[-f \int_{\text{ray}} \frac{s(x, y, z)}{Q(x, y, z)} \, dl \right]. \tag{6.5}$$

Taking the logarithm, we obtain

$$-\frac{1}{f} \ln A(f) = \int_{\text{ray}} \frac{s(x, y, z)}{Q(x, y, z)} \, dl. \tag{6.6}$$

Equation (6.6) has the same form as equation (6.1) with $-[\ln A]/f$ and s/Q replacing T and s, respectively, and the formalism used to solve equations (6.2)–(6.4) can be applied in this case as well. Thus, incorporating amplitude data, the Q structure of the rockmass can be determined if the velocity structure is known.

It should be noted, however, that amplitude tomography is much more difficult to use in practice than travel-time tomography. The simple formulation of the problem is possible only in such cases where amplitude dissipation is influenced mainly by absorption of the rock and not by interference effects. This is valid in the case of absorbing anomalies larger than the seismic wavelength, if they are not connected with strong-velocity anomalies. Amplitude decay caused by scattering, diffraction or reflection effects at rock boundaries, such as faults, are not taken into account in our simple formulation of the line integral (6.5) (Neumann-Denzau and Behrens, 1984; Ivansson, 1987). In addition, it is difficult to account for the source spectrum, source and receiver directivity, and geometric spreading. Attenuation tomography, on the other hand, is well worth the effort, since amplitude data carry additional information about the state of the rockmass that complements the information derived from travel time data.

6.2 Effects of Seismic Ray Bending

From mathematical principles of tomography, described in the previous section, it follows that the problem we face is identification of an algorithm that can reconstruct source-receiver propagation paths, given that the ray paths are not necessarily straight lines, as they are dependent on the velocity distribution within the rockmass, and are not known in advance.

In most cases of seismic tomography applications in mines (especially in the early days of development), ray paths were approximated by straight lines. It is obvious that when velocity structure is not homogeneous, this may result in serious artifacts, leading to the necessity of using highly complex and computationally expensive two-point ray tracing technique. Comprehensive reviews of the ray tracing techniques can be found in a number of publications, such as those of Červeny (1987), Nolet (1987), Thurber and Aki (1987), Vidale (1988, 1990), and Virieux (1991).

Despite numerous applications, detailed analysis of the significance of ray bending in tomographic imaging so far have not been done. Several investigations have shown, however, that straight-line approximation is valid when seismic velocity variations within the rock do not depart from the average value by more than 10–15 percent, and when the size of disturbance is smaller than the seismic wavelength (Dines and Lyttle, 1979; Ramirez, 1986; Ivansson, 1987; Roberts et al., 1989). An extensive analysis and comparison of straight-line and bent-ray paths in mining tomography was performed by Kasina (1988). The analysis includes the evaluation of applicability of straight-line ray paths for selected models of medium, such as faults, low-velocity anomalies, and layered rockmass structures. An interesting analysis of different sources of distortion in tomographic velocity images, based on synthetic and field data, was performed by Dyer and Worthington (1988). They concluded that since the greater the velocity contrast, the greater is the ray bending and the possibility of uneven ray coverage, it is difficult to define any model independent guidelines concerning the effectiveness of straight ray as compared to ray bending inversion. They also found that the errors in locations of shots and receivers can lead to significant distortion in the tomographic velocity image, distortion that can be much more significant than that in the straight line ray approximation, despite the occurrence of pronounced ray bending.

A less ambitious, but very efficient, way to take into account the ray-bending effects is the technique used in the past, for example, by Chander (1977), Julian and Gubins (1977), and also Aki et al. (1977) for velocity inversion employing P arrival times from distant earthquakes. The procedure takes into account the ray bending effects by approximating integral (6.2) by a sum of straight paths within each block and applies Snell's law at the block interfaces. The original one-dimensional approach of Aki et al. (1977) was extended to three dimensions and extensively discussed by Koch (1985b). The method was found to be highly useful for imaging of three-dimensional velocity structures and has been applied to various geophysical problems (e.g., Waltham, 1988; Engell-Sørensen, 1991a, 1991b). An alternative procedure, taking into account the ray-bending effects but still using the straight-line paths, was described by Bates and McKinnon (1979) and Ivansson (1985). The procedure is based on the evaluation of the straight-line intervals of the true slowness model by adding, to the recorded times, the appropriate time differences calculated from the current velocity model.

An entirely different incorporation of the ray bending effects was proposed by Devaney (1984). As the input data, he used Fourier spectra of the recorded signals. It was shown that this technique is capable of reconstructing the velocity distribution in weakly inhomogeneous rock formations.

In addition to the errors associated with the ray path modeling, tomographic velocity images can be distorted by several other factors. One of them, connected with the effect of seismic wave diffraction and recently discovered by Wieland (1987), seems to play a major role. Wieland raised the question as to whether we can safely neglect the effect of seismic-wave diffraction in heterogeneous structures. Since, from the observer's point of view, there is no difference between direct and diffracted waves, the diffracted waves arriving before direct waves would be read as the early arrivals, while the diffracted waves arriving later would be generally ignored. In order to investigate this problem, Wieland (1987) considered a geometrically simple case: a spherical inclusion in a homogeneous space, for which he derived an analytical solution. Originally, he assumed that the anomalous inclusion has a 100-km diameter, the velocity anomaly varies between ± 10 percent, and the homogeneous space velocity is 8 km/s. Two cases have been analyzed. The first one for the incident signal with the dominant frequency of 1.5 Hz and cutoff frequency of 3 Hz, and the second one with the dominant frequency of 5 Hz and cutoff frequency of 10 Hz. The important fact is that identical results can be obtained with all linear dimensions and all velocities m times smaller or greater, where m is any positive number. The same applies if we leave the velocity unchanged, take the linear dimensions and the path length m times smaller or greater, and choose the characteristic frequencies m

times greater or smaller. It is easy to show that all Wieland's qualitative conclusions can be applied to distances and frequencies used, for example, in crosshole transmission tomography in mines.

Wieland's most significant result is that there is a strong asymmetry affecting positive and negative velocity anomalies. The ray approximation works well for positive velocity anomalies, but in the presence of a negative anomaly, the observed travel times are likely to be the travel times of the fastest path around the anomaly, even at relatively short distances. Since we are not able to distinguish direct and diffracted waves, and the first one is used for tomographic modeling, negative anomalies are unlikely to manifest themselves in travel time data. If such anomalies exist, they are likely to be underestimated in a tomographic inversion. Similar analysis performed for positive velocity anomalies shows that a fast inclusion will appear geometrically larger than it actually is. These disquieting phenomena, known as Wieland's effects (Nolet, 1987), have not been recognized so far, but it is likely that they have influenced many published tomographic results.

6.3 Inversion Techniques

Once the seismic rays have been traced through a rock, a potentially very large system of equations [(equations (6.4)] must be solved for the unknown slowness or the attenuation perturbation vector δs. The number of unknowns can be of the order of 10^3, and the system is poorly conditioned. In addition, the problem of storage and large matrix manipulation in a reasonable computational time must also be solved. Furthermore, since a given seismic ray intersects only a small portion of the rock, most of the elements of L matrix in that row are equal to zero. Thus the matrix L is sparse, with 99 percent or even more of the elements equal to zero for large problems. The classic least-squares solution of equations (6.4) is equivalent to the system of normal equations $L^T L = L^T \delta s$, which require an inversion of the $(m \times m)$ matrix $L^T L$. Since the number m of unknowns can be very large, the construction, storage, and direct inversion of $L^T L$ is not possible. In addition, direct solution methods based on the transformations of L matrix, such as the singular value decomposition (SVD) method, are also not attractive since they require too much computer time and lead to large, dense, intermediate matrices. Although the original matrix L is sparse, it fills up during the computational process. In most real cases, therefore, the solution of equations (6.4) cannot be obtained by techniques that require the use of explicit L or $L^T L$ matrices.

The solution of the problem can be achieved by an approach in which an access to only one row of the matrix L is required at a time and L may reside in a secondary storage. Two different classes of such one-row-action algorithms have been used so far in seismic tomography: the stationary iterative backprojection, and conjugate gradient methods. The stationary iterative backprojection methods are the most popular. They form a large family of reconstruction algorithms, also known under the name of *algebraic reconstruction techniques* (ARTs). The ART for image reconstruction was first introduced into the open literature by Gordon *et al.* (1970). Almost the same algorithm has been proposed in a patent description by Hounsfield (1972), originally filed in 1968. The simplest variant of the ART procedure, however, was known much earlier (Karczmarz, 1937) and applied for solving the systems of consistent linear equations. An extensive discussion of different aspects of the method, including historical notes, is given by Herman (1980).

The original ART technique, introduced by Gordon *et al.* (1970), is based on applying corrections to all blocks contributing to a single ray and repeating the procedure for each ray, taking into account the previous changes. The simplest version of the ART algorithm involves the following steps:

(1) Guess initial approximations δs_j^* $(j = , \ldots, m)$ for each seismic ray i $(i = 1, \ldots, n)$.

(2) Compute all ith ray segments l_{ij} (if they differ from the previous iteration) and residuals $r_i^* = \delta T_i - \Sigma l_{ij} \cdot \delta s_j^*$, where the sum is taken for all blocks passed by the ith ray.

(3) Adjust the δs_j values according to the formula

$$\delta s_j = \delta s_j^* + l_{ij} r_i^* / \sum_k l_{ik}^2. \tag{6.7}$$

These values are applied now in the next ray, until all rays have been used.

(4) Repeat steps 2 and 3 until some termination criteria are met. At this point we set $\hat{s} = s_0 + \delta s^*$ as a solution for the slowness or attenuation model s.

It should be noted that the above iterative scheme updates the solution along a single ray. This causes the final solution to depend on the order in which the rays are considered. In addition, the procedure converges only if $L\delta = s = \delta T$ has an exact solution (Van der Sluis and Van der Vorst, 1987). In practice, the perturbation vector δs oscillates as a result of model approximations and errors involved in data. The convergence of the procedure can be improved by computing the corrections for all rays first, keeping the residuals fixed, and averaging these corrections before updating the approximation for δs. In this way, formula (6.7) should be replaced by

$$\delta s_j = \delta s_j^* + \frac{1}{n_j} \left(\sum_i l_{ij} \cdot r_i^* \Big/ \sum_k l_{ik}^2 \right), \tag{6.8}$$

where n_j denotes the number of rays passing through block j. Methods in which the solution is updated only after all equations (rays) have been processed are called simultaneous iterative reconstruction techniques (SIRTs). Note that the SIRT solution is independent on the order in which the rays are supplied. More details, comparison, and extensive discussion of numerical procedures for solving the large sparse matrix systems arising from tomographic problems are given by Van der Sluis and Van der Vorst (1987). Physical interpretation of the SIRT procedure and its similarity with the generalized least squares inversion are discussed by Trampert and Leveque (1990).

Different variants of the backprojection technique have been successfully applied to invert seismic travel-time data for the velocity structure from large sets of earthquakes (e.g., Humphreys *et al.*, 1984; Nakanishi, 1985; Humphreys and Clayton, 1988). One of the most impressive applications of this technique was performed by Clayton and Comer (1984) and Comer and Clayton (1984). In their study of laterally heterogeneous structure of the whole Earth, 50,000 blocks were distinguished and over one million observations were used. A row action technique was required, since the straight inversion of a 50,000 × 50,000-element matrix is, at present, not possible. Some other applications of the ART technique have been carried out for exploration geophysics and in mines (e.g., McMechan, 1983; Neumann-Denzau and Behrens, 1984; Scales, 1987; Dubinski and Dworak, 1989; Roberts *et al.*, 1989).

The second class of the procedures employed for the solution of large and sparse systems of equations generated by seismic tomography is based on a classic conjugate gradient algorithm (Hestenes and Stiefel, 1952). The computational scheme of the conjugate gradient procedure for least squares can be described as follows (Scales, 1987):

(1) Guess trial values of the perturbation vector δs^*.
(2) Compute $r = \delta T - L\delta s^*$, $u^* = p = L^T r^*$, $q^* = Lp^*$, and $\alpha^* = (u^*, u^*)/(q^*, q^*)$, where (x, x) denotes an inner product equal to $x^T x$.
(3) Adjust vector δs to $\delta s = \delta s^* + \alpha p^*$. The adjusted values of δs become now the new trial values of the perturbation vector δs.
(4) Adjust vectors r, u, p, q and scalar α to $r = r^* - \alpha q^*$, $u = L^T r$, $p = u + (u, u)/(u^*, u')$, $q = Lp$, and $\alpha = (u, u)/(q, q)$.
(5) Repeat steps 3 and 4 until some termination criteria are met. At that point we set $\hat{s} = s_0 + \delta s^*$ as the solution for s.

The conjugate gradient method is highly effective because it is possible to employ normal equations $L^T L \delta s = L^T \delta T$ without explicitly forming the product $L^T L$. This is important, since, as it was shown in Chapter 4, the implicit solution of equation (6.4) is more stable than that based on the explicit solution of normal equations. If the matrix L in equation (6.4) is ill-conditioned (which happens very often in tomography problems), then the matrix

$L^T L$ in the system of normal equations will be much more ill-conditioned, as the condition number of $L^T L$ is the square of the condition number of L. An additional advantage of the conjugate gradient method is the fact that it consists of a few simple vector operations and that the correction vectors **p** are calculated recursively and are not stored.

The conjugate gradient method was introduced to seismic tomography by Nolet (1985). In a numerical experiment, in which he compared the performance of different inversion techniques, he used the version of the conjugate gradient method developed by Page and Saunders (1982), known under the name of LSQR. The LSQR procedure is similar to the inversion technique based on the singular value decomposition (SVD) method, discussed in Chapter 4. For the large-size matrix L, however, it is not possible to calculate the eigenvectors of $L^T L$ in any reasonable computation time. Page and Saunders (1982) proposed, instead of diagonalizing $L^T L$, to tridiagonalize it using a simple scheme in which the columns of the transformation matrix turn out to be orthogonal. A simple scheme of the LSQR algorithm is given by Nolet (1987). Van der Sluis and Van der Vorst (1987) show that the first iteration solutions obtained by the LSQR do not contain components belonging to the very small eigenvalues of $L^T L$. The smallest eigenvalues eventually contribute to the solution more and more as the iteration proceeds.

Several studies have shown that the LSQR procedure is an extremely useful technique for solving large and sparse systems of equations generated by seismic tomography. The method is sufficiently fast and accurate. Spakman (1985, 1986) has used the LSQR for a large-scale upper-mantle tomography. An extensive investigation of how well the SIRT and LSQR algorithms are able to solve tomographic problems, in which many model parameters and a large amount of input data are used, was performed by Nolet (1985) and Spakman and Nolet (1988). Lees and Crosson (1989) used the LSQR procedure for a three-dimensional velocity structure inversion in a small block (2 km horizontally and 2 km vertically) in the vicinity of the Mount St. Helens Volcano. The classic conjugate gradient algorithm of Hestenes and Stiefel (1952) has been successfully applied in exploration geophysics (e.g., Koehler and Taner, 1985; Gerszterkorn et al., 1986). An interesting numerical experiment simulating mining conditions and the LSQR seismic-wave velocity inversion was performed by Scales (1987). He also reviewed the performance of several different inversion techniques and concluded that the algorithms based on the conjugate gradient method are fast and accurate and can be easily adapted to take the advantage of the sparsity of tomographic problems.

To demonstrate the effectiveness of seismic tomography applied under mining conditions, the results of velocity reconstruction in a hard-rock pillar at a Sudbury mine in Canada, performed by Young et al. (1987), are briefly described. The investigation was conducted in an 100×200-m pillar, 760 m

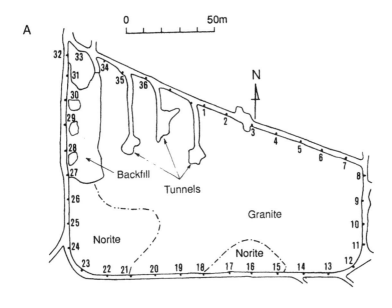

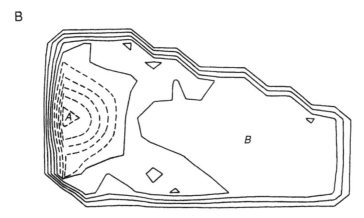

Figure 6.2 Seismic wave velocity reconstruction in a hard-rock pillar at a Sudbury mine, Canada. (A) Plan of the pillar showing 36 source and receiver positions and the backfilled area. (B) Results of tomographic inversion showing low-velocity anomaly in the backfilled area of the pillar and relatively homogeneous high velocity area away from the backfill. The estimated P-wave velocity at point A is 2500 m/s and at point B in the center of the pillar it is 6500 m/s. [From Young *et al.*, 1987 Fig. 3); reprinted from: Herget, G. & S. Vongpaisal (eds.), Proceedings: 6th congress of the International Society for Rock Mechanics-Comptes-rendus: 6èmé congrés de la Societé Internationale de Méchanique des Roches/Berichte: 6er Kongress der Internationalen Gesellschaft für Felsmechanik, Montreal, Canada, 1987. 1987-90. 1854 pp., 3 volumes, Hfl.850. A.A. Balkema, P.O. Box 1675, 3000 BR Rotterdam.]

below the surface. An array of 36 single-component accelerometers was fixed to the pillar walls and the sensors surrounded the pillar (Fig. 6.2A). Shots were successively fired at each sensor location and recorded at 36 points. Direct P-wave arrival times were picked up from over 1200 records and used as the input data in the ART-based reconstruction algorithm. The results indicate the presence of a low-velocity zone, approximately 2500 m/s, in the backfilled part of the pillar (Fig. 6.2B). Most of the pillar's central area is characterized by a high-velocity zone, approximately 6500 m/s, suggesting fairly homogenous intact rockmass. The edges of the pillar are distressed and the P-wave velocity there was found to be about 5500 m/s.

Other applications of seismic tomography in mines can be found in the work of Ivansson (1987). Aspects of seismic tomography, such as its potential and limitations, ray coverage, resolution in the image reconstruction, and identification of artifacts, are discussed in detail by Gustavsson *et al.* (1986), Ivansson (1986), Menke (1989), Phillips and Fehler (1991), and many other authors.

Chapter 7 | Stress-Induced Anisotropy and the Propagation of Seismic Waves

The propagation of seismic waves in a perfectly elastic and isotropic medium is described in Chapter 3. A perfectly elastic material is aniso-tropic if it deviates from the directionally regular elastic behavior of an isotropic material; the elastic properties of a medium have different values for different directions. Anisotropy in rocks is a result of many factors, such as variation in material properties, distribution of crystals and cracks, and various stress effects. It has two major effects on the propagation of seismic waves in the Earth's crust. The velocities of both P and S waves vary with the direction of propagation, and S waves split into two or more phases with fixed velocities and polarizations that propagate through the region of anisotropy (e.g., Crampin, 1987b). The first effect is small and difficult to recognize, whereas the second effect is easily recognized whenever suitable three-component records of S waves are obtained and analyzed in a digital form.

Seismic anisotropy is a new field undergoing rapid expansion in recent years. Observations of shear-wave splitting above small earthquakes in Turkey led Crampin *et al.* (1984b) to suggest that stress-aligned fluid-filled micro-racks surround the zone of seismicity, and these they called *extensive-dilatancy anisotropy* (EDA). As a result of further extensive research, it is now recognized by many seismologists, although not by all, that the azimuthal anisotropy associated with S-wave splitting in most rocks in the Earth's crust is caused by preferentially oriented fluid-filled inclusions, aligned by the contemporary stress field acting on the rockmass.

Until recently, most seismic studies in the crust were confined to P waves, and liquid-filled EDA cracks have very little effect on P waves. Liquid-filled EDA cracks have also little effect on S-wave velocity, but they cause their splitting, which is likely to be the most reliable indicator of anisotropy in the crust (Crampin, 1987a). Since distributions of cracks and stresses are of crucial importance in mines, monitoring of shear waves could become an important part of seismic research at mines, provided that three-component shear-wave motion is recorded digitally at high sampling rates.

7.1   Stress–Strain Relations

The most general stress–strain relations are described by formula (3.5). From general considerations of strain energy in a perfectly elastic homogeneous and anisotropic medium it follows that relations of the form $A_{ijkl} = A_{klij}$ hold between the elastic constants in (3.5). From this it further follows that not more than 21 of these coefficients can be independent. Thus 21 elastic constants are required to describe the elastic behavior of the general anisotropic material. The properties of symmetry provide further relations between the elastic constants for other specific types of the material.

For rock materials, the most important is orthorhombic symmetry, in which there are three mutually perpendicular planes of symmetry. In this case there are nine independent elastic constants. It is often convenient in anisotropic analysis to describe the stress–strain relations (3.5) in the form

$$p_i = c_{ij} e_j, \qquad (i, j = 1, \ldots, 6). \tag{7.1}$$

Then the stress–strain relations for orthorhombic symmetry, with the axes of symmetry taken along the x_1, x_2, and x_3 directions, are

$$\begin{aligned}
p_1 &= c_{11} e_1 + c_{12} e_2 + c_{13} e_3, \\
p_2 &= c_{12} e_1 + c_{22} e_2 + c_{23} e_3, \\
p_3 &= c_{13} e_1 + c_{23} e_2 + c_{33} e_3, \\
p_4 &= c_{44} e_4, \\
p_5 &= c_{55} e_5, \\
p_6 &= c_{66} e_6,
\end{aligned} \tag{7.2}$$

with nine independent constants c_{ij}.

Young's modulus for uniaxial stress in the x_1 direction is

$$\frac{p_1}{e_1} = \frac{D}{\left(c_{22} c_{33} - c_{23}^2\right)}, \tag{7.3}$$

where

$$D = \begin{vmatrix} c_{11} & c_{12} & c_{13} \\ c_{21} & c_{22} & c_{23} \\ c_{31} & c_{32} & c_{33} \end{vmatrix}, \tag{7.4}$$

and there are different Poisson's ratios for the x_2 and x_3 directions, since

$$\frac{e_2}{e_1} = \frac{c_{23} c_{13} - c_{12} c_{33}}{c_{22} c_{33} - c_{23}^2} \quad \text{and} \quad \frac{e_3}{e_1} = \frac{c_{12} c_{23} - c_{22} c_{13}}{c_{22} c_{33} - c_{23}^2}. \tag{7.5}$$

Thus each direction of stress has its own Young's modulus and its own variation of Poisson's ratio.

If there are three orthogonal planes of symmetry at a point, the strain–energy function [see relation (3.14)] may be written in the form

$$W = \tfrac{1}{2}\big(Ae_{11}^2 + Be_{23}^2 + Ce_{33}^2 + 2Fe_{22}e_{33} + 2Ge_{33}e_{11} \\ + 2He_{11}e_{22} + Le_{23}^2 + Me_{31}^2 + Ne_{12}^2 \big),$$

(7.6)

involving nine independent coefficients.

A further simplification occurs when there is a single axis of symmetry, and the properties of the material are the same in all directions at right angles to the axis. Rocks at great depths have been compressed by the overburden, and this may affect their elastic properties in the vertical and horizontal directions. Their elastic constants for deformation at right angles to the vertical direction might be equal but different from those for deformations parallel to it. This is also the case of sedimentary rocks with the vertical axis perpendicular to the bedding. Such materials for which the stress–strain relations are symmetric about a fixed axis are called *transversely isotropic*. It is recommended that the term *transverse isotropy* should be used only for hexagonal symmetry system when the symmetry axis is normal to the free surface, and anisotropy or azimuthal anisotropy be reserved for all other symmetry systems and for other orientations of hexagonal systems (e.g., Crampin *et al.*, 1984a). For the vertical axis of symmetry, the stress–strain relations are

$$p_1 = c_{11}e_1 + (c_{11} - 2c_{66})e_2 + c_{13}e_3,$$
$$p_2 = (c_{11} - 2c_{66})e_1 + c_{11}e_2 + c_{13}e_3,$$
$$p_3 = c_{13}e_1 + c_{13}e_2 + c_{33}e_3,$$
$$p_4 = c_{44}e_{23},$$
$$p_5 = c_{55}e_{31},$$
$$p_6 = c_{66}e_{12},$$

(7.7)

involving five independent elastic constants $A = c_{11}$, $F = c_{12}$, $L = c_{44}$, and two others, C and N. Thus the number of elastic constants increases from two for the isotropic medium to five for the transversely isotropic material.

7.2 Equations of Motion

There are many similarities between the behavior of seismic waves in isotropic and transversely isotropic media. In both cases, wave motion in the vertical plane (*P-*, *SV-* and Rayleigh-wave motion), called often the *sagittal plane* when referred to through the direction of phase propagation, decouples from the horizontal transverse motion (*SH-* and Love-wave motion) in plane-layered structures. The propagation of seismic waves in such structures can be described by simple analytical expressions. The behavior of seismic waves in general anisotropic media with azimuthal variations of elastic properties is much more complicated. Here the sagittal and horizontal–transverse motions are not decoupled and their properties must be calculated numerically.

The equations of motion without body force in general anisotropic, homogeneous, and perfectly elastic, media are

$$\rho \frac{\partial^2 u_i}{\partial t^2} = A_{ijkl} \frac{\partial^2 u_k}{\partial x_j \partial x_l}. \tag{7.8}$$

The analysis of the fourth-order tensor A_{ijkl} of 21 independent elastic constants is rather complex and matrix notation is needed. The Cartesian system in equations (7.8) can be always rotated to have the direction of phase propagation along the x_1 axis, with the x_3 axis being vertical.

Substitution of the equation for plane waves propagating in the x_1 direction with the phase velocity c

$$u_i = a_i \exp[i\omega(t - x_1/c)], \tag{7.9}$$

where a_i is the amplitude vector defining the polarization of particle motion, into the equations of motion (7.8) provides three linear equations

$$\rho c^2 a_1 = A_{1111}a_1 + A_{1121}a_2 + A_{1131}a_3,$$

$$\rho c^2 a_2 = A_{2111}a_1 + A_{2121}a_2 + A_{2131}a_3, \tag{7.10}$$

$$\rho c^2 a_3 = A_{3111}a_1 + A_{3121}a_2 + A_{3131}a_3,$$

where the common multiplier $(-i\omega)^2 \exp[i\omega(t - x_1/c)]$ has been omitted. These equations may be solved in a variety of ways. The preferred technique for numerical solution is to write the equations as a linear eigenvalue problem (e.g., Crampin, 1984a)

$$(\mathbf{M} - \rho c^2 \mathbf{I})\mathbf{a} = 0, \tag{7.11}$$

where $\mathbf{M}$ is the 3×3 matrix with elements A_{i1k1}, $\mathbf{I}$ is the 3×3 identity matrix, and $\mathbf{a}$, with elements a_i, is the amplitude vector of the displacement. The matrix $\mathbf{M}$ is real, symmetric, and positive definite. Consequently, the eigenvalue problem (7.11) has three real positive roots for ρc^2 with orthogonal eigenvectors.

The three real roots of equations (7.11) can be identified with three body waves in every direction of phase propagation with orthogonal polarizations and with velocities, in general, different and varying with direction. These waves correspond to a quasi-P wave, denoted by qP, with approximately longitudinal particle polarization, and two quasi–S waves, denoted by $qS1$ and $qS2$, with approximately transverse particle motion. The waves are therefore not, in general, purely dilatational and rotational, and their particle motions are not parallel to the displacements of P, SV, and SH waves in isotropic horizontally layered solids.

A consequence of the dependence of the phase velocity on direction in anisotropic media is that the wavenumber κ is a vector rather than scalar quantity of isotropic media. Body waves, surface waves, and group velocities are all affected. The group velocity $C = \partial\omega/\partial\kappa$ in isotropic media now also becomes a vector

$$C = (\partial\omega/\partial\kappa_1, \partial\omega/\partial\kappa_2, \partial\omega/\partial\kappa_3),   (7.12)$$

and the wave energy is not transported parallel to the phase velocity, except in a few restrictive cases. As a result, description of reflection and refraction of plane body waves in anisotropic media is very difficult. Since the three plane waves traveling in the same direction are not generally radial and transversal, anomalous conversions may occur between pairs of qP and qS waves at the interfaces of different media.

7.3 Anisotropic Symmetry Systems

The general triclinic material is described by the full 21 independent elastic constants and has no planes of symmetry. The other extreme case is the isotropic material with two elastic constants, in which all planes are symmetry planes. All others anisotropic symmetry systems are readily classified by the relative arrangement of planes of symmetry into eight distinct systems, described in detail by Crampin (1984a). All the more common symmetry systems have one or more symmetry planes with the spatial orientations shown in Fig. 7.1, reproduced from Crampin (1984a).

The orientation of the anisotropic symmetry planes to the free surface and to the direction of propagation is a major factor for the propagation of seismic waves. Symmetry planes have two major effects on body waves. There are two cases when P and SV motions are decoupled from SH motion: (1) when the propagation direction is in a vertical symmetry plane and (2) when the propagation direction is in a symmetry plane, and the polarization of the P wave and one of the S waves is parallel and the polarization of the other S wave is at right angles to the symmetry plane (Crampin, 1984a).

There are particularly simple relations between the elastic constants c_{ij} [$i,j = 1,\dots,6$; see relation (7.1)] and the velocities of body waves propagating in symmetry planes in weakly anisotropic solids. The three body-wave phase velocities in the plane $x_3 = 0$ in the x_1, x_2, and x_3 coordinate system, where

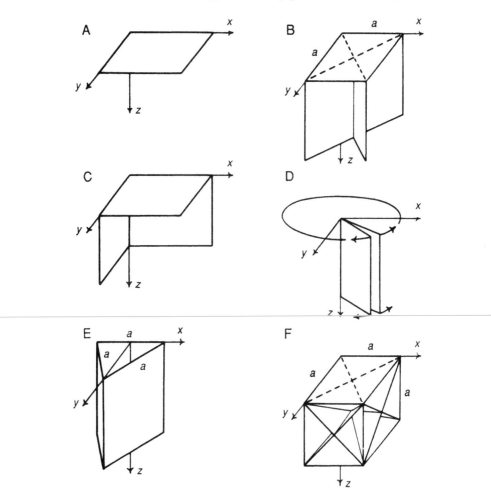

Figure 7.1 Orientation of symmetry planes of the more common anisotropic symmetry systems: (A) monoclinic; (B) tetragonal; (C) orthorhombic; (D) hexagonal; (E) trigonal; (F) cubic. [From Crampin (1984a, Fig 2), copyright by Royal Astronomical Society]

x_1, x_2, and x_3 are not necessarily principal axes, are

$$\rho V_P^2 = A_1 + A_2 \cos 2\theta + A_3 \sin 2\theta + A_4 \cos 4\theta + A_5 \sin 4\theta,$$
$$\rho V_{SP}^2 = B_1 + B_2 \cos 4\theta + B_3 \sin 4\theta, \tag{7.13}$$
$$\rho V_{SR}^2 = C_1 + C_2 \cos 2\theta + C_3 \sin 2\theta,$$

where V_P, V_{SP}, and V_{SR} are the velocities of the wave qP and the two quasi–shear waves qSP and qSR polarized parallel and at right angles, respectively, to the symmetry plane; θ is the azimuth measured from the x_1 direction; and

$$A_1 = \tfrac{1}{8}[3(c_{11} + c_{22}) + 2(c_{12} + 2c_{66})],$$
$$A_2 = \tfrac{1}{2}(c_{11} - c_{22}), A_3 = (c_{61} + c_{62}),$$
$$A_4 = \tfrac{1}{8}[c_{11} + c_{22} - 2(c_{12} + 2c_{66})],$$
$$A_5 = \tfrac{1}{2}(c_{61} - c_{62}), \tag{7.14}$$
$$B_1 = \tfrac{1}{8}[c_{11} + c_{22} - 2(c_{12} - c_{66})],$$
$$B_2 = -A_4, B_3 = -A_5,$$
$$C_1 = \tfrac{1}{2}(c_{53} + c_{44}), C_2 = \tfrac{1}{2}(c_{55} - c_{44}), C_3 = c_{45}.$$

The P-wave equation in system (7.13) was derived by Backus (1965); and the shear-wave equations, by Crampin (1977). The equations are correct to the first order in the difference between the anisotropic and isotropic elastic constants. They become especially simple when the azimuth θ is measured from a direction of sagittal symmetry (when the sagittal plane is a plane of mirror symmetry). Then the coefficients of the sine terms in equations (7.13) vanish, and the reduced equations with cosine terms only are obtained (Crampin, 1977):

$$\rho V_P^2 = A_1 + A_2 \cos 2\theta + A_4 \cos 4\theta,$$
$$\rho V_{SP}^2 = B_1 + B_2 \cos 4\theta, \tag{7.15}$$
$$\rho V_{SR}^2 = C_1 + C_2 \cos 2\theta,$$

where the coefficients are specified by relations (7.14).

7.4 Wave Propagation in Cracked Solids

The presence of aligned inclusions, such as cracks, pores, or impurities, is probably the most common cause of effective anisotropy in the Earth (e.g., Crampin, 1981). Wave propagation in a medium containing a uniform weak concentration of aligned cracks can be simulated by wave propagation in a purely elastic anisotropic solid that has the same velocity variations with direction as the cracked solid (Crampin, 1978). It is always possible to make this approximation for dilute concentrations of inclusions, to avoid their interactions, when the dimensions of the concentrations are small in comparison with the wavelength of seismic waves. A key step in this procedure is the use of effective elastic constants for the cracked material describing wave propagation in anisotropic media.

Several theories have been developed to calculate the effective elastic constants of materials containing aligned circular cracks or ellipsoidal inclusions. Crampin (1978) obtained effective elastic constants by modeling the variation of wave velocity in a cracked solid derived by Garbin and Knopoff (1973, 1975a, 1975b) in their first-order approximation theory. A more general approach for calculating the elastic constants of cracked media, including first- (no mutual interactions between the cracks) and second-order (crack–crack interactions accounted for) interactions between the scattering inclusions, was developed by Hudson (1981). His theory, extended and used by Crampin (1984b), provides one of the most popular models employed for the explanation of observed anisotropy. These theories are based on the scattering of long-wavelength waves by the cracks, under the assumptions that the cracks are in dilute concentration and have small aspect ratios. The aspect ratio d of an ellipsoidal inclusion with rotational symmetry and with semiaxes a, a, and c is $d = c/a$. The models assuming $c \ll a$ correspond to circular cracks and are occasionally called "flat-crack models" (Douma, 1988). The other limiting case corresponds to spherical inclusions with the aspect ratio $d = 1$.

The aspect ratio of inclusions in the Earth is not necessarily small (Crampin et al., 1986), and it is probably the most important crack parameter affected by stress changes (Peacock et al., 1988). Anderson et al. (1974) and Nishizawa (1982) presented models for calculation of the effective elastic constants of solids containing aligned ellipsoidal inclusions, which are valid for all values of the aspect ratio of these inclusions. Unlike the model of Hudson (1981), which is based on the scattering of elastic waves, both the models of Nishizawa (1982) and Anderson et al. (1974) are based on a static approach to calculate the effective elastic constants proposed by Eshelby (1957).

A comparison of Nishizawa's ellipsoid model with Hudson's circular crack model in terms of elastic constants and group velocities was carried out by Douma (1988) for dry and liquid-filled inclusions. Different crack densities ϵ were considered, defined as $\epsilon = Nr^3/V$, where N is the number of cracks of radius r in a volume V in an isotropic solid. The group velocity as a function of the angle between the ray direction and the symmetry axis of the effective medium for the both models is shown in Fig. 7.2 for liquid-filled inclusions and in Fig. 7.3 for dry inclusions. Both figures are reproduced from Douma (1988). His numerical study shows that for a large range of aspects ratios both Hudson's (1981) and Nishizawa's (1982) models give similar results. Significant differences between the models arise only for large aspect ratios $d > 0.3$ and for high crack densities.

A solid containing aligned circular cracks or ellipsoidal inclusions can be considered as a homogeneous transversely isotropic material. Once the effective elastic constants of such a material are obtained, using, for example, one of the described theoretical models, the characteristics of wave propagation in the medium can also be calculated (see, e.g., Keith and Crampin,

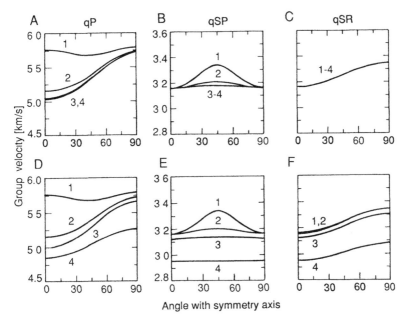

Figure 7.2 Group velocity of the three body waves in an isotropic medium containing liquid-filled inclusions as a function of the angle (in degrees) between the ray direction and the symmetry axis of the effective medium. The top figures correspond to Hudson's model of cracked media and the bottom figures, to that of Nishizawa. The curves numbered 1–4 were calculated for four aspect ratios of the inclusions equal to 0.0001, 0.001, 0.1, and 0.5, respectively. The adopted crack density is 0 05 [From Douma (1988, Fig. 2.2).]

1977a–1977c; Crampin, 1981). The five independent elastic constants describing transverse isotropy can be reduced to four constants by introducing dimensionless parameters. Furthermore, for weak transverse isotropy the approximate relations with only three dimensionless parameters were derived by Thomsen (1986).

The three parameters δ, ε, and γ are expressed by elastic constants as follows (Thomsen, 1986):

$$\delta = \frac{(c_{13} + c_{44})^2 - (c_{33} - c_{44})^2}{2c_{33}(c_{33} - c_{44})},$$

$$\varepsilon = \frac{c_{11} - c_{33}}{2c_{33}}, \tag{7.16}$$

$$\gamma = \frac{c_{66} - c_{44}}{2c_{44}}.$$

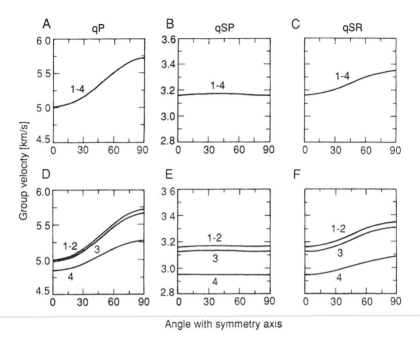

Figure 7.3   The same as in Fig. 7.2, except for an isotropic medium with dry inclusions [From Douma (1988, Fig. 2.3).]

Then the linearized phase velocities of the three body waves qP, qSP, and qSR, assuming that the symmetry axis is directed along the x_3 axis, are

$$V_{qP} \cong V_P(1 + \delta \sin^2 \theta \cos^2 \theta + \varepsilon \sin^4 \theta),$$

$$V_{qSP} \cong V_S\left(1 + \frac{V_P^2}{V_S^2}(\varepsilon - \delta)\sin^2 \theta \cos^2 \theta\right), \tag{7.17}$$

$$V_{qSR} \cong V_S(1 + \gamma \sin^2 \theta),$$

where $V_P = (c_{33}/\rho_e)^{1/2}$ and $V_S = (c_{44}/\rho_e)^{1/2}$ are the velocities along the symmetry axis, in which ρ_e is the density of the effective medium, and θ is the angle between the normal to the wavefront and the symmetry axis.

The three anisotropy parameters are all equal to zero for isotropic materials, and their deviation from zero can be regarded as a measure of a degree of anisotropy. From equations (7.16) it follows that the parameter ε

represents the relative difference between the qP-phase velocities perpendicular and parallel to the axis of symmetry, whereas the parameter γ represents this difference for the qSR waves (Thomsen, 1986)

$$\varepsilon \cong \frac{V_{qP}(\pi/2) - V_P}{V_P} \qquad (7.18)$$

and

$$\gamma \cong \frac{V_{qSR}(\pi/2) - V_S}{V_S}. \qquad (7.19)$$

the parameter δ can be obtained from measurements carried out at the angles θ equal to 0°, 45°, and 90°

$$\delta \cong 4\left(\frac{V_{qP}(\pi/4)}{V_{qP}(0)} - 1\right) - \left(\frac{V_{qP}(\pi/2)}{V_{qP}(0)} - 1\right). \qquad (7.20)$$

For weak to moderate anisotropy the parameters δ, ε, and γ are smaller than 0.2. If the parameter δ is equal to ε, elliptical anisotropy is observed (Thomsen, 1986). Such a type of anisotropy implies that the fronts of qP waves are ellipsoidal, whereas the fronts of qSP waves become spherical (Rudzki, 1911).

7.5   Shear-Wave Splitting Induced by Anisotropy

Shear-wave splitting occurs when shear waves propagate through solids that are effectively anisotropic, such as rocks containing aligned cracks or ellipsoidal inclusions. The behavior of shear waves crossing an anisotropic region is illustrated schematically in Fig. 7.4, reproduced from Crampin (1987a). On entering a region of aligned cracks, a shear wave splits into two (or more) components with different polarizations and different velocities. These components separate in time and provide distinct characteristics in the three-dimensional particle motion when displayed in mutually orthogonal cross sections of the particle displacements called polarization diagrams (Crampin, 1978). The splitting phenomenon is also called *shear-wave birefringence* and *shear-wave double refraction*.

Two quantities are readily determined from seismograms displaying shear-wave splitting. These are the time delay between the two shear-wave arrivals and the polarization of the first (faster) shear wave (e.g., Crampin, 1981). The delay between the two split shear waves is directly proportional to the length of the path in the anisotropic region and the relative difference in velocity between the two shear waves for a given direction of propagation.

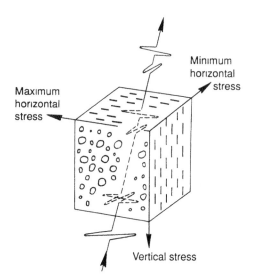

Figure 7.4 Schematic illustration of shear-wave splitting in cracked rock. When a shear wave enters a cracked medium, the component of motion with displacement normal to the crack face encounters an effectively lower shear modulus and consequently travels more slowly than the component with displacement parallel to the cracks. [From Crampin (1987a, Fig. 2), copyright by Royal Astronomical Society.]

The polarization of faster split shear wave is controlled by the symmetry of crack geometry along the wave path (Crampin, 1981).

Shear waves displaying splitting with characteristic patterns of polarization were first observed above a swarm of small earthquakes near the North Anatolian fault in Turkey (e.g., Crampin and Booth, 1985). It was found that the leading split shear wave was polarized in a direction parallel or subparallel to the local or regional maximum horizontal principal stress that has been inferred from fault-plane solutions and related evidence. Figure 7.5, reproduced from Chen *et al.* (1987), shows some examples of seismograms of small earthquakes recorded near the North Anatolian fault in Turkey. Although shear-wave splitting is difficult to recognize on records displayed in the form of conventional three-component time-series seismograms, the characteristic abrupt change of direction and distinct patterns of particle motion are clearly visible on polarization diagrams. It was found that the polarizations of the leading split shear waves are consistent with parallel, vertical, water-filled cracks, and the observed patterns of particle displacement in polarization diagrams are similar to those in synthetic seismograms corresponding to the wave propagation in cracked structures (Crampin and Booth, 1985). In general, the initial polarization of shear waves recorded at the surface above

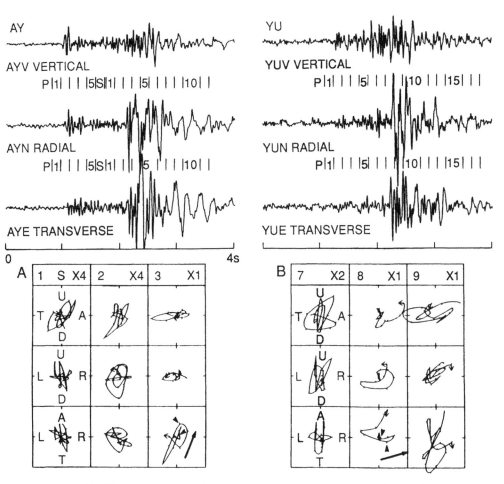

Figure 7.5 Three-component seismograms, rotated into vertical, radial, and transverse components, recorded in the shear-wave window above two small earthquakes in Turkey at two stations The stations, focal depth and epicentral distance (in kilometers), and source to station azimuth are (A) AY, 7.1, 5 5, 96°; (B) YU, 7.5, 2.8, 196°. The numbered polarization diagrams correspond to the time intervals marked above the shear-wave portion of the seismograms They show three mutually orthogonal cross sections of the particle motion labeled "Up," "Down," "Toward," "Away from the epicenter," and "Left and right of the radial direction from the epicenter." Ticks on the polarization diagrams are 0.01 s, and the relative gain is shown above each set of diagrams. Shear-wave splitting is seen in the lower horizontal cross sections, where the onsets of the split shear waves are marked by arrowheads, and the bar indicates the vector of polarization of the leading shear wave. [From Chen et al. (1987, Fig A1), copyright by Royal Astronomical Society.]

seismic sources is approximately parallel for all azimuths and angles of incidence within the shear-wave window at all recording sites. The second arriving shear wave can often be resolved approximately orthogonal to the first split shear-wave arrival (e.g., Leary et al., 1990).

The other rockmass property that may be inferred from observations of shear-wave splitting is the average density of flaws, cracks, and microcracks. The density of cracks ϵ was defined by O'Connell and Budiansky (1974) in terms of a flat crack of radius r embedded in a volume V. Considering the energy of elastic deformation associated with the crack in an uniform medium, and if there are N such cracks in the volume V, the crack density is

$$\epsilon = Nr^3/V. \tag{7.21}$$

When different crack populations N_t of radii r_t are present in the volume V, the total crack density is the sum of the densities ϵ_t given by relation (7.21). Then the crack density is described in terms of the energy of an average crack deformation $\langle r^3 \rangle = \Sigma N_t \epsilon_t / \Sigma N_t$ and relation (7.21) becomes

$$\epsilon = N\langle r^3 \rangle / V, \tag{7.22}$$

where N is now the total number of cracks. Hudson (1981) described polarized shear-wave velocities in terms of the average crack density. His relations provide the useful approximate relation between the crack density ϵ, shear-wave velocity V_S or travel time T_S, and differential shear-wave velocity ΔV_S or travel time ΔT_S for vertically propagating seismic rays in parallel vertical microcracks (Leary et al., 1990)

$$|\Delta V_S/V_S| \cong |\Delta T_S/T_S| \cong \epsilon. \tag{7.23}$$

Values of crack density ranging between 0.01 and 0.05 were found to be typical in various geologic and tectonic regions (Leary et al., 1990).

The calculation of synthetic seismograms is one of the major techniques for the interpretation of wave propagation. The separation of the phase and group velocity during body-wave propagation in anisotropic media means that the propagation depends critically on whether the waves have plane or curved (often called spherical) wavefronts. Synthetic seismograms of plane waves traveling in plane-layered anisotropic media are comparatively simple to compute, but have limited applications. Nevertheless, plane waves have demonstrated the significance of shear-wave polarization anomalies for recognizing and estimating anisotropic structures. Applications of plane waves to some simple anisotropic structures can be found in the review by Crampin (1981).

Shear-wave splitting has also been observed in shear-wave vertical seismic profiles (VSPs), and effective technique in exploration seismology, and in most shear-wave reflection surveys made for exploration of hydrocarbon reservoirs. The reflection surveys and VSP are interpreted as indicating the orientation of subsurface fractures (e.g., Crampin, 1987b). The VSP measurements are carried out in boreholes, away from the free surface, close to the source, and below the layer of interest if necessary. Such conditions are also available in deep underground mines, and experience gained in VSP studies might be relevant to some situations encountered in mines.

In order to demonstrate the effects of shear-wave splitting in VSP, Crampin (1985) has calculated synthetic seismograms of plane shear waves propagating at a range of angles of incidence and a range of azimuths through the simple model shown in Fig. 7.6. The model consists of a horizontal layer of vertical parallel saturated cracks striking in the x_2 direction, with a horizontal symmetry axis in the x_1 direction. Figure 7.7 shows synthetic seismograms of a 40-Hz shear wave with a linear polarization bisecting SV- and SH-wave polarizations propagating at a range of angles of incidence through a 100-m-thick layer, recorded 100 m below the layer. Figure 7.7A shows propagation through an isotropic layer; and Fig. 7.7B, propagation through an anisotropic layer for planes of incidence normal to the strike of the vertical cracks. On the right of each three-component trace, a polarization diagram is drawn. The polarization diagrams of the shear waves traveling through the isotropic layer show linear motion, whereas the diagrams of shear waves through the crack-induced anisotropic layer display

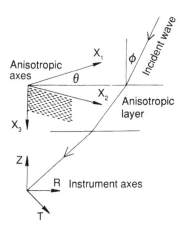

Figure 7.6 The structural model of the wave path for synthetic seismograms and arrangement of axes. Z, R, and T are vertical, radial and transverse axes of a receiver. X_1, X_2, and X_3 are axes fixed in the anisotropic medium, which simulates vertical cracks parallel to the X_2, X_3 plane (shaded) so that X_1 is the horizontal direction of the symmetry axis. θ is the azimuth from the symmetry axis X_1 (the angle of the plane of incidence to the strike of the cracks), and Φ is the angle of incidence of the plane shear wave to the layer [From Crampin (1985, Fig. 5).]

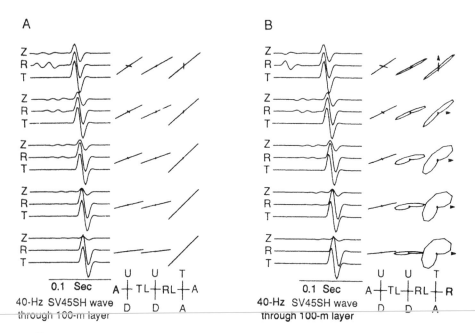

Figure 7.7 Synthetic plane-wave seismograms and polarization diagrams for 40-Hz shear waves polarized *SV*45*SH* (with a linear polarization bisecting *SV*- and *SH*-wave polarizations) with incident angles $\Phi = 25°$, $20°$, $15°$, $10°$, and $5°$ (from the top of each figure) onto a layer, recorded 100 m below the layer, for (A) 100-m-thick isotropic layer ($\rho = 2.6$ g/cm^3, $V_p = 5.04$ km/s, $V_s = 2.92$ km/s) and (B) 100-m-thick layer simulating vertical cracks at right angles to the plane of the incident waves ($\theta = 0°$). The polarization diagrams are orthogonal cross sections of the particle displacements, with axes up (U), down (D), toward (T), and away (A) from the source, and left (L) and right (R) toward the source. The polarizations of the initial shear-wave onsets are marked by arrowheads on the horizontal sections. [From Crampin (1985, Fig. 6)]

elliptical motion, which varies with the angle of incidence, that is, with the delay between the split shear waves, and includes a 90° change in the direction of the first motion polarity.

These synthetic seismograms have been calculated for plane shear waves. The direction of propagation of the surface of constant phase, however, may differ substantially from the direction of group propagation along a ray in anisotropic rock (Crampin, 1981). For the particular crack-induced anisotropy considered by Crampin (1985), this deviation of group and phase propagation on a spherical wavefront will not make a great difference in the delay between the split waves, because the differences between the group and phase velocities are small. Shear-wave polarizations, however, are sensitive to small variations of the model, and the greatest effects would appear in polarization diagrams. Thus the exact behavior of shear-wave splitting must be modeled by synthetic seismograms for the waves with spherical wavefronts propagating from appropriate point sources. There are at present two principal techniques for computing synthetic seismograms with spherical wavefronts in isotropic media: the reflectivity method (Fuchs, 1968; Fuchs and Müller, 1971) and the ray method (Babich and Alekseev, 1958; Červeny *et al.*, 1977). The reflectivity technique, for calculating synthetic seismograms in plane uniform anisotropic layers, is especially suited to the analysis of multilayered media, as it makes use of the anisotropic propagator matrices. The ray method is more suited to calculating wave propagation in continuously varying media. The description of these techniques is beyond the scope of this book; an elegant review of the involved problems has been given by Crampin (1981).

Experience shows that only two parameters can be used in practice to characterize the shear-wave splitting observed at a receiver, although in general four parameters might be available in ideal circumstances (Chen *et al.*, 1987). The two parameters are the polarization vector of the first split shear wave and the time delay between the split shear waves. The other two possible parameters are the second split shear-wave arrivals and the relative attenuation between the two split waves. The measurements of polarizations and time delays of split shear waves is of great significance for proper interpretation of anisotropic effects, and the procedure followed in this respect by Chen *et al.* (1987) is briefly described here.

The first step in determination of the first shear-wave polarization vector is to rotate the horizontal seismograms into components that are radial and transverse with respect to the line between an epicenter and a receiver. Since phase changes in the shear wavetrain are often difficult to recognize on seismograms, it is necessary to display the data in polarization diagrams. These diagrams are three mutually perpendicular orthogonal sections of the particle displacement plotted for successive time intervals along the three-component wavetrains (see Fig. 7.5). The vertical sections are used only to distinguish between the shear-wave motion and *P*- or possible *S*-to-*P*-converted wave motions. The horizontal section is the most important for shear waves and it contains most of the shear-wave energy. The most important step is the identification of the onset of the first shear-wave arrival; the polarization direction of the first arrival in the horizontal plane is then determined from the corresponding polarization diagrams. The first motion of the shear wave is usually sufficiently linear for identifying the polarization direction. When slightly elliptical motion is observed, the average polarization direction can be chosen. Diagrams showing strongly elliptical or circular polarization should be discarded. Whenever possible, vector polarizations should be determined by assessing the direction (polarity) of the first motion.

The delay between the split shear waves is much more difficult to measure than the first-arrival polarization. To measure the time delay, identification of the onset of a second split shear-wave arrival is needed. There are several difficulties in this respect. Each type of rock along a given ray path may have different crack densities and dimensions and different matrix velocities, and small changes at internal interfaces will cause further splitting of each split wave propagating through them. There may, therefore, be a multiple choice of possible secondary split shear waves, and no consistent second arrivals may be identified. Second split shear waves may be obscured by signal-generated coda and background noise. Finally, the shear waves radiated from the source may not excite the two possible polarizations equally, and then either split shear wave could be very small or even absent along any given ray path (Chen *et al.*, 1987).

Despite these difficulties, consistent arrivals can still be chosen. Synthetic seismograms show that the onset of second split wave is marked by abrupt changes in the direction of the particle motion of the first shear wave (Crampin, 1981). If the observed shear-wave splitting is caused by distributions of vertical, parallel saturated cracks, corresponding to an elastically transversely isotropic medium with a horizontal symmetry axis, then the polarization of the second split shear wave is likely to be approximately orthogonal to the polarization of the first shear wave (Crampin and Booth, 1985). Thus the onset of the second split shear wave can be identified by a change of polarization of the first shear wave, with a significant component of energy in the direction perpendicular to the first polarization direction (Chen *et al.*, 1987).

Numerous studies of shear-wave particle motions from earthquakes have shown anomalous behavior corresponding to seismic anisotropy in the crust in a wide variety of geologic and tectonic regions (e.g., Crampin *et al.*, 1980, 1986, 1990; Booth *et al.*, 1985, 1990; Buchbinder, 1985; Crampin and Booth, 1985; Kaneshima *et al.*, 1987, 1989; Peacock *et al.*, 1988; Iannoccone and Deschamps, 1989; Savage *et al.*, 1990; Kaneshima, 1990; Shih and Meyer, 1990). Results from vertical seismic profiles in sedimentary, metamorphic and igneous rocks are also consistent with shear-wave splitting (e.g., Peacock and Crampin, 1985; Daley *et al.*, 1988; Majer *et al.*, 1988). Five possible causes of effective anisotropy in the crust have been suggested (Crampin *et al.*, 1984a): aligned crystals, aligned fabric such as aligned grains, periodic thin layers, direct stress-induced anisotropy, and aligned cracks or microcracks. Crampin (1987a, 1987b) believes that azimuthal anisotropy is almost universal in the shallow crust and that it is caused by the presence of fluid-filled cracks and particularly microcracks, which are aligned by stress into parallel vertical, or nearly vertical, orientations, striking parallel to the direction of maximum horizontal stress. These distributions of aligned cracks are EDA (extensive-dilatancy anisotropy, defined earlier) cracks.

Typical dimensions of EDA cracks are likely to be a few micrometers in metamorphic and igneous rocks, a few millimeters in sedimentary rocks, and up to a few meters in fractured beds (Crampin, 1990). Large cracks may leave permanent relics, but our understanding of microcracks in the crust is based mostly on speculation. Since cracks in the in situ rockmass are effectively inaccessible, observations and interpretation of shear waves penetrating the remote rockmass may well be the only way to deduce the in situ crack geometry. It seems that much of the effective anisotropy of EDA cracks results from the distribution of microcracks or aligned pore space pervading most intact rocks in the crust, and that their geometry may be comparatively mobile and easily modified by changes in stress (Crampin, 1990).

Thus, the most valuable aspect of shear-wave polarization observations for earthquake studies seems to be their use to determine stress orientation at depth (e.g., Aster *et al.*, 1990). Several studies have confirmed that initial shear-wave polarization directions and regional stress orientations are correlated, and this result is consistent with models of vertical cracks aligned parallel to the direction of maximum compressive stress. The use of the other shear-wave splitting parameter, the time delay between split waves, used in earthquake studies (e.g., for earthquake prediction), is less clear. Temporal variations have been reported in the behavior of split shear waves in a seismic gap at Anza, California (Peacock *et al.*, 1988). Furthermore, Crampin *et al.* (1990) reported that shear-wave splitting delay times at station KNW of the Anzac seismic network exhibit temporal variations that can be correlated with the occurrence of the $M_L = 5.6$ North Palm Springs earthquake of July 8, 1986. These results, however, have not been supported by another analysis of the same data reported by Aster *et al.* (1990).

Although there are unpublished reports that shear-wave splitting caused by EDA cracks has been observed in mines (Crampin, 1987b), no comprehensive study on this matter has been published as yet. Investigating such cracks by monitoring shear waves by three-component instruments with high dynamic range is likely to provide the detailed crack geometry and the corresponding stress variations in the source area of seismic events and rockbursts in mines. The possible application of the observations of time delays between the split shear-wave arrivals for prediction of large seismic events is briefly discussed in Chapter 13.

7.6 Shear-Wave Splitting in Isotropic Media

Any single observation of shear-wave splitting can be simulated by propagation through a combination of isotropic discontinuities. In the absence of anisotropy, the observed polarization of shear waves should be that radiated from the source, modified by interaction with internal interfaces and free surface topography. Shear waves observed at a free surface may be seriously distorted by interaction with a free surface if the angle of incidence is greater than the critical angle (e.g., Evans, 1984). This angle defines a shear-wave window, in which the shear waveforms recorded at the surface are not distorted.

Recent studies have shown that shear-wave polarizations may also be distorted by internal interfaces in isotropic media, leading to difficulties in interpretation of shear-wave splitting induced by anisotropy (Cormier, 1984; Douma and Helbig, 1987). The change in polarization of a plane shear wave transmitted through an interface between half-spaces of sandstone and halite was calculated by Douma and Helbig (1987) for an range of angles of incidence. They found that the greatest deviation is of a few degrees, and they suggested that such effects are cumulative and might have serious implications for the study of shear-weave splitting induced by anisotropy.

This problem was considered by Liu and Crampin (1990), who calculated synthetic seismograms for shear waves, with plane and curved wavefronts, incident on internal isotropic-to-isotropic and isotropic-to-anisotropic interfaces and assessed the effects of such interfaces on observations of anisotropy-induced shear-wave splitting. Their results are briefly recounted here.

Liu and Crampin (1990) considered plane shear waves SH and SV with relative amplitudes A_{SH} and A_{SV}, respectively, incident on a plane boundary from medium 1 to medium 2, with parameters taken from Douma and Helbig (1987) and corresponding to sandstone and halite, respectively. The polarization angle ψ, defined in Fig. 7.8, of the transmitted shear wave in the plane parallel to the plane of constant phase is related to the amplitude and transmission coefficients by (e.g., Douma and Helbig, 1987)

$$\psi = \tan^{-1}(B|R|), \qquad (7.24)$$

where $B = A_{SH}/A_{SV}$ and $R = T_{SH}/T_{SV}$, in which T_{SH} and T_{SV} are the transmission coefficients of SH and SV waves. The ratio B specifies the polarization angle of the incident wave and the transmission coefficients are functions of the properties of media 1 and 2 (ρ_1, V_{P1}, V_{S1} and ρ_2, V_{P2}, V_{S2}, respectively) and angle of incidence (Aki and Richards, 1980).

The polarization angle and phase difference of the transmitted shear wave, for an incident wave with equal amplitude of SH and SV waves ($\psi = 45°$ and $B = 1$), as functions of the angles of incidence are shown in Fig. 7.9, reproduced from Liu and Crampin (1990), for the sandstone to halite (low-to-high-velocity) interface and for the halite to sandstone (high-to-low-velocity) interface. At normal incidence, the polarization and the phase of the transmitted wave are unchanged. The deviation of the polarization and phase of the transmitted wave increases with increasing angle of incidence, reaching 3° at the smallest critical angle $\arcsin(V_{S1}/V_{P2})$ for the low-to-high-

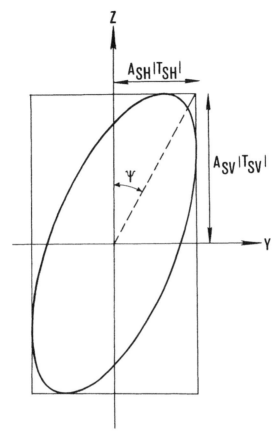

Figure 7.8 Definition of polarization angle ψ of a shear wave in the plane parallel to the plane of constant phase. A_{SH} and A_{SV} are the amplitudes of the SH and SV components of the incident shear wave, and T_{SH} and T_{SV} are the transmission coefficients of SH and SV waves. [From Liu and Crampin, (1990, Fig. 1).]

velocity interface, and $2°$ at the angle $\arcsin(V_{S2}/V_{P2})$ for the high-to-low-velocity interface.

Beyond these critical angles, the incident wave is totally reflected and there appears an interface wave with complex transmission coefficients. A complex transmission coefficient means that waves, called *inhomogeneous waves*, with an imaginary vertical component of the normal to the wavefront are generated in medium 2. The resultant inhomogeneous interface waves become elliptically polarized, carrying energy parallel to the interface at the phase velocity of the incident shear wave along the interface, and decaying

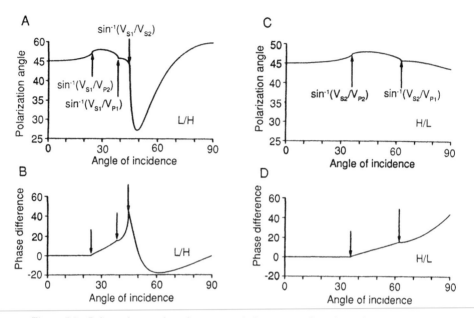

Figure 7.9   Polarization angles of transmitted shear waves (top figures) and phase differences (bottom figures) as functions of incident angles (in degrees) for sandstone-to-halite interface (A, B) and halite-to-sandstone interface (C, D). The ray paths at critical angles of incidence are marked by arrows. [From Liu and Crampin (1990, Fig. 2).]

exponentially with distance from the interface (Hudson, 1980; Kennett, 1983). The polarization ellipse in the plane containing SH and SV components is determined by the following relation (Smith and Ward, 1974)

$$\sin^2(\Delta\Phi) = \frac{U_{SH}^2}{\left(A_{SH}|T_{SH}|\right)^2} + \frac{U_{SV}^2}{\left(A_{SV}|T_{SV}|\right)^2} - 2\frac{U_{SH}U_{SV}\cos(\Delta\Phi)}{A_{SH}A_{SV}|T_{SH}T_{SV}|}, \quad (7.25)$$

where $\Delta\Phi = \Phi_{SH} - \Phi_{SV}$ is the phase difference and Φ_{SH} and Φ_{SV} are the phase angles of SH and SV waves, and U_{SH} and U_{SV} are the displacement components of SH and SV waves, respectively.

Thus, a linearly polarized plane shear wave, transmitted through an isolated isotropic-to-isotropic interface within the innermost window, preserves its initial phase and linear motion. The particle motion becomes elliptical only for the angles of incidence greater than the smallest critical angle (Liu and Crampin, 1990).

Particle motions related to plane and curved wavefronts of shear waves transmitted through an isotropic-to-isotropic interface have different characteristics. On curved wavefronts, the effects of the critical angle are spread over a range of angles, the exact behavior for any particular geometry depending on the curvature of the wavefront and the frequency of the incident wave (Liu and Crampin, 1990).

The behavior of shear waves at an isotropic-to-anisotropic interface, where the anisotropy is caused by distributions of vertical parallel fluid-filled micro-cracks, depends on the pathlength through the anisotropy. The effects of the anisotropy are visible only when the path is long enough to separate the two shear waves. As a result, the polarizations of shear waves immediately below the interface are very similar to those for the isotropic-to-isotropic interface, and a change from linear to elliptical motion is observed at the critical angle of incidence. The observed shear-wave splitting shows several differences from the elliptical polarizations associated with isotropic-to-isotropic inter-faces. The splitting is distinct within the shear-wave window, and the polar-ization of the leading shear wave is controlled by the orientation of the anisotropy and not by the polarization of the incident wave. Thus, the important result of the study by Liu and Crampin (1990) is that the initial polarization of the shear waves in polarization diagrams is controlled by the orientation of EDA cracks. The small changes in orientation of the wave caused by the interface may modify the details of the pattern in polarization diagrams, but generally will not alter the polarization direction of the leading split shear wave.

Chapter 8 | Attenuation and Scattering of Seismic Waves

In all problems so far considered, perfectly elastic media were assumed. In such media wave motion will continue indefinitely, once it has been initiated. It is common experience, however, that as a wave is propagated through real materials, wave amplitudes attenuate as a result of different processes responsible for energy dissipation, which can be summarized as internal friction. The strains and stresses associated with propagating waves can cause irreversible changes in the crystal structures of the medium, and work could also be done on grain boundaries within the medium. Such media are called *anelastic*, since the configuration of material particles is dependent on the history of applied stresses. There is no satisfactory theory of internal friction, although several mechanisms have been suggested. The gross effect of internal friction is described by the dimensionless quantity Q, known as the *quality factor* of a given system, which can be defined in various ways.

Wave attenuation by internal friction, often called *intrinsic attenuation*, is a very large subject. Its theoretical aspects are described only briefly here. A more detailed treatment of this subject can be found in a number of textbooks, such as the works of Ewing *et al.* (1957), Aki and Richards (1980), Ben-Menahem and Singh (1981), and Bullen and Bolt (1985).

Apart from intrinsic attenuation, elastic waves propagating in heterogeneous media are attenuated by scattering effects. Deflections of a portion of wave energy occur when elastic waves encounter an obstacle and then an interfering scattered wave will spread out from the obstacle in all directions. The modification of seismic waves caused by three-dimensional heterogeneities is broadly called *seismic-wave scattering*. Scattering attenuation is not an energy dissipation mechanism, but only an energy redistribution in space and time, which is a geometric effect. Under the single scattering approximation, the scattering attenuation cannot be separated from the intrinsic attenuation. To separate these two attenuation effects, a multiple scattering theory is needed. Although there is no general solution for the multiple scattering theory, several special cases have been studied.

Seismic-wave scattering is extremely complex phenomenon. Not surprisingly, this new field in research undergoes very rapid expansion, and numerous papers on this subject are published each year. Here observational techniques and methods are emphasized, especially those relevant to obser-

128

vations of scattering phenomena in underground mines. The theory, once again, is presented in a concise form. Excellent reviews of various aspects of seismic-wave scattering can be found in recently published three special issues of *Pure and Applied Geophysics*, edited by Wu and Aki (1988a, 1989, 1990).

8.1 Anelastic Effects

There is no internal energy loss in perfectly elastic media described by constitutive equations (3.7). In real media, however, the attenuation of seismic energy is an established fact, and it is necessary to introduce damping into the elastic constitutive relations. Anelastic damping depends on a number of physical mechanisms, which cannot be represented by a single modification of the constitutive equations. The simplest description of anelasticity is a superposition of two mechanisms of resistance to deformation: linear elasticity and shear viscosity of an ideal viscous fluid or Stokes' viscosity. A material behaving in such a way is called a *viscoelastic material*. If stress components at a given time are related linearly to strain components and the principle of linear superposition is observed, then the medium is said to be linearly viscoelastic. The principle of linear superposition means that the strain output from a combination of two different stress inputs applied at different times is equal to the sum of the strain outputs resulting from these stresses acting separately.

In viscoelastic bodies there are various strain responses to constant stress. Elastic strain is instantaneous and reversible but not necessarily linear. The stress above which the behavior is no longer elastic is called the *elastic limit*. The strain does not disappear after the removal of the stress greater than the elastic limit. For viscoelastic behavior a number of new concepts is introduced. Creep is the slow continuous deformation of a material under constant stress. When the stress is removed, the strain gradually decreases during a process known as *recovery* (Fig. 8.1). If the recovery is complete, elastic creep occurs, for stresses smaller than the strength of the material. If the recovery is partial, flow is present. Flow in which the strain rate is linear with the stress is a viscous flow, and flow in which the strain rate is nonlinear with the stress is a plastic flow. A creep that occurs at an increasing rate may terminate in rupture. In viscoelastic materials under constant strain, the stress gradually decreases during a process known as *relaxation*.

Mechanical viscoelastic analogs are often used to describe the simplest one-dimensional models of viscoelastic materials. The simplest representative of linear elasticity is a linear spring with the condition that its extension is proportional to the applied force, showing instantaneous elasticity and recovery. The mechanical analog to a viscous fluid is the linear dashpot element,

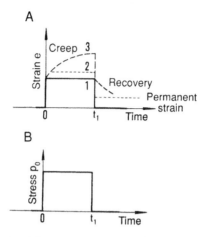

Figure 8.1   Strain behavior (A) under constant stress (B) for various substances: (1) instantaneous elastic response; (2) plastic response with a permanent residual strain; (3) viscoelastic creep and recovery.

used to represent Stokes' linear viscosity. Its rate of extension is proportional to the applied force, and when a step of constant stress is applied, the dashpot will deform continuously at a constant rate. From these two elements models of principal viscoelastic materials can be composed.

8.2   Viscoelastic Constitutive Relations

The stress–strain relations for a perfectly elastic isotropic body in equilibrium are uniquely described by equations (3.12). To enter the time t into the stress-strain relations, it is convenient to introduce the tensors P_{ij} and E_{ij} defined by (Bullen and Bolt, 1985)

$$P_{ij} = p_{ij} - \tfrac{1}{3} p_{ll} \delta_{ij},$$ (8.1)

$$E_{ij} = e_{ij} - \tfrac{1}{3} e_{ll} \delta_{ij}.$$ (8.2)

P_{ij} and E_{ij} are called the *deviatoric* (or *distortional*) stress tensor and *deviatoric strain tensor*, respectively. For $i \neq j$, $P_{ij} = p_{ij}$ and $E_{ij} = e_{ij}$. From relations (3.12) and (3.13) it can be deduced that

$$\tfrac{1}{3} p_{ll} = k\theta.$$ (8.3)

Similarly, from relations (8.1)–(8.3) and (3.12) and (3.13) it follows that

$$P_{ij} = 2\mu E_{ij} \tag{8.4}$$

for all i and j. For perfect elasticity, relation (8.3) describes behavior under a symmetric stress, and relation (8.4) describes the effects of any departures from symmetry. The main imperfections of elasticity that are observed arise only under stresses that are not fully symmetric. Thus, to describe these imperfections, relation (8.4) should be modified, while formula (8.3) is kept unchanged.

The simplest model displaying a deviation from perfectly elastic behavior is an ideal viscous fluid. In this case instead of relation (8.4) we have

$$P_{ij} = 2\gamma \frac{d}{dt} E_{ij}, \tag{8.5}$$

where γ is a new parameter denoting the shear viscosity of the fluid.

By combining the perfect elasticity relations (8.4) with viscous fluid relations (8.5), one might obtain a constitutive relation relevant to imperfectly elastic behavior of a solid

$$P_{ij} = 2\mu E_{ij} + 2\gamma \frac{dE_{ij}}{dt}. \tag{8.6}$$

These relations correspond to Kelvin–Voigt or firmoviscous material. The presence of the term with γ in formula (8.6) implies exponential delay in reaching the full strain under a given deviatoric stress and in recovery of an initial configuration after the removal of stress. It also implies that in viscoelastic solids the rigidity can no longer be described by the single parameter μ; it depends on γ and on the form of applied stress.

Relations (8.6) are not representative for observed elastic afterworking, when creep follows application or removal of a deviatoric stress. Furthermore, following a sudden change in the deviatoric stress, the creep is often preceded by some immediate change of strain, which is not accounted for in formula (8.6). These effects can be taken into consideration by adding to relations (8.6) a term with dP_{ij}/dt in analogy to the term $2\gamma E_{ij}/dt$ (Bullen and Bolt, 1985)

$$P_{ij} + \tau \frac{dP_{ij}}{dt} = 2\mu E_{ij} + 2\gamma \frac{dE_{ij}}{dt}, \tag{8.7}$$

where τ is a further parameter. It can be shown that under very slowly changing stresses, the terms with τ and γ in formula (8.7) are not important and the behavior is close to that of a perfectly elastic solid of rigidity μ. Under rapidly changing stresses, the terms with τ and γ predominate and the behavior is similar to that of a perfectly elastic solid of rigidity γ/τ. Under stresses changing intermediately, significant damping and dissipation of energy occur.

If the parameter μ is taken equal to zero, relations (8.7) become

$$2\gamma \frac{dE_{ij}}{dt} = P_{ij} + \tau \frac{dP_{ij}}{dt}. \tag{8.8}$$

They describe the elastic behavior of a medium called the *Maxwell* or *elasticoviscous substance*. It can be shown that under deviatoric stresses changing very slowly, the first term on the right-hand side of formula (8.8) is predominant and the behavior approximates to that of a fluid of viscosity γ. Under rapidly changing stresses, the second term predominates and the Maxwell substance is close to a perfectly elastic solid of rigidity γ/τ.

A linear viscoelastic solid may be defined as a material for which the stress–strain relations can be expressed as linear differential equations that involve only the stress, the strain, and their derivatives with respect to time. The general stress–strain relation in differential form for a single component of stress p and corresponding strain e is (e.g., Ben-Menahem and Singh, 1981)

$$Rp(t) = Se(t), \tag{8.9}$$

where R and S are linear differential operators of the form

$$R = \sum_0^L r_L \frac{\partial^L}{\partial t^L} \quad \text{and} \quad S = \sum_0^M s_M \frac{\partial^M}{\partial t^M}, \tag{8.10}$$

in which the constants r_L and s_M are material constants. This form of representation of the stress–strain relation corresponds directly to the element constants of mechanical analogs, defined as m for the springs and η for the dashpots. For the spring we may write

$$p = me, \tag{8.11}$$

and for the dashpot,

$$p = \eta\dot{e}, \tag{8.12}$$

where a dot denotes differentiation with respect to time.

A spring and a dashpot in series give equation (8.8) or a Maxwell model. In this case the force acting on both elements is the same and the total extension is the sum of that in each element. The model is shown in Fig. 8.2. In this material, a given stress generates at the same time an elastic strain $e_1 = (1/m)p$ and a strain rate $\dot{e}_2 = (1/\eta)p$. The total rate of strain is

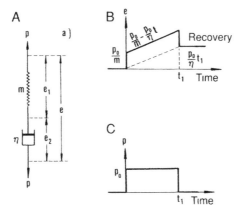

Figure 8.2 Mechanical model of Maxwell substance (A) and its strain behavior in the form of creep and recovery (B) under constant stress (C).

therefore

$$\dot{e} = \dot{e}_1 + \dot{e}_2 = \frac{1}{m}\dot{p} + \frac{1}{\eta}p. \tag{8.13}$$

Defining the relaxation time $\tau_0 = \eta/m$, the integration of equation (8.13) leads to

$$p = m\exp(-t/\tau_0)\left(e(0) + \int_0^t \dot{e}(t)\exp(t/\tau_0)\,dt\right). \tag{8.14}$$

Integration by parts provides an alternative expression

$$\frac{p}{m} = e - \frac{1}{\tau_0}\int_0^t e(\tau)\exp[-(t-\tau)/\tau_0]\,d\tau. \tag{8.15}$$

The strain as a function of stress is also given by a direct integration of equation (8.13):

$$e(t) = \frac{1}{m}p(t) + \frac{1}{\eta}\int_0^t p(\tau)\,d\tau. \tag{8.16}$$

We examine the stress–strain behavior in a Maxwell material for a few selected cases. A step of stress is described as $p = p_0 H(t)$, where $H(t)$ is the Heaviside step function

$$H(t) = \begin{cases} 0, & t < 0 \\ 1, & t \geq 0 \end{cases} \quad \text{and} \quad \frac{d}{dt}H(t) = \delta(t), \tag{8.17}$$

and $\delta(t)$ is the Dirac delta function. The strain response to a stress step follows directly from equation (8.16)

$$e(t) = \frac{p_0}{m}\left(1 + \frac{t}{\tau_0}\right)H(t) = p_0\phi(t)H(t), \tag{8.18}$$

where the function $\phi(t) = 1/m + t/\eta$ (for $t > 0$), describing the strain response to a constant force, is called the *creep function*. Similarly, for a strain step $e = e_0 H(t)$, equation (8.15) renders

$$p(t) = me_0 \exp(-t/\tau_0) = e_0\psi(t), \tag{8.19}$$

where $\psi(t) = m \exp(-mt/\eta)$ is called the *relaxation function*. Equation (8.19) describes the stress relaxation phenomenon under constant strain, shown in Fig. 8.2. From equation (8.16) it follows that for a harmonic stress cycle $p = p_0 \sin \omega_0 t$, the strain response is

$$\begin{aligned} e(t) &= \frac{p_0}{m}\left(\sin \omega_0 t - \frac{1}{Q}\cos \omega_0 t + \frac{1}{Q}\right) \\ &= \frac{p_0}{m}\left[\left(1 + \frac{1}{Q^2}\right)^{1/2}\sin(\omega_0 t - \chi_0) + \frac{1}{Q}\right], \end{aligned} \tag{8.20}$$

where $\tan \chi_0 = 1/Q$ and $Q = \omega_0\tau_0 = \omega_0\eta/m$. Thus the strain lags behind the activating stress by the angle $\tan^{-1}(1/Q)$, which is close to $1/Q$ when $Q \gg 1$.

The Kelvin–Voigt model is represented by the spring and the dashpot connected in parallel (Fig. 8.3). In this case the total stress is the sum of two separate stresses: an elastic stress $p_1 = me$ and a viscous stress $p_2 = \eta\dot{e}$. Thus

$$p = p_1 + p_2 = me + \eta\dot{e}. \tag{8.21}$$

In the Maxwell model the force is the same on both elements. Here the force is different but the strain is the same on both elements at all times. The strain is obtained, for a given stress, by solving equation (8.21):

$$e(t) = e(0)\exp(-t/\tau_0) + \frac{1}{m}\int_0^t p(\tau)\exp[-(t-\tau)/\tau_0]\,d\tau. \tag{8.22}$$

The creep and relaxation functions are

$$\phi(t) = [1 - \exp(-mt/\eta)]/m \tag{8.23}$$

and

$$\psi(t) = mH(t) + \eta\delta(t), \tag{8.24}$$

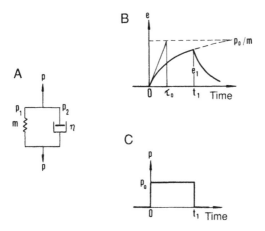

Figure 8.3 Mechanical model of Kelvin–Voigt substance (A) and its creep and recovery (B) under constant stress (C).

respectively. The Kelvin–Voigt model does not show a time-dependent relaxation. The presence of the delta function follows from the fact that an infinite stress is needed to accommodate an abrupt change of strain in the dashpot.

Neither the Maxwell nor the Kelvin–Voigt model is sufficient to account for the behavior of most viscoelastic materials. The Kelvin–Voigt model cannot accommodate an abrupt change in strain and does not show residual strain after unloading, and the Maxwell model has no creep features. A more satisfactory model for the behavior of rocks in the Earth's crust, under stresses and strains associated with seismic vibrations, is the *standard linear solid*, also called the *three-element elastic model* or the *generalized Kelvin–Voigt substance*. It is composed of a Kelvin–Voigt element (m_1, η_1) connected in series with a spring m_2 (see Fig. 8.4). The equations describing this system are

$$p = m_1 e_1 + \dot{\eta}_1 e_1 = m_2 e_2 \quad \text{and} \quad e = e_1 + e_2. \tag{8.25}$$

By eliminating e_1 and e_2 from these equations we have

$$p + \tau_p \dot{p} = m_r (e + \tau_e \dot{e}), \tag{8.26}$$

where

$$\tau_p = \frac{\eta_1}{m_1 + m_2}, \qquad \tau_e = \frac{\eta_1}{m_1}, \qquad m_r = \frac{m_1 m_2}{m_1 + m_2}. \tag{8.27}$$

The parameter τ_p is known as the stress relaxation time under constant

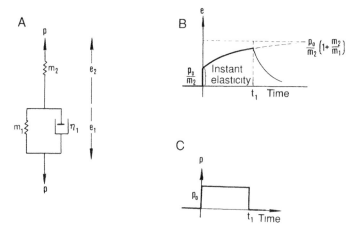

Figure 8.4 Mechanical model of standard linear solid (A) and its creep and recovery (B) under constant stress (C).

strain, and τ_e is the strain relaxation time under constant stress. The solution of equation (8.26) gives

$$e(t) = \frac{1}{m_r}\left[p(t) - \left(1 - \frac{\tau_p}{\tau_e}\right)\int_0^t \exp[-(t-\tau)\tau_e]\,\dot{p}(\tau)\,d\tau\right]. \quad (8.28)$$

The substance exhibits both creep and relaxation. For a step stress $p = p_0 H(t)$, the strain $e(t)$ is

$$\begin{aligned}
e(t) &= \frac{p_0}{m_r}\left[1 - \left(1 - \frac{\tau_p}{\tau_e}\right)\exp(-t/\tau_e)\right]H(t) \\
&= \frac{p_0}{m_2}\left\{ 1 + \left(\frac{\tau_e}{\tau_p} - 1\right)[1 - \exp(-t/\tau_e)]\right\}H(t),
\end{aligned} \quad (8.29)$$

with the limiting values $e(\infty) = p_0/m_r$ and $e(0) = p_0/m_2$. The strain as a function of time is shown in Fig. 8.4. The final value of the stress/strain ratio is m_r which is known as the *relaxed elastic modulus*, and the initial value of the stress/strain ratio is m_2, known as the *unrelaxed elastic modulus*. Their ratio is $m_2/m_r = \tau_e/\tau_p$. For a strain step $e = e_0 H(t)$, the stress relaxation with time is

$$p(t) = m_r e_0\left[1 + \left(\frac{\tau_e}{\tau_p} - 1\right)\exp(-t/\tau_p)\right]H(t), \quad (8.30)$$

with the limiting values $p(0) = m_2/e_0$ and $p(\infty)/m_r/e_0$. For sinusoidal disturbance $p = p_0 \exp(i\omega t)$, $e = e_0 \exp(i\omega t)$, the basic equation (8.26) leads to

$$p_0(1 + i\omega\tau_p) = e_0(1 + i\omega\tau_e), \tag{8.31}$$

and

$$\frac{p_0}{e_0} = m_r \frac{1 + i\omega\tau_e}{1 + i\omega\tau_p} = m_2 \frac{m_1 + i\omega\eta_1}{(m_1 + m_2) + i\omega\eta_1} = K \exp(i\delta), \tag{8.32}$$

where

$$\tan \delta = \frac{\omega(\tau_e - \tau_p)}{1 + \omega^2 \tau_e \tau_p} = \frac{1}{Q}. \tag{8.33}$$

Similarly as in the Maxwell model, δ provides again a measure of the lag of the strain behind the stress and of the system damping.

There are a number of other mechanical analogs of viscoelastic materials, based on various combinations of springs and dashpots. Some of them were recently described by Qaisar (1989). Another class of models of anelastic behavior is suggested by experimental studies. Instead of fitting complicated systems of springs and dashpots, the strain increase can be described as a power or logarithmic function of time. Jeffreys (1958) proposed a composite form for the creep function in the Earth

$$\phi(t) = 1/m_0 + q\left[(1 + at)^\alpha - 1\right]/a, \tag{8.34}$$

where m_0, q and a are positive constants and $0 \le a \le 1$. A value of $\alpha = 0.25$ provides an effective model for many problems related to the propagation of seismic waves.

8.3 Intrinsic Attenuation and Its Quality Factor Q

The usual way that attenuation is introduced into perfectly elastic isotropic wave motion is by specifying the complex elastic moduli, complex wave velocities, and complex wavenumbers. For real values of the medium density ρ and angular frequency ω, the complex elastic moduli $\bar{\mu}$ and $\bar{k}$ are (e.g., Ben-Menahem and Singh, 1981)

$$\bar{\mu}(\omega) = \mu + \int_0^\infty \dot{\mu}(t)\exp(-i\omega t)\,dt = \mu + i\mu^*, \tag{8.35}$$

$$\bar{k}(\omega) = k + \int_0^\infty \dot{k}(t)\exp(-i\omega t)\,dt = k + ik^*. \tag{8.36}$$

The functions $\bar{\mu}(\omega)$ and $\bar{k}(\omega)$ are known as the *dynamic shear modulus* and *dynamic bulk modulus*, respectively. The complex wave velocities are defined as

$$\bar{\alpha}(\omega) = \left[\frac{\bar{k}(\omega) + 4\bar{\mu}(\omega)/3}{\rho} \right]^{1/2} = \alpha + i\alpha^*, \tag{8.37}$$

$$\bar{\beta}(\omega) = \left[\frac{\bar{\mu}(\omega)}{\rho} \right]^{1/2} = \beta + i\beta^*. \tag{8.38}$$

The dynamic wave numbers are defined by

$$\bar{\kappa}_\alpha = \frac{\omega}{\bar{\alpha}(\omega)} = \frac{\omega}{c_\alpha(\omega)} - i\chi_\alpha(\omega) = \kappa_\alpha - i\kappa_\alpha^*, \tag{8.39}$$

$$\bar{\kappa}_\beta = \frac{\omega}{\bar{\beta}(\omega)} = \frac{\omega}{c_\beta(\omega)} - i\chi_\beta(\omega) = \kappa_\beta - i\kappa_\beta^*, \tag{8.40}$$

where

$$c_\alpha = \frac{1}{\mathrm{Re}\left\{\rho/\left[\bar{k}(\omega) + 4\bar{\mu}(\omega)/3\right]\right\}^{1/2}}, \qquad c_\beta(\omega) = \frac{1}{\mathrm{Re}\left[\rho/\bar{\mu}(\omega)\right]^{1/2}} \tag{8.41}$$

are the phase velocities, and

$$\chi_\alpha(\omega) = -\omega\,\mathrm{Im}\left[\frac{\rho}{\bar{k}(\omega) + 4\bar{\mu}(\omega)/3}\right], \qquad \chi_\beta(\omega) = \omega\,\mathrm{Im}\left[\frac{\rho}{\bar{\mu}(\omega)}\right]^{1/2} \tag{8.42}$$

are the anelastic attenuation coefficients.

It can be shown that any formal solution of the equations in the theory of linear elasticity offers a corresponding solution for a linear viscoelastic body if the elastic moduli that occur in the elastic solution are replaced by the corresponding complex moduli. This is known as the *correspondence principle*. The other important principle is the principle of causality, which simply states that in a physical system, no output can occur before the input. Formulated in terms of signals that carry information across the medium, this principle states that a signal originated at a point $x = 0$ at time $t = 0$ cannot arrive at the distance $x = ct$ before time t, where c is a finite characteristic velocity of the medium. Relations between the real (Re) and imaginary (Im) parts of the complex propagation functions, described by formulas (8.35)–(8.42), predict causal dispersion resulting from anelasticity, regardless of the physical mechanism involved in absorption.

Velocity dispersion in a standard liner solid is a good example of dispersion effects caused by anelasticity. The theory of viscoelasticity is embodied in Boltzmann's aftereffect equation, which can be written as

$$e(t) = \int_{-\infty}^{t} \dot{p}(\tau)\phi(t - \tau)\, d\tau, \tag{8.43}$$

where $\phi(t)$, the creep function, is determined by the physical mechanism of attenuation (or rheology of the material). This equation assumes that the strain $e(t)$ at time t is caused by a linear superposition of the total stress history $p(t)$ up to the time t, and therefore it incorporates both the superposition and the causality principles. The Boltzmann aftereffect equation for a standard linear solid, with $p(-\infty) = 0$ and $t - \tau = \theta$, is (e.g., Liu *et al.*, 1976)

$$e(t) = \frac{p(t)}{m_r} - \frac{1}{m_r}\left(1 - \frac{\tau_p}{\tau_e}\right)\int_0^{\infty} \exp(-\theta/\tau_e)\dot{p}(t - \theta)\, d\theta. \tag{8.44}$$

For sinusoidal disturbance

$$e(t) = \frac{p(t)}{m_r}\left[1 - \frac{\omega^2 \tau_e^2}{1 + \omega^2 \tau_e^2}\left(1 - \frac{\tau_p}{\tau_e}\right) - i\frac{\omega(\tau_e - \tau_p)}{1 + \omega^2 \tau_e^2}\right] = \frac{p(t)}{m_r}(A - iB),$$

$$\tag{8.45}$$

where

$$A = 1 - \frac{\omega^2 \tau_e^2}{1 + \omega^2 \tau_e^2}\left(1 - \frac{\tau_p}{\tau_e}\right)$$

and

$$B = \omega(\tau_e - \tau_p)/(1 + \omega^2 \tau_e^2).$$

Defining the complex modulus m_c as

$$m_c = \frac{m_r}{A - iB}, \tag{8.46}$$

we have

$$\frac{\bar{\kappa}(\omega)}{\omega/c_r} = \left[\frac{m_r}{m_c(\omega)}\right]^{1/2}, \tag{8.47}$$

where $c_r = (m_r/\rho)^{1/2}$ is the phase velocity associated with the relaxed elastic modulus. For a plane wave propagating in a linear viscoelastic solid, the wavenumber $\bar{\kappa}$ is defined by

$$\bar{\kappa}(\omega) = \frac{\omega}{c(\omega)} - i\chi(\omega), \tag{8.48}$$

where $c(\omega)$ is the phase velocity and $\chi(\omega)$ is the attenuation factor. Combining equation (8.47) with equation (8.48), we obtain the following relations:

$$c(\omega) = \frac{c_r}{B} \left\{ 2A \left[\left(1 + \frac{B^2}{A^2} \right)^{1/2} - 1 \right] \right\}^{1/2}, \qquad (8.49)$$

$$\chi(\omega) = \frac{\omega}{c_r} \left\{ \frac{A}{2} \left[\left(1 + \frac{B^2}{A^2} \right)^{1/2} - 1 \right] \right\}^{1/2}. \qquad (8.50)$$

The internal friction coefficient $\tan \delta$ is

$$\tan \delta = \frac{\mathrm{Im}(m_c)}{\mathrm{Re}(m_c)} = \frac{\omega(\tau_e - \tau_p)}{1 + \omega^2 \tau_e \tau_p} = \frac{\tau_e - \tau_p}{\tau_0} \left(\frac{\omega \tau_0}{1 + \omega^2 \tau_0^2} \right) = \frac{1}{Q}, \qquad (8.51)$$

where $\tau_0 = (\tau_e \tau_p)^{1/2}$. Equation (8.15) is the same as relation (8.33).

The phase velocity dispersion and the frequency variation of $1/Q$ for the standard linear solid are shown in Fig. 8.5. The phase velocity variation with frequency has zero slope at low and high frequencies. The coefficient $1/Q$ has an ω dependence at low frequencies and $1/\omega$ dependence at high frequencies, and it has a single peak at $\omega = 1/\tau_0$. The observed Q in the Earth, however, is flat over a wide range of frequency. This relative constancy of Q can be described by a distribution of relaxation mechanisms. Superimposing many such mechanisms, each one shifted somewhat relative to the other, a flat dependence of $1/Q$ over a given finite frequency interval can be obtained. A linear superposition of creep elements, each with the characteristics of a standard linear solid, leads to the generalization of equation (8.44) either for a finite number or for a continuous distribution of relaxation mechanisms (Liu et al., 1976; Ben-Menahem and Singh, 1981).

The dimensionless dissipation parameter or the quality factor Q, already used to describe the response of a Maxwell substance [relation (8.20)] and of a standard linear solid (relations (8.45)–(8.51)) to a harmonic stress cycle, is an important seismological quantity, and its more formal introduction is needed. If a volume of material is cycled in stress at a frequency ω, then a dimensionless measure of the internal friction or the anelasticity is given by (e.g., Aki and Richards, 1980)

$$\frac{1}{Q(\omega)} = \frac{\Delta W}{2\pi W}, \qquad (8.52)$$

where W is the peak strain energy stored in the volume and ΔW is the energy lost in each cycle as a result of elastic imperfections of the material. For small attenuation, that is, assuming $Q \gg 1$, the maximum strain energy

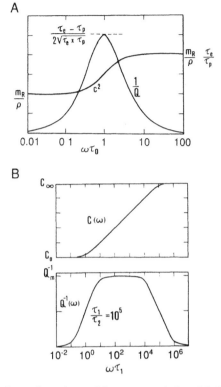

Figure 8.5 Phase velocity dispersion and frequency variation of $1/Q$ for the standard linear solid (A) and for a model with continuous band-limited relaxation times (B).

$W \cong 2\overline{W}$, where $\overline{W}$ is the average stored energy, and a standard definition of the intrinsic Q recommended by O'Connell and Budiansky (1978) is

$$Q = \frac{4\pi\overline{W}}{\Delta W}. \tag{8.53}$$

The attenuation for a given wave type (P or S) is defined as the inverse of the quality factor Q and is related to other parameters by

$$\frac{1}{Q} = \frac{\chi c}{\pi f} = \frac{D}{\pi}, \tag{8.54}$$

where χ is the anelastic attenuation coefficient, c is the wave velocity, f is the frequency, and D is the logarithmic decrement. The logarithmic decrement is a measure of the amplitude decay $D = \ln|A(t_n)/A(t_{n+2})|$ during a cycle with the period $T = t_{n+2} - t_n$. The dimension of the anelastic attenuation coefficient χ is generally given as decibels (dB) per unit length or nepers per unit length (e.g., Toksöz et al., 1988). The relation between the two is χ (dB/unit length) = 8.68χ [nepers (Np; also known as napiers)/unit length].

For a medium with linear stress–strain relations and small attenuation, wave amplitude A is proportional to $W^{1/2}$. Hence

$$\frac{1}{Q(\omega)} = \frac{\Delta A}{\pi A},$$ (8.55)

from which the amplitude variations as a result of attenuation can be found (Aki and Richards, 1980). The attenuation effects can be observed either as the temporal decay of amplitude in a standing wave at fixed wavenumber or the spatial decay in a propagating wave at fixed frequency. In the first case, the amplitude $A = A(t)$, with an initial value $A = A_0$ and decreasing a fraction π/Q at successive times $n(2\pi/\omega)$, is

$$A(t) = A_0(1 - \pi/Q)^n \quad \text{for } t = n(2\pi/\omega).$$ (8.56)

Thus

$$A(t) = A_0\left(1 - \frac{\omega t}{2Qn}\right)^n \Rightarrow A_0 \exp\left(-\frac{\omega t}{2Q}\right) \quad \text{for large } n.$$ (8.57)

This relation defines the value of a temporal Q obtained from observations of exponentially decaying values of $A(t)$ (Aki and Richards, 1980). For the spatial decay, the form $A = A(x)$ for distance x can be considered, assuming that the direction of maximum attenuation is along the x axis that is also the direction of wave propagation. Then $\Delta A = (dA/dx)l$, where $l = 2\pi c/\omega$ is the wavelength, and equation (8.55) becomes $dA/dx = (\omega/2cQ)A$ with the exponentially decaying solution (Aki and Richards, 1980)

$$A(x) = A_0 \exp\left(-\frac{\omega x}{2cQ}\right).$$ (8.58)

This relation defines the value of a spatial Q obtained from observations of exponentially decaying values of $A(x)$. The values of Q measured from relations (8.57) and (8.58) will be the same in homogeneous media only. The presence of dispersion effects caused by heterogeneity of the material will result in different values of temporal and spatial measures of the quality factor Q.

For small attenuation, the quality factor Q can be expressed in terms of stress–strain relations

$$\frac{1}{Q} = \frac{m^*}{m} = \tan \delta \cong \delta,$$ (8.59)

where m^* and m are the imaginary and real parts of the appropriate elastic modulus $\bar{m} = m + im^*$, and δ is the so-called loss angle, the phase angle by which the strain lags behind the stress. It can be shown for large values of the quality factor Q_α for P waves and the quality factor Q_β for S waves that (e.g., Ben-Menahem and Singh, 1981)

$$\frac{1}{Q_\alpha} = \frac{2\kappa_\alpha^*}{\kappa_\alpha} = \frac{k^* + (4/3)\mu^*}{k + (4/3)\mu} = \frac{2\alpha^*}{\alpha}, \tag{8.60}$$

$$\frac{1}{Q_\beta} = \frac{2\kappa_\beta^*}{\kappa_\beta} = \frac{\mu^*}{\mu} = \frac{2\beta^*}{\beta}. \tag{8.61}$$

It is convenient to define the bulk dissipation parameter Q_k

$$\frac{1}{Q_k} = \frac{k^*}{k}, \tag{8.62}$$

which is connected with the longitudinal and shear dissipation factors by the relation

$$\frac{1}{Q_\alpha} = \frac{N}{Q_\beta} + \frac{1-N}{Q_k}, \tag{8.63}$$

where $N = (4/3)(\beta/\alpha)^2$.

It follows from observations that P waves are in general less attenuated than S waves, implying that bulk losses are smaller than shear losses in the Earth's crust. If there are no losses in pure compression, and several studies indicate that shear mechanisms such as grain boundary sliding dominate the attenuation processes, then from relation (8.63) it follows that $Q_\beta/Q_\alpha = (4/3)(\beta/\alpha)^2$, and for $\lambda \cong \mu$ along the path $Q_\alpha \cong (9/4)Q_\beta$. The attenuation of body waves can also be represented by the quantity t^*, introduced by Futterman (1962), which is defined as the integral over time along the path of inverse Q. The approximate but more convenient form is $t^* = T/Q$, where T is the total travel time of the wave. The parameter t^*, sometimes called the *attenuation operator*, is given the subscript α for P waves and β for S waves, the same as for the quality factor Q. Thus for $\lambda \cong \mu$ along the path $t_\beta^* = 4t_\alpha^*$. For this case, therefore, the problem is reduced to a single unknown, either Q or t^*.

By accepting so far the density to be real, the possibility of losses from imperfect inertia was ignored (Ben-Menahem and Singh, 1981). Introducing the complex density $\bar{\rho} = \rho + i\rho^*$ and assuming $0 < \rho^*/\rho \ll 1$, we have

$$\frac{\mu + i\mu^*}{\rho + i\rho^*} \cong \frac{\mu}{\rho}\left[1 + i\left(\frac{\mu^*}{\mu} - \frac{\rho^*}{\rho}\right)\right], \tag{8.64}$$

and

$$\frac{2\kappa_\beta^*}{\kappa_\beta} = \frac{\mu^*}{\mu} - \frac{\rho^*}{\rho}. \tag{8.65}$$

Similary for P waves

$$\frac{2\kappa_\alpha^*}{\kappa_\alpha} = \frac{k^* + (4/3)\mu^*}{k + (4/3)\mu} - \frac{\rho^*}{\rho}. \tag{8.66}$$

The last two equations can also be written in the form

$$\frac{1}{Q_\beta} = \frac{1}{Q_\mu} - \frac{1}{Q_\rho} \quad \text{and} \quad \frac{1}{Q_\alpha} = \frac{N}{Q_\mu} + \frac{1-N}{k} - \frac{1}{Q_\rho}, \tag{8.67}$$

where

$$\frac{1}{Q_\mu} = \frac{\mu^*}{\mu} \quad \text{and} \quad \frac{1}{Q_\rho} = \frac{\rho^*}{\rho}. \tag{8.68}$$

Equations (8.67) represent the total attenuation in terms of imperfections in rigidity, bulk modulus, and density.

Although physical definitions of the quality factor Q are given by formulas (8.52)–(8.54), in practice Q is usually measured by methods based either on equations (8.57) or (8.58). In such an approach different results may be obtained from different methods of measurement even in simple cases (Aki and Richards, 1980). The most important case is the influence of scattering. Without scattering, the quality factor measured using relations (8.57) or (8.58) will represent anelastic attenuation only. In reality, however, a pulse traveling through a medium is affected by anelasticity and scattering and the influence of both effects cannot be separated unless some models for the Q components are used. Before discussing these problems, however, a brief review of the attenuation of body waves in anisotropic media is needed.

8.4   Attenuation in Anisotropic Media

Attenuation may be introduced into anisotropic wave motion by specifying imaginary parts to the elastic constants. This problem has been discussed by Crampin (1981), and our outline here follows his approach. Substituting the displacement equation of plane wave propagating in the x_1 direction with the complex velocity $\bar{c} = c + ic^*$

$$u_t = a_t \exp[i\omega(t - x_1/\bar{c})] \tag{8.69}$$

into the equation of motion (7.8), we obtain an eigenvalue problem similar to (7.11)

$$\left(\overline{\mathbf{M}} - \rho\bar{c}^2\mathbf{I}\right)\mathbf{a} = 0, \tag{8.70}$$

where $\overline{\mathbf{M}}$ is the 3×3 matrix with complex elements $\bar{A}_{t1k1} = A_{t1k1} + iA^*_{t1k1}$. The complex $\overline{\mathbf{M}}$ matrix results in three complex eigenvalues of the three body waves qP, qSP, and qSR. For each wave the following relation can be written

$$\rho\bar{c}^2 = \varepsilon + i\varepsilon^*, \tag{8.71}$$

where $\bar{\varepsilon} = \varepsilon + i\varepsilon^*$ is one of the three eigenvalues of $\overline{\mathbf{M}}$. Expanding relation (8.71) and neglecting squares of $\varepsilon^*/\varepsilon$, the velocity becomes

$$\bar{c} = (\varepsilon/\rho)^{1/2}[1 + i(\varepsilon^*/\varepsilon)/2], \tag{8.72}$$

and the quality factor Q for each wave is expressed as

$$\frac{1}{Q} = \frac{\varepsilon^*}{\varepsilon}. \tag{8.73}$$

The approximate equations (7.13), describing the velocity variation in planes of mirror symmetry in perfectly elastic anisotropic media, can be altered for anelastic media by replacing real elastic constants by complex elastic constants. The coefficients in equations (7.13) are linear functions of the elastic constants [relations (7.14)]. Consequently, replacing real elastic constants c_{ij} by complex $\bar{c}_{ij}$, the real and imaginary parts separate, and the imaginary parts satisfy similar equations as the real parts in equations (7.13)

$$\varepsilon^*_P = \mathrm{Im}\left(\rho\overline{V}_P^2\right) = A^*_1 + A^*_2\cos 2\theta + A^*_3\sin 2\theta + A^*_4\cos 4\theta + A^*_5\sin 4\theta,$$

$$\varepsilon^*_{SP} = \mathrm{Im}\left(\rho\overline{V}_{SP}^2\right) = B^*_1 + B^*_2\cos 4\theta + B^*_3\sin 4\theta, \tag{8.74}$$

$$\varphi^*_{SR} = \mathrm{Im}\left(\rho\overline{V}_{SR}^2\right) = C^*_1 + C^*_2\cos 2\theta + C^*_3\sin 2\theta,$$

where the coefficients A^*_m, B^*_n, and C^*_n (where $m = 1,\ldots,5$ and $n = 1,2,3$) are the same linear combinations of the imaginary elastic constants c^*_{ij} as the coefficients A_m, B_n, and C_n are the combinations of the real elastic constants described by relations (7.14).

The quality factor Q_{qp} for quasi—longitudinal waves qP can be written as

$$\frac{1}{Q_{qP}} = \frac{\varepsilon^*_P}{\varepsilon_P} = \frac{A^*_1 + A^*_2\cos 2\theta + A^*_3\sin 2\theta + A^*_4\cos 4\theta + A^*_5\sin 4\theta}{A_1 + A_2\cos 2\theta + A_3\sin 2\theta + A_4\cos 4\theta + A_5\sin 4\theta}. \tag{8.75}$$

Similar expressions can also be written for the quality factors Q_{qSP} and $Q_{qS}R$ for quasi–S waves. Since the relations A_{m-1}/A_1, B_{n-1}/B_1, and C_{n-1}/C_1

are small, the full equations for the quality factors can be expressed approximately in the form

$$\frac{1}{Q_{qP}} = \frac{A_1^* + A_2^* \cos 2\theta + A_3^* \sin 2\theta + A_4^* \cos 4\theta + A_5^* \sin 4\theta}{A_1},$$

$$\frac{1}{Q_{qSP}} = \frac{B_1^* + B_2^* \cos 4\theta + B_3^* \sin 4\theta}{B_1} \qquad (8.76)$$

$$\frac{1}{Q_{qSR}} = \frac{C_1^* + C_2^* \cos 2\theta + C_3^* \sin 2\theta}{C_1}.$$

When the angle θ is measured from a direction of sagittal symmetry, the full equations (8.76) for the quality factors contract to the reduced equations, similarly as the full equations (7.13) for velocity variations contract to the reduced equations (7.15). The reduced equations for the dissipation coefficients are

$$\frac{1}{Q_{qP}} = \frac{A_1^* + A_2^* \cos 2\theta + A_4^* \cos 4\theta}{A_1},$$

$$\frac{1}{Q_{qSR}} = \frac{B_1^* + B_2^* \cos 4\theta}{B_1}, \qquad (8.77)$$

$$\frac{1}{Q_{qSR}} = \frac{C_1^* + C_2^* \cos 2\theta}{C_1},$$

where the constants are the same as in the previous equations.

From equations (8.76) and (8.77) it follows that the quality factors have similar dependence on the azimuth as that shown for the velocities. The equations can be used for modeling attenuation in the same way as equations (7.13) and (7.15) are used to model variations of velocity. If the variation of attenuation with direction in a given medium is known, from either observations or theoretical considerations, then effective complex elastic constants can be estimated from equations (8.76) and (8.77). The real and imaginary parts of the constants can be separated and transformed independently (Crampin, 1981).

8.5 Scattering Effects and Coda Waves

One of the most conspicuous features of seismograms of local earthquakes are the scattered waves that follow all primary waves. For local microearthquakes and mine tremors, whose source durations could be as short as a fraction of a second, seismic energy is often recorded several or even tens of seconds after the direct shear-wave arrivals. Thus the entire seismogram can be considered as composed by primary waves and scattered waves. The primary waves would constitute the whole seismogram if the medium inhomogeneities were not present. The scattered waves are generated by the interactions between primary waves and inhomogeneities. Media without and with inhomogeneities are often called *unperturbed* and *perturbed media*. Heterogeneities in the medium can be considered as continuous and discontinuous. Discontinuous heterogeneities are inclusions in the medium. The media inside and outside of the inclusions are homogeneous, and sharp discontinuities occur on the boundaries. The scattering problem for a single inclusion can be formulated as a boundary value problem. For a medium with numerous inclusions a multiple scattering theory, based on the solution of the single scattering problem, can be used. Another approach is the perturbation method often used in seismology for both discontinuous and continuous media (e.g., Wu, 1989).

There are three important lengths involves in any scattering phenomenon: the linear dimension L of the inhomogeneous region, the scale length a of the inhomogeneity, and the wavelength. The medium-extent length L is also called "travel distance" as the distance in the medium where scattering takes place. The relation between the wavelength and the linear dimension L affects the characteristics of scattering phenomena. These variables are connected by the parameter κL, where κ is the wavenumber. The inhomogeneity scale length a is also called the *correlation distance*, as it expresses the separation for which fluctuations of variables in a random medium become uncorrelated in statistical studies. Usually a is assumed to be much smaller than the length of the medium extent L, that is, $a \ll L$. Another important parameter is κa. When $\kappa a \geq 1$, the effect of the shape of a scatterer must be considered. The inclusion shape is not important for the case $\kappa a \ll 1$, which is called *Rayleigh scattering*. If $1 < \kappa a < 10$, the waves travel in a medium with heterogeneity scale similar to their wavelength and scattering becomes strong. When the wavelength and the obstacles are of similar size, that is, $\kappa a \cong 1$, scattering can become particularly strong and is called *Mie scattering* [see, e.g., a review by Herraiz and Espinosa (1987)].

In the perturbation method the heterogeneous medium is decomposed into an unperturbed (reference) medium and the perturbations. Assuming weak heterogeneity (discontinuous medium), the scattering problem is transferred into a radiation problem by considering the response of the perturbations to the primary waves as the excitation of secondary sources. The unperturbed medium is a homogeneous, isotropic, unbounded body. The equation of motion for displacement **u** in a general inhomogeneous, isotropic, elastic body is (Aki and Richards, 1980)

$$\rho \frac{\partial^2 u_i}{\partial t^2} = \frac{\partial}{\partial x_i}(\lambda \nabla \cdot \mathbf{u}) + \frac{\partial}{\partial x_j}\left[\mu\left(\frac{\partial u_i}{\partial x_j} + \frac{\partial u_j}{\partial x_i}\right)\right], \tag{8.78}$$

where $\nabla \cdot \mathbf{u} = \partial u_j / \partial x_j$. This is the ith component of the vector wave equation describing the behavior of seismic waves in a given phase of the medium. The density ρ and Lamé's constants λ and μ in the perturbed medium are expressed as

$$\rho = \rho_0 + \delta\rho,$$
$$\lambda = \lambda_0 + \delta\lambda, \tag{8.79}$$
$$\mu = \mu_0 + \delta\mu,$$

where ρ_0, λ_0, and μ_0 are the density and elastic constants in the unperturbed homogeneous medium, and $\delta\rho$, $\delta\lambda$, and $\delta\mu$ are functions of space with magnitudes assumed to be much smaller than the corresponding unperturbed values, that is $|\delta\rho/\rho_0|, |\delta\lambda/\lambda_0|, |\delta\mu/\mu_0| \ll 1$.

Substituting relations (8.79) into equations (8.78) and arranging the unperturbed terms on the left-hand side and the perturbed terms on the right-hand side, the following equations are obtained (Aki and Richards, 1980):

$$\rho \frac{\partial^2 u_i}{\partial t^2} - (\lambda_0 + \mu_0)\frac{\partial}{\partial x_i}(\nabla \cdot \mathbf{u}) - \mu_0 \nabla^2 u_i$$

$$= -\delta\rho \frac{\partial^2 u_i}{\partial t^2} + (\delta\lambda + \delta\mu)\frac{\partial}{\partial x_i}(\nabla \cdot \mathbf{u}) + \delta\mu \nabla^2 u_i \tag{8.80}$$

$$+ \frac{\partial(\delta\lambda)}{\partial x_i}\nabla \cdot \mathbf{u} + \frac{\partial(\delta\mu)}{\partial x_j}\left(\frac{\partial u_i}{\partial x_j} + \frac{\partial u_j}{\partial x_i}\right).$$

The solution $\mathbf{u}$ for the total displacement field can be considered as the sum of the primary field $\mathbf{u}^0$ in the reference medium and the scattered field $\mathbf{u}^1$

$$\mathbf{u} = \mathbf{u}^0 + \mathbf{u}^1. \tag{8.81}$$

This displacement $\mathbf{u}^0$ is the solution for the unperturbed medium satisfying the equation [see formula (3.21)]

$$\rho_0 \frac{\partial^2 u_i^0}{\partial t^2} - (\lambda_0 + \mu_0)\frac{\partial}{\partial x_i}(\nabla \cdot \mathbf{u}^0) - \mu_0 \nabla^2 u_i^0 = 0. \tag{8.82}$$

Substituting relation (8.81) into equation (8.80), substracting equation (8.82) from equation (8.80), accepting that $|u_i^1| \ll |u_i^0|$, and neglecting higher-order terms, the following equation for $\mathbf{u}^1$ is obtained:

$$\rho_0 \frac{\partial^2 u_i^1}{\partial t^2} - (\lambda_0 + \mu_0) \frac{\partial}{\partial x_i} (\nabla \cdot \mathbf{u}^1) - \mu_0 \nabla^2 u_i^1 = Q_i, \tag{8.83}$$

where

$$Q_i = -\delta\rho \frac{\partial^2 u_i^0}{\partial t^2} + (\delta\lambda + \delta\mu) \frac{\partial}{\partial x_i} (\nabla \cdot \mathbf{u}^0) + \delta\mu \nabla^2 u_i^0$$

$$+ \frac{\partial(\delta\lambda)}{\partial x_i} \nabla \cdot \mathbf{u}^0 + \frac{\partial(\delta\mu)}{\partial x_j} \left(\frac{\partial u_i^0}{\partial x_j} + \frac{\partial u_j^0}{\partial x_i} \right). \tag{8.84}$$

Equation (8.83) is the equation of motion in a homogeneous, unbounded, isotropic, elastic medium with body force $\mathbf{Q}$ defined by relation (8.84).

We neglected so far body forces in our considerations of the propagation of seismic waves. They will be discussed in Chapter 9, which is devoted to the focal mechanism of seismic events.

The solution of equations (8.83) can be found in the book of Aki and Richards (1980). In practice, however, the usual deterministic approach to inhomogeneities is very difficult because of the large number of parameters involved, and statistical techniques become appropriate. Under the new probabilistic approach, a small number of statistical parameters are sufficient to describe the heterogeneities of the Earth. Its application to high-frequency records was pioneered by Aki (1969), whose attention was focused on the parts of the seismograms after all the direct waves such as P, S, and surface waves have arrived, and where the backscattered waves were most likely to be found. In order to consider scattered waves and their modeling in some detail, however, further procedures used to solve the wave propagation problems should be introduced.

Most signals studied in seismology are transient; that is, they have in practice a finite length, such as records of seismic events. For such a transient signal $f(t)$, the Fourier transform $F(f)$ exists, defined as

$$f(t) \rightarrow F(f) = \int_{-\infty}^{\infty} f(t) \exp(-\imath ft) \, dt, \tag{8.85}$$

where t is the time and f is the frequency. This is the transformation from the time domain into the frequency domain. The function $F(f)$ is called a *complex spectral density function*. The function $f(t)$ can be recovered from its complex spectral density function $F(f)$ by the inverse Fourier transform

$$F(f) \rightarrow f(t) = \int_{-\infty}^{\infty} F(f) \exp(ift) \, df. \tag{8.86}$$

The kernel of the inverse transform [i.e., $\exp(ift)$] is almost identical to the kernel of the direct Fourier transform but for the positive sign of the exponent. The relationship between the function of time $f(t)$ and the function of frequency $F(f)$ is often stressed by saying that $f(t)$ and $F(f)$ constitute a Fourier transform pair. In some seismological applications the sign convention for Fourier transforms is different. The positive sign of the exponent is taken for the direct transform, and the negative sign is taken for the inverse Fourier transform.

The amplitude spectral density or the amplitude spectrum $A(f)$ is defined as the absolute value of $F(f)$

$$A(f) = |F(f)| = \left\{(\mathrm{Re}[F(f)])^2 + (\mathrm{Im}[F(f)])^2\right\}^{1/2}, \qquad (8.87)$$

and the phase spectrum $\varphi(f)$ as

$$\varphi(f) = \tan^{-1}\frac{\mathrm{Im}[F(f)]}{\mathrm{Re}[F(f)]}. \qquad (8.88)$$

The Fourier transform can be expressed by the amplitude and phase spectra as

$$F(f) = A(f)\exp[i\varphi(f)]. \qquad (8.89)$$

The squared amplitude spectrum defines the energy density spectrum $E(f)$

$$E(f) = A^2(f), \qquad (8.90)$$

which is, obviously, independent of the phase $\varphi(f)$.

In seismological applications, the angular frequency $\omega = 2\pi f$ instead of the frequency f is often used.

To describe stationary stochastic processes, such as ambient seismic ground noise or scattered waves, the power spectral density $P(\omega)$ is introduced, which is the Fourier transform of the autocorrelation function $p(\tau)$, defined as (Aki and Richards, 1980)

$$p(\tau) = \langle f(t)f(t + \tau)\rangle, \qquad (8.91)$$

where the symbols $\langle \cdot \rangle$ indicate averaging over time t. Thus, the power spectral density is

$$P(\omega) = \int_{-\infty}^{\infty} p(\tau)\exp(-i\omega\tau)\,d\tau, \qquad (8.92)$$

which is independent of the phase.

The spatial Fourier transform is used to transform the space variables x_l to the wavenumbers κ_l,

$$f(x_l) \rightarrow F(\kappa_l) = \int_{-\infty}^{\infty} f(x_l)\exp(-i\kappa_l x_l)\,dx_l. \tag{8.93}$$

This is the transformation from the space domain into the wavenumber domain, expressed for the function of a single coordinate. Similar relations can be written for a function $f(x_1, x_2, x_3)$ involving triple transformation. In general, if a function $f(x_1, x_2, x_3, t)$ describes some propagating physical variable of interest, its Fourier transform is the following quadruply transformed function

$$F(\kappa_1, \kappa_2, \kappa_3, \omega) = \int_{-\infty}^{\infty} dx_1 \int_{-\infty}^{\infty} dx_2 \int_{-\infty}^{\infty} dx_3 \int_{-\infty}^{\infty} dt f(x_1, x_2, x_3, t)$$
$$\cdot \exp\left[-i(\kappa_1 x_1 + \kappa_2 x_2 + \kappa_3 x_3 - \omega t)\right], \tag{8.94}$$

presented here as an illustration.

The scattered waves recorded after the passage of primary waves from local earthquakes are called *coda waves*. This meaning was used in the first studies of scattered waves from local earthquakes (Aki, 1969). When only body waves are considered, as in mining seismology, a distinction is made between P and S codas. The P coda waves form a wavetrain between direct P and S phases, while the S coda waves follow the direct S waves. The S coda is dominant on the records of local earthquakes and is most often studied. The beginning of the coda was originally assigned at the point where the decay of amplitudes becomes regular. The accepted procedure now, originally proposed by Rautian and Khalturin (1978), is to take as the coda beginning twice the travel time of direct waves measured from the origin time of the seismic event, usually called *lapse time*. If this rule is too demanding, the coda beginning can be taken closer to the arrival of direct waves, but it is necessary to avoid contamination with the contribution from primary waves. The end of the coda is usually placed where the signal-to-noise ratio attains a given arbitrary value.

There are many theories proposed to interpret and model the coda waves: the diffusion theory (Aki and Chouet, 1975), the single-scattering theory (Aki, 1969; Aki and Chouet, 1975; Sato, 1977a), the multiple-scattering theory (Kopnichev, 1977; Gao et al., 1983), the energy transport theory (Wu, 1985), and the energy flux theory (Frankel and Wennerberg, 1987). The single-scattering and multiple-scattering models are based on the ray theory approach, the energy transport theory is based on the balance of energies carried by the primary and scattered waves, and the energy flux model of coda considers the energy balance between the direct and scattered waves.

Recently Zeng *et al.* (1991) have shown that both the single and multiple-scattering formulas and the energy transport formulas can be unified in an integral equation describing the energy of scattered waves.

Since Aki's (1969) paper it became popular to model coda waves as backscattered waves from more or less uniformly distributed scatterers in the Earth's crust. *Forward scattering* is the process in which most of the scattered energy is pumped forward in the incident primary-wave direction. The opposite result is called *backward scattering*. In general, impedance fluctuations, which are the medium density times the wave velocity fluctuations, tend to generate mainly backward scattering, while velocity perturbations create forward scattering, although the result depends strongly on many other factors.

Very often weak scattering, as opposed to strong scattering, is assumed in coda studies. In weak scattering the fluctuations of perturbed parameters are small in comparison with their corresponding mean values. The fractional energy loss from primary waves, measured by the ratio $\Delta I/I$, where ΔI is the loss of energy by scattering and I is the energy of primary waves, is also very small, and Born's approximation is acceptable. The Born approximation neglects both the loss of energy from primary waves and multiple scattering. Although it violates the energy conservation law, it has been used in various physical problems, and is often used in high-frequency coda wave analyses. In single scattering only one wave scatterer encounter is considered inside a homogeneous medium with discrete scatterers, and weak scattering is required. Multiple scattering is associated with strong scattering. If the scattering is very strong, it can be considered as a diffusion process. This theory has been adopted to explain the scattering found on lunar seismograms characterized by very large and very long coda waves. It has been shown, however, that the diffusion theory is not applicable to terrestrial coda waves (Kopnichev, 1977; Dainty and Toksöz, 1981), although it has occasionally been used to model coda waves for large lapse times.

The first observation of coda waves, relevant to the backscattering model, revealed that the total duration of local seismic waves is independent of epicentral distance (Soloviev, 1965). Coda waves were found to be insensitive to the nature of direct path and to have similar amplitudes and frequency contents for different stations and for a given source (Aki, 1969). This independence from path implied that the separation of source and path effects is possible in the coda power spectra (Aki and Chouet, 1975)

$$P(\omega|t) = S(\omega)C(\omega|t),\qquad (8.95)$$

where $P(\omega|t)$ is the coda power spectrum for frequency ω at lapse time t, $S(\omega)$ includes the source parameters, and $C(\omega|t)$ represents the regional effect and is independent of path between the source and the station. If the factor $C(\omega|t)$ is common to all seismic events in a given area and two events are considered, relation (8.95) becomes

$$\frac{P_1(\omega|t)}{P_2(\omega|t)} = \frac{S_1(\omega)}{S_2(\omega)}, \tag{8.96}$$

which allows for separate source and path effects.

Relation (8.95) has been confirmed in various seismic areas (e.g., Aki and Chouet, 1975; Rautian and Khalturin, 1978; Roecker et al., 1982; Pulli, 1984; Phillips et al., 1988; Woodgold, 1990). Its application to coda wave analysis requires relating the power spectrum $P(\omega|t)$ and coda amplitudes, and finding an expression for $S(\omega)C(\omega|t)$, for which models of coda generation are needed. Several excellent reviews of various problems related to coda waves have been published recently (e.g., Toksöz et al., 1988; Wang and Herrmann, 1988; Frankel, 1989; Wu, 1989; Chouet, 1990; Gao and Li, 1990; Sato, 1990). The most comprehensive review of coda waves, summarizing the work done in this subject up to 1986, has been published by Herraiz and Espinosa (1987). Their approach to coda wave modeling and analysis is followed here, with several modifications resulting from more recent works.

8.6 Single-Scattering Models of Coda Generation

A random medium is the medium with average characteristics in which deviations from the mean values of velocity, density, or Lamé parameters produce random heterogeneities. Aki and Chouet (1975) derived a formula for coda amplitudes for body waves, assuming isotropic single scattering from discrete randomly distributed heterogeneities, and the source and the receiver located at the same point in an infinite medium. This is a valid assumption for coda waves that arrive at the receiver long after the passage of the primary P or S waves. Aki and Chouet (1975) introduced the single-backscattering model and considered that body waves could be responsible for coda waves and not necessarily surface waves, which were accepted in the initial surface-wave model of Aki (1969). This was a turning point in the explanation of coda waves of local earthquakes.

Taking into account the geometric spreading and the anelastic attenuation of body waves, relation (8.95) can be written as

$$P(\omega|t) = S(\omega)t^{-2}\exp(-\omega t/Q), \tag{8.97}$$

where $S(\omega)$ sums up the effect of both the primary- and secondary-wave sources, t^{-2} is the geometric spreading factor in which t is the lapse time, and Q represents both intrinsic and scattering attenuation effects. It is this separation of the source and path terms that makes coda wave measurements a powerful tool in seismic studies. The source term can be expressed as

$$S(\omega) = |\Omega(\omega|r_0)|^2 (8\pi r_0^4 n_t / c), \tag{8.98}$$

where the factor $|\Omega(\omega|r_0)|$ represents the amplitude spectra of the backscattering wavelet from a single scatterer located at a reference distance r_0 from the source and receiver, n_t, is the number of scatterers per unit of volume, and c is the wave velocity. Thus the second factor includes the total number of scatterers and for a given r_0 is constant. The first factor represents the intensity of a single secondary wave leaving a scatterer at a reference distance r_0. Its relation to the primary source is given by

$$|\Omega(\omega|r_0)| = M_0 |\Omega_0(\omega|r_0)|, \tag{8.99}$$

where M_0 is the seismic moment described in Chapter 9. Under the assumption that scatterers are uniformly distributed in space, the term $|\Omega_0(\omega|r_0)|$ is constant in a given area, and the variation in the source term $S(\omega)$ from different seismic events is caused by differences in their seismic moment.

Now the power spectra and the recorded amplitudes of coda waves need to be related. The recorded amplitudes can be approximated by their root-mean-square values, defining coda envelopes. Mean values, given by the envelope $\langle f^2(t) \rangle$, are related to power spectra by [see equations (8.91) and (8.92)]

$$p(t,\tau) = \langle f(t)f(t+\tau) \rangle = \frac{1}{2\pi} \int_{-\infty}^{\infty} P(\omega|t)\exp(i\omega t)\, d\omega. \tag{8.100}$$

For zero lag $(\tau = 0)$, $p(t,0) = \langle f^2(t) \rangle$, and relation (8.100) becomes

$$\langle f^2(t) \rangle = \frac{1}{2\pi} \int_{-\infty}^{\infty} P(\omega|t)\, d\omega. \tag{8.101}$$

For a narrow-bandpass-filtered signal it is possible to have $P(\omega|t) = P$ constant within a given frequency band and $P(\omega|t) = 0$ otherwise. Then

$$\langle f^2(t) \rangle = 2P(\omega|t)\, \Delta f, \tag{8.102}$$

where Δf is the bandwidth of the filter with a center frequency at ω. Then the root-mean-square amplitude $A(\omega|t)$ of the coda waves is

$$A(\omega|t) = [2P(\omega|t)\Delta f]^{1/2}. \tag{8.103}$$

Thus the mean-square amplitude is approximately equal to the product of the power spectral density and the bandwidth. If the peak-to-peak amplitude is measured, then it corresponds to roughly twice the root-mean-square value of the signal, and the right-hand side of relation (8.103) should be multiplied by a factor of 2. Substituting the last equation (8.103), into equation (8.97), we obtain the relation

$$A(\omega|t) = C(\omega)t^{-1}\exp(-\omega t/2Q_c), \tag{8.104}$$

where $C(\omega) = [2S(\omega)\Delta f]^{1/2}$ is constant for a given event and frequency and is often called the *coda source factor*, and the factor Q_c is used instead of Q to specify the attenuation of coda waves.

The coda wave theory of Aki and Chouet (1975) is valid for the collocated source and receiver. This approximation is acceptable for coda waves that arrive after twice the S-wave lapse time. In some cases, especially for small seismic events, it is necessary to measure the coda waves close to the S-wave arrival, and then the source–receiver distance must be taken into account. Such a separation is included in the single scattering model of Sato (1977a). His model, called the single isotropic scattering model or *SIS* model, assumes a three-dimensional infinite and perfectly elastic medium in which the scatterers are homogeneously and randomly distributed. The distribution is characterized by the mean free path l. This is a parameter that controls the energy transferred from the primary to the scattered waves throughout the traveled path. The scatterers reduce the mean energy flux density of the incident plane wave by a factor $\exp(-x/l)$, where x is the distance along the direction of propagation. If l is greater than the travel distance, no multiple scattering needs be considered. The medium is characterized by the constant wave velocity c, and only S waves generated from a point source within a short time are considered.

The sum of the energy scattered by the inhomogeneities on the surface of an expanding ellipsoid whose foci are the source and receiver is

$$E(\omega|t) = \frac{W(\omega)}{4\pi l r^2} K\left(\frac{t}{t_s}\right)^2 \quad \text{for } t > t_s, \tag{8.105}$$

where $W(\omega)$ is the total energy radiated by the source within a unit frequency band, r is the source–receiver distance, t_s is the S-wave travel time, and the function $K(x)$ for $x = t/t_s$ is given by

$$K(x) = \left|\frac{1}{x}\ln\left(\frac{x+1}{x-1}\right)\right|^{1/2} \quad \text{for } x > 1. \tag{8.106}$$

Using relation (8.103) between the amplitude and the power spectrum of coda waves and the relation between the energy density and the power spectrum given by Aki and Chouet (1975), Pulli (1984) obtained the following equation from equation (8.105):

$$A(\omega|t) = \frac{1}{\omega} \left| \frac{W(\omega)\Delta f}{2\pi\rho l} \right|^{1/2} \frac{K(t/t_s)}{r} \exp(-\omega t/2Q_c), \qquad (8.107)$$

where ρ is the medium density, and the right-hand side of the equation is multiplied by the coda decay term $\exp(-\omega t/2Q_c)$, originally not considered by Sato (1977a). This is consistent with the expression for coda decay from the single-scattering model given by Aki (1980a) and later by Sato (1988). A simplified form of this equation, separating source and path terms, is

$$A(\omega|t) = C(\omega)r^{-1}K(t/t_s)\exp(-\omega t/2Q_c). \qquad (8.108)$$

This type of equation is often used in practical analyses of coda waves (e.g., Pulli, 1984; Jin and Aki, 1986; Woodgold, 1990).

The quality factor Q_c from coda waves is a combination of both anelastic or intrinsic attenuation described by the factor Q_i and scattering expressed by the factor Q_s. This fact is summed up by the relation (Dainty and Toksöz, 1981)

$$\frac{1}{Q} = \frac{1}{Q_c} = \frac{1}{Q_i} + \frac{1}{Q_s}, \qquad (8.109)$$

where Q is the quality factor obtained from measurements of the decay of seismic waves in general. The effects of anelastic attenuation and scattering cannot be separated directly, since both physical processes are described by the same mathematical form of the exponential function. Dainty (1981) has suggested, however, the expression

$$\frac{1}{Q_c} = \frac{1}{Q_i} + \frac{c}{\omega l} = \frac{1}{Q_i} + \frac{gc}{\omega} \qquad (8.110)$$

to explain Q_c observations dependent on frequency in terms of both anelastic attenuation and scattering. The scattering coefficient g evaluates the capacity of the medium to originate scattering. If the medium contains discrete heterogeneities, then the coefficient g can be obtained by multiplying the density of scatterers by their scattering cross section, which is the reciprocal of the mean free path. The scattering coefficient is also called turbidity coefficient, defined by Chernov (1960) as $g = \Delta I/IL$, where ΔI is the loss of energy by scattering when a wave of energy I passes through a layer of thickness L. The relative importance of these factors in a given area determines the characteristics of seismograms of local earthquakes. Small absorption and strong scattering produce seismograms with long duration, whereas high absorption diminishes the scattering process and shortens duration of the records.

The single isotropic scattering model of coda waves was improved by Sato (1977b) by including $P \to S$ and $S \to P$ conversions, and applied to the records of local earthquakes in the Kanto district, Japan, to estimate the mean free path (Sato, 1978). Further improvement in modeling of the excitation of coda waves was introduced by taking into account nonisotropic scattering and nonspherical source radiation (Sato, 1982). This approach is particularly suitable for a stope environment in mines, since it allows the introduction of an inhomogeneous distribution for the scatterers, and was used by Cichowicz and Green (1989) to study scattering effects at Western Deep Levels gold mine in South Africa. The single-scattering model was also used to study the scattering problem in a three-dimensional random medium with a vertical Q structure made of flat layers and a heterogeneity size that may vary with depth (Chouet, 1990).

In general, the single-scattering model of coda waves is acceptable for short travel times, whereas multiple scattering becomes important for long travel times evaluated along the seismic coda (Gao et al., 1983; Sato, 1984). The single-scattering hypothesis, however, is not valid in some areas, where indications of strong scattering were found for coda from local events (Dainty et al., 1987).

8.7 Multiple-Scattering Models

The first attempt to solve multiple scattering problems was made by Kopnichev (1977), who assumed a random, uniform, and isotropic medium and studied isotropic double and triple scattering of surface and body waves, supposing that the primary and scattered waves are of the same nature.

Gao et al. (1983) extended the single-scattering model of Aki and Chouet (1975) to the multiple-scattering case in a two-dimensional elastic medium in which scatterers with a cross section σ are uniformly distributed with a density n_s. They calculated the power spectrum $P(\omega|t)$ for double, triple, and quadruple scattering, showing that it can be decomposed into two parts

$$P(\omega|t) = P_s(\omega|t) + P_m(\omega|t), \qquad (8.111)$$

where $P_s(\omega|t)$ and $P_m(\omega|t)$ are the contributions from single and multiple scattering, respectively. These terms can be approximated by the numerical result obtained for the seventh-order multiple scattering

$$P_s(\omega|t) = \frac{n_s \sigma S(\omega)}{T} \exp\left[-\omega t \left(\frac{1}{Q_s} + \frac{1}{Q_i}\right)\right] \qquad (8.112)$$

and

$$P_m(\omega|t) = (n_s\sigma)^2 cS(\omega)\exp\left[-\omega t\left(\frac{0.74}{Q_s} + \frac{1}{Q_i}\right)\right],\qquad(8.113)$$

where $S(\omega) = r_0|\Omega(\omega|r_0)|^2$ is the source term representing the amplitude spectrum of the primary waves at the reference distance r_0 multiplied by this distance.

Relations (8.112) and (8.113) indicate that at short lapse time t, the power spectrum density of coda waves is mainly formed by the contributions from single scattering, while as t increases, multiple scattering becomes more important. The described results can be better understood by introducing two additional parameters $\bar{Q} = Q/Q_s$ and $\gamma = 2n_s\sigma r$, where $r = ct/2$ is the general traveled distance, and presenting the coda power spectra for single and multiple scattering as functions of γ for different values of $\bar{Q}$, which are shown in Fig. 8.6, reproduced from Gao et al. (1983). The parameter γ can be regarded as a normalized time since it represents time in terms of the mean free travel time l/c, which is necessary for a wave with velocity c to travel the mean free path l, and $\gamma = ct/l$. For small $\bar{Q}$, that is, when the scattering loss is smaller than the absorption loss, the coda attenuation before the multiple scattering is important. The intersection of $P_s(\omega|t)$ and $P_m(\omega|t)$ curves indicates that for the value of γ of about 0.8, the multiple scattering starts to be dominant. Neglecting this effect implies that for large lapse time, the values of Q_s are overestimated by a factor of 0.74.

Another approach to the multiple scattering of coda waves is the energy transport theory of Wu (1985). He derived the normalized energy distribution of seismic waves in a random but statistically uniform and isotropic medium for an isotropic point source with unit energy, assuming the isotropic scattering as the approximation of Rayleigh scattering (the size of the scatterers is smaller or nearly equal to the wavelength)

$$4\pi r^2 E(r) = \eta_e P_d r\exp(-\eta_e d_0 r) + \eta_e r\int_0^1 f(\xi, B_0)\exp\left(-\frac{\eta_e r}{\xi}\right)\frac{d\xi}{\xi^2}\qquad(8.114)$$

$$= 4\pi r^2(E_d + E_c),$$

where r is the distance between the source and the receiver, $E(r)$ is the average energy density proportional to the intensity at the observation point, $\eta_e = \eta_a + \eta_s$ is the extinction coefficient of the medium, η_a is the absorption coefficient of the medium equal to twice the anelastic attenuation coefficient χ given in equation (8.54) and η_s is the scattering coefficient defined as the total scattered power by a unit volume of random medium per unit incident

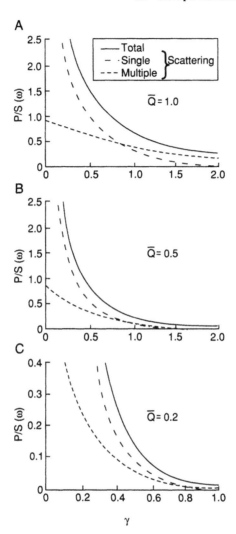

Figure 8.6   The ratio of coda power spectrum P over source factor $S(\omega)$ as a function of γ (explained in the text) for three values of $\overline{Q} = Q/Q_s$ and for the case of total, single, and multiple scattering. [From Gao *et al.* (1983, Fig. 2).]

power flux density, and

$$B_0 = \frac{\eta_s}{\eta_e} = \frac{n_s}{\eta_s + \eta_a} \tag{8.115}$$

is the seismic albedo of the medium, P_d and $f(\xi, B_0)$ are defined as

$$P_d = \frac{2d_0^2(1 - d_0^2)}{B_0(d_0^2 + B_0 - 1)}, \tag{8.116}$$

in which d_0 is the diffuse multiplier determined by

$$\frac{B_0}{2d_0} \ln\left(\frac{1 + d_0}{1 - d_0}\right), \tag{8.117}$$

and

$$f(\xi, B_0) = \left[\left(1 - B_0\xi \tanh^{-1}\xi\right)^2 + \left(\frac{\pi}{2}B_0\xi\right)^2\right]^{-1}. \tag{8.118}$$

The first term in relation (8.114) is the diffuse term $4\pi r^2 E_d$, contributed from the multiple scattering, and the second term is the coherent term $4\pi r^2 E_c$, representing the coherent part of the energy density.

The normalized energy distribution as a function of distance is shown in Fig. 8.7 (from Wu and Aki, 1988b), in which the distance D_e is normalized by the extinction length L_e of the medium, which in turn is the reciprocal of the extinction coefficient η_e

$$D_e = r/L_e = \eta_e r. \tag{8.119}$$

The shape of the curves in Fig. 8.7 depends strongly on the seismic albedo B_0 of the medium. For a purely absorbing medium $B_0 = 0$, and the energy decreases following a straight line in the semilogarithmic coordinates. As a result of the scattering process, however, the energy distribution curve becomes dependent on the seismic albedo B_0. For the case of large albedo, when the medium is strongly heterogeneous and scattering is significant, the curves are of arch shape and their maxima depend on the extinction coefficient η_e. It is possible therefore to obtain B_0 and η_e from the energy–distance curves and to separate the scattering effect from the intrinsic attenuation. This theory has been applied to local earthquakes in the Hindu Kush region (Wu and Aki, 1988b).

In the energy-flux model of coda waves (Frankel and Wennerberg, 1987), the energy balance between the direct waves and coda waves is considered and no assumptions concerning single or multiple scattering is made. The conservation of energy principle is combined with the observation of the

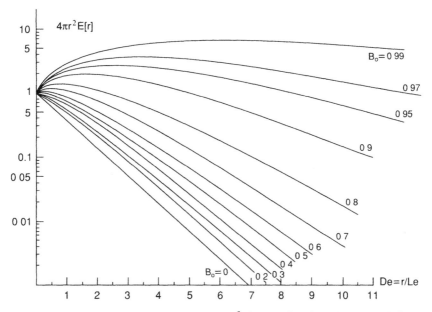

Figure 8.7 Normalized energy distributions $4\pi r^2 E(r)$ as a function of normalized distance $D_c = R/L_c$, where L_c is the extinction length defined in the text, for different values of the seismic albedo B_0. [From Wu and Aki (1988b, Fig. 2).]

spatial homogeneity of the coda energy to provide a formula for the coda amplitude in the time domain, where the values of the scattering and intrinsic quality factors are explicitly separated. The expression in two-dimensional media is

$$A_c(t,\omega) \propto \left[E_0(\omega)/\pi \right]^{1/2}(ct)^{-1} \exp(-\omega t/2Q_i)\left[1 - \exp(-\omega t/Q_s) \right]^{1/2},$$
$$(8.120)$$

where $E_0(\omega)$ is the elastic energy radiated by the source at frequency ω. The theoretical coda decays in two-dimensional media from the single-scattering model of Aki and Chouet (1975) and the energy-flux model of Frankel and Wennerberg (1987) are shown in Fig. 8.8, reproduced from Frankel and Wennerberg (1987). The coda decay predicted by the single-scattering model is much steeper than the decay observed in the synthetics, when Q_s measured for direct waves in each numerical simulation is smaller than about 200. The coda decays derived from the energy-flux model, on the other hand, correspond to the observed coda decays in all simulations.

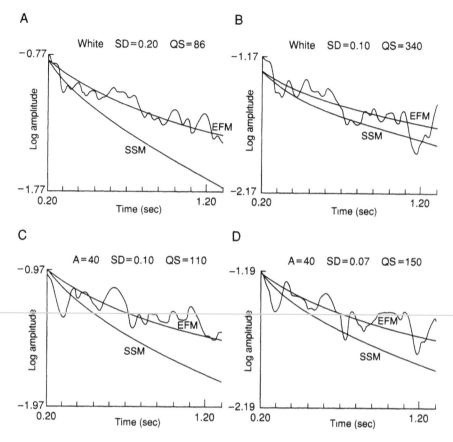

Figure 8.8   Average coda envelopes of synthetic seismograms (30 Hz), for a source–receiver distance of 180 m and for four random media, as a function of lapse time exceeding twice the direct wave travel time. Smooth curves represent the theoretical decays calculated from the energy flux model (EFM) and the single-scattering model (SSM), based on the scattering Q measured for each medium. [From Frankel and Wennerberg (1987, Fig. 8)].

Frankel and Wennerberg (1987) used the energy-flux model of coda waves to separate the scattering and intrinsic quality factors, based on analyses of coda amplitude and decay. This method was tested on synthetics from finite-difference simulations. They found that the coda decay was much more sensitive to the intrinsic attenuation than to the scattering attenuation. The intrinsic Q_i, therefore, can be estimated from the coda decay, and the scattering Q_s can be determined from the ratio of the coda amplitude to the direct wave amplitude. Applications of the finite-difference and finite-element methods to numerical studies of seismic wave scattering in both simple and complex velocity models have been recently reviewed by Frankel (1989).

8.8   Frequency Dependence of Q and Backscattering Coefficient

Several studies have found that the quality factor Q is frequency-dependent, often expressed in the form

$$Q(f) = qf^n, \tag{8.121}$$

where q is a constant, with the exponent n ranging in various regions from 0.4 to 1.1 in the frequency interval from 1 to 30 Hz. The first estimates of Q_c, made by Aki and Chouet (1975) from Japan and California, have shown a distinct increase of Q_c with frequency and a remarkable variation in its value at 1 Hz among different regions. Rautian and Khalturin (1978) observed the same increase of Q_c with frequency for the Garm region with the exponent $n = 0.5$. Since 1978 many measurements of the coda quality factor Q_c have been carried out in various parts of the world (e.g., Herrmann, 1980; Console and Rovelli, 1981; Roecker *et al.*, 1982; Rodriquez *et al.*, 1983; Biswas and Aki, 1984; Pulli, 1984; Jin *et al.*, 1985; Steensma and Biswas, 1988; Chapman and Rogers, 1989; Kvamme and Havskov, 1989; Woodgold, 1990). It was found that both q and n show regional variation, often related to tectonic features.

To study the attenuation of high-frequency S waves, Aki (1980a) applied a single-station method and used direct body waves and coda waves to eliminate the source effect. The amplitude spectrum of primary S waves can be expressed as

$$A_s(\omega, r) = S(\omega, \theta) r^{-1} \exp(-\omega r / 2\beta Q_\beta), \tag{8.122}$$

where r is the distance from the source to the receiver; $S(\omega, \theta)$ represents the source term, including the radiation pattern effect, which is a function of the observation azimuth θ; and β is the unperturbed S-wave velocity. From the coda decay observations it follows that the coda amplitude spectrum at time t_0 can be written as

$$A_c(\omega|t_0) = S(\omega)C(\omega|t_0), \tag{8.123}$$

where $C(\omega|t_0)$ is the path term independent of distance. Thus the ratio at a given time t is

$$\frac{A_s(\omega, r)}{A_c(\omega|t)} = \frac{S(\omega, \theta)}{S(\omega)r} \frac{\exp(-\omega r / 2\beta Q_\beta)}{C(\omega|t)}. \tag{8.124}$$

Taking natural logarithms of both sides and averaging the left-hand side over many events from a limited distance range $r \mp \Delta r$, it is possible to remove any systematic variations with distance

$$\left\langle \ln\left(\frac{rA_s(\omega,r)}{A_c(\omega|t)}\right)\right\rangle_{r \mp \Delta r} = a - br, \qquad (8.125)$$

where a is independent of r and

$$b = \frac{\omega}{2\beta Q_\beta} \qquad (8.126)$$

is the slope coefficient of the averaged logarithmic ratio of the spectra against the distance r.

This method was applied by Aki (1980a) to data from the Kanto region in Japan, who obtained a clear dependence of Q_β on frequency of the same form as that for coda waves (8.121), with the exponent n between 0.6 and 0.8. A similar agreement between the two quality factors Q_β and Q_c was also found in other areas (e.g., Roecker *et al.*, 1982; Del Pezzo *et al.*, 1985; Rebollar *et al.*, 1985). These results strongly support the hypothesis that the coda waves are composed of S to S backscattered waves (Aki, 1980b). Further observations clarified the apparent frequency dependence of Q_β, especially for frequencies between 1 and 30 Hz (e.g., Console and Rovelli, 1981; Frankel, 1982; Roecker *et al.*, 1982; Rovelli, 1983). It was found that the frequency exponent n ranges from 0.5 to 0.9 for frequencies between 1 and 30 Hz. The estimation of Q_β at lower frequencies (below 0.5 Hz) is based mostly on surface waves. Aki (1980a) combined the estimates of $1/Q_\beta$ by different methods and published by various authors and conjectured that $1/Q_\beta$ has a peak around 0.5 Hz and decreases on both sides. This peak was considered by Sato (1982) in his theoretical work. The fit of observational results by the curves describing the frequency dependence of Q for a given model is a decisive test for the model quality to explain attenuation and scattering phenomena.

A number of attenuation mechanisms have been proposed to explain the dependence of Q_β on frequency. They were critically reviewed by Aki (1980a) and Sato (1984). It seems at present that no other model than scattering generated by randomly distributed inhomogeneities in an elastic medium is able to explain the presence of S coda waves on the seismic records (Sato, 1990).

The medium heterogeneity is usually estimated by the scattering coefficient g, introduced in relation (8.110), which measures the intensity of scattering. It is the reciprocal of the mean free path and has a dimension of the reciprocal of length. The coefficient g is the average of the differential scattering coefficient $g(\theta)$ over the angle θ measured from the direction of the incident primary wave ($\theta < \pi/2$ represents the forward direction). The coefficient $g(\theta)$ is defined as 4π times the fractional loss of energy by scattering per unit travel distance of the primary wave, and per unit solid angle at the radiation direction θ. For an isotropic scattering, $g(\theta) = g$. The backscattering coefficient g_π is the value of $g(\theta)$ for $\theta = \pi$. Physically, $g(\theta)$ in this case expresses 4π times the fractional energy lost by backscattering into a unit solid angle around θ for every kilometer traveled by the primary wave (see the review by Herraiz and Espinosa, 1987).

For the single backscattering model, Aki (1981b) related coda power spectrum $P(\omega|t)$ and the backscattering coefficient g_π by the following formula

$$P(\omega|t) = \frac{\beta}{2} g_\pi(\omega) |S(\omega)|^2 \left(\frac{\beta t}{2}\right)^{-2} \exp(-\omega t/Q_c). \qquad (8.127)$$

The backscattering coefficient g_π may be calculated from the following formula, which can be obtained from relation (8.127):

$$\exp(a) = \frac{(\Delta f)^{1/2} \beta t_0}{2(\beta g_\pi)^{1/2}} \exp(\omega t_0/2Q_c), \qquad (8.128)$$

where the exponent a can be found for each frequency band Δf from a linear fit to the natural logarithm of the amplitude ratios A_s/A_c (S to coda waves) for different distances, and Q_c may be estimated by an analysis of coda waves. The method requires several events to obtain the average value of g_π. Aki (1980a, 1980b) found $g_\pi = 0.02$ km^{-1} for the frequencies 1.5 and 3 Hz in the Dodaira and Tsukuba regions of Japan. This means that S waves with frequencies of 1.5–3 Hz lose 2 percent of their original energy by scattering for every traveled kilometer if $g(\theta) = g_\pi$ for all azimuths θ. This value of g_π is consistent with those obtained for the same area by Sato (1978), who found the values from 0.006 to 0.04 km^{-1} for the frequency range of 1 to 30 Hz.

The described method provides only an average value of the coefficient g_π, which gives an indirect estimation of the heterogeneity by expressing the energy loss by backscattering. To study the influence of heterogeneities on the wave velocity and elastic parameters of the medium, the evaluation of $g(\theta)$ for all values of the azimuth θ is needed. For such an approach, more detailed statistical models are used, which can be found in the relevant literature (e.g., Wu, 1982; Sato, 1982, 1990; Herraiz and Espinosa, 1987).

There is some confusion in the literature concerning the frequency dependence of attenuation in the crust at frequencies above 1 Hz, recently pointed out by Wennerberg and Frankel (1989). Dainty (1981) assumed that anelastic crustal Q is independent of frequency, whereas scattering Q is proportional to frequency. These assumptions are usually made to interpret measurements of coda Q, whose variation with frequency is supposed to reflect the effect of scattering. The assumption made in advance that anelasticity is not dependent on frequency is difficult to justify, and scattering, on the other hand, may be frequency-independent, as demonstrated numerically by Frankel and Clayton (1986). Wennerberg and Frankel (1989) pointed out basic parallels between theories of anelastic and scattering attenuation. They noted that the frequency dependence of Q can be related to a distribution of scales of physical properties of the medium. The frequency dependence of anelastic Q is related to the distribution of relaxation times in exactly the same manner as the frequency dependence of scattering Q is related to the distribution of scatterer sizes. Thus, without additional information or assumptions, seismic observations alone of the frequency dependence of Q cannot distinguish energy loss from scattering effects.

As a seismic pulse propagates through a heterogeneous elastic medium or through an anelastic medium, its energy is lost either to other parts of the wavefield, in the case of scattering, or to heat, in the case of anelasticity. As a consequence, the pulsewidth is broadened and its amplitude decreases with distance more rapidly than as a result of geometric spreading alone (Fig. 8.9, reproduced from Sato, 1984). The point is that identical descriptions of pulse attenuation can be made using either energy loss or medium heterogeneity as a model and then scattering cannot be separated from anelastic attenuation without additional assumptions (Wennerberg and Frankel, 1989).

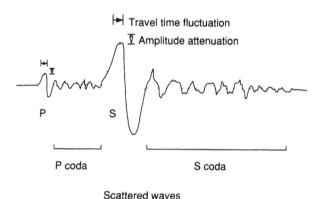

Figure 8.9 Example of seismograms in a homogeneous (dotted curve) and in an inhomogeneous (solid curve) medium. [From Sato (1984, Fig. 1).]

8.9   Methods of Calculation of Quality Factor Q

Although several approaches to the calculation of the quality factor Q from short period seismic records were described to some extent in the previous sections, these and other methods are presented here in a more systematic manner. Attenuation at low frequencies of direct P or S waves can be estimated from displacement spectra, assuming that the departure from a constant spectral level at frequencies lower than the corner frequency are caused by the travel path attenuation only. When no prior information is available on frequency-dependent effects for direct P or S waves, a frequency-independent Q is usually assumed. In this case the spectral amplitude $A(f)$ is given by

$$A(f) = A_0(f)t^{-1}\exp(-\pi ft/Q), \qquad (8.129)$$

where $A_0(f)$ is the source spectrum, t is the travel time, and f is the frequency. Taking the natural logarithm of both sides of relation (8.129), the source factor becomes a constant in the following equation:

$$\ln[tA(f)] = \ln A_0 - \pi ft/Q. \qquad (8.130)$$

Relation (8.130) describes a straight line as a function of frequency. The best-fit line can be determined by least-squares technique. If an average value of Q is required, the records of several events can be used. Although this method is the simplest one, its application is often difficult, leading to unstable results due to the heterogeneity of the rockmass in mines or in active tectonic areas in general.

The method of spectral ratio, which is based on the spectral ratio of signals recorded at two different stations of the same event, is probably the most widely used method for determining Q from direct P or S waves (e.g., Rebollar, 1984; Campillo et al., 1985; Fletcher et al., 1986; Chael, 1987; Blakeslee et al., 1989; Kvamme and Havskov, 1989). The spectrum of the body-wave group at station 1 can be written as (e.g., Kvamme and Havskov, 1989)

$$A_1(f) = S(f,\theta_1)t_1^{-1}\exp(-\pi ft_1/Q)I_1(f)C(f,\theta_1), \qquad (8.131)$$

where t_1 is the travel time measured from the origin time, θ_1 is the direction (azimuth) of the direct path relative to the station, $I_1(f)$ is the instrumental response, and $C_1(f,\theta_1)$ is the correction for local site effects. Considering another record of the same event from a different station 2 and taking the spectral ratio of the two records, the following relation is obtained:

$$\frac{A_2(f)}{A_1(f)} = C(f)\left(\frac{t_1}{t_2}\right)\exp\left[-\pi(t_2 - t_1)f/Q\right], \qquad (8.132)$$

under additional assumptions that the azimuths of the two stations are approximately equal, and that $C(f)$ describes a correction for possible differences in instrumental response and local site effects. The quality factor Q then can be calculated accepting either its dependence or independence of frequency. Taking the natural logarithm of both sides of relation (8.132), the following equation is obtained:

$$\ln\left[t_2 A_2(f)/t_1 A_1(f)\right] = \ln C(f) - \pi(t_2 - t_1)f/Q, \qquad (8.133)$$

where $\ln C(f)$ should be close to zero if local site effects between the two stations are similar. If the quality factor Q is independent of frequency, then relation (8.133) is again an equation of a straight line with the slope related to Q. An assumption, on the other hand, that $Q = qf^n$ leads to the exponential form of equation (8.133), which can fit the observed values of the spectral ratio.

An interesting technique of measuring seismic attenuation within an active fault zone was developed by Blakeslee *et al.* (1989). Using a pair of stations and pairs of earthquakes, spectral ratios are performed to isolate attenuation associated by wave propagation within the fault zone. This empirical approach eliminates common source, propagation, instrumental, and near-surface site effects and can be highly useful in mining applications. The observed spectrum $A(f)$ of direct body waves can be described in a manner slightly different from that in relation (8.131)

$$A(f) = S(f)P(f)F(f)R(f)I(f), \qquad (8.134)$$

where the propagation path is divided into three distinct segments each with its own filtering effect: the country rock path $P(f)$, the fault zone $F(f)$, and the site effect $R(f)$. The geometry of a geneal one-dimensional case involving two stations and two earthquakes is shown in Fig. 8.10, reproduced from Blakeslee *et al.* (1989). For this geometry, the two earthquakes EQ_a and EQ_b are colinear between the two stations ST_1 and ST_2, and they will produce the following four spectra:

$$\begin{aligned}
A_{1b} &= S_b P_1 F_1 F_x R_1 I_1, \\
A_{2a} &= S_a P_2 F_2 F_x R_2 I_2, \\
A_{1a} &= S_a P_1 F_1 R_1 I_1, \\
A_{2b} &= S_b P_2 F_2 R_2 I_2.
\end{aligned} \qquad (8.135)$$

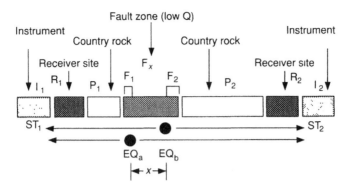

Figure 8.10   One-dimensional geometry of sources and stations, which allows isolation of the filtering effect of the fault zone. [From Blakeslee *et al.* (1989, Fig 2).]

These four spectra can be combined as follows

$$\frac{A_{1b}}{A_{1a}}\frac{A_{2a}}{A_{2b}} = \frac{S_b F_x}{S_a}\frac{S_a F_x}{S_b} = F_x^2. \tag{8.136}$$

Thus, fault-zone attenuation is effectively separated from country rock and near-site losses.

The quality factor Q_F in the fault zone can be determined in a similar way as it is done by the standard method of spectral ratio, described by relation (8.133)

$$\ln[\, A_{1b}A_{2a}/A_{1a}A_{2b}] = \text{const} - \pi f d/cQ_F, \tag{8.137}$$

where d is the difference in propagation distance resulting from the spatial separation between the two events, and c is either the *P*- or *S*-wave velocity in the fault zone.

Another method of studying attenuation is based on acceleration spectra. From the study of accelerograms from Californian earthquakes, Anderson and Hough (1984) and Hough *et al.* (1988) found that the asymptote of the high-frequency acceleration spectrum is typically described by exponential decay

$$A(f,r) = A_0 \exp(-\pi\chi f), \tag{8.138}$$

where χ is the spectral decay parameter representing the contribution from a frequency-independent contribution to Q (Hough *et al.*, 1989)

$$\chi(r) = \int \frac{dr}{Q_i c}, \tag{8.139}$$

where r is the distance, Q_i is the frequency-independent component of attenuation, and c is the wave velocity. The factor A_0 in relation (8.138) incorporates source, geometric spreading, and frequency-dependent attenuation effects

$$A_0 = S(f)\frac{1}{r}\exp(-\pi r/qc), \tag{8.140}$$

where $S(f)$ is the source acceleration spectrum and $Q_d = qf$ is the frequency-dependent component of the total Q_t given by

$$\frac{1}{Q_t} = \frac{1}{Q_i} + \frac{1}{Q_d}, \tag{8.141}$$

provided that the source acceleration spectrum is flat at frequencies above the corner frequency. The acceptance of the exponent $n = 1$ [see relation (8.121)] is justified here by the fact that, over a limited frequency band, any frequency dependence of Q that can be parameterized by a fractional power between zero and one can also be parameterized by the appropriate choice of Q_i and q (Hough *et al.*, 1988).

A least-squares regression is performed to fit an exponential decay to the logarithm of the spectrum between a frequency chosen to be above the corner frequency and the frequency at which a spectrum is dominated by noise. These regressions, performed over a number of spectra obtained at various distances, provide estimates of $\chi(r)$ and $A(0, r)$, which are functions of hypocentral distance, site, and source characteristics (Hough *et al.*, 1988). It was found that the spectral decay parameter $\chi(r)$ increased slowly from a finite intercept of a linear approximation

$$\chi(r) = \chi_0 + mr, \tag{8.142}$$

where the intercept χ_0 is suggested to represent a site effect caused by attenuation in the weathered layer and the slow increase with distance measured by the slope m to be a regional effect that describes whole-path attenuation during lateral propagation of seismic waves (Anderson and Hough, 1984). The extrapolated intercept $A(0, r) = A_0$ of equation (8.138) depends on the source size. To avoid the uncertainty in correcting for source size, the accepted practice is to study a set of earthquakes of similar magnitude, assuming that their high-frequency spectral levels are identical.

Attenuation of direct P waves can also be measured in the time domain, although the accuracy here seems to be lower than that in the frequency domain. An example of the attenuation effect on the shape of the displacement and velocity pulse in the time domain is shown in Fig. 8.11, reproduced from O'Neill (1984). The pulse is calculated from the source model of Sato and Hirasawa (1973) for the attenuation operator $t^* = 0$ and 0.06 s. The

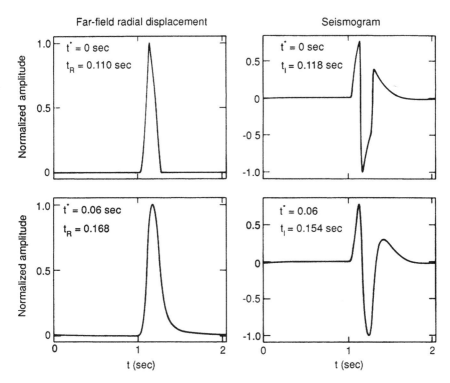

Figure 8.11 Attenuation effect on a displacement pulse and the corresponding short-period siesmogram, calculated from the source model of Sato and Hirasawa (1973) for the attenuation operator $t^* = 0$ and 0.06 s. The rise time is denoted by t_R and the initial pulse width by t_I. [From O'Neill (1984, Fig 1).]

initial pulse width computed from the attenuated seismogram is considerably different from that obtained when the attenuation effect is neglected. Both the pulse duration τ_D and the pulse rise time τ_R can be used as a measure of attenuation. They are defined in Fig. 8.12, modified from Hauksson *et al.* (1987), where an example of their dependence on travel time is also shown. To calculate the quality factor Q_α from *P*-wave pulse durations, the following relation can be used (Ohtake, 1987)

$$\tau_D = \tau_I + Ct/Q_\alpha, \tag{8.143}$$

where τ_I is the initial pulse width, C is a constant (often assumed to be equal to one), and t is the travel time. A similar approach is used to determine Q_α from the pulse rise time (Kjartansson, 1979)

$$\tau_R = \tau_0 + Ct/Q_\alpha, \tag{8.144}$$

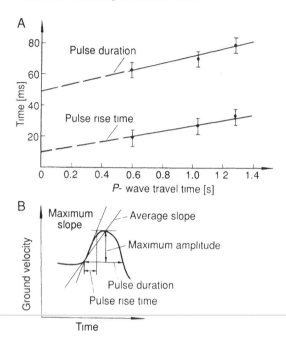

Figure 8.12 *P*-wave pulse duration and rise time as a function of travel time (A) and schematic illustration of how maximum slope, average slope, pulse duration. and pulse rise time are defined (B). [From Hauksson *et al.* (1987, Fig 14).]

where τ_0 is the pulse rise time at the source, t is the travel time, and $C = 0.485$ for Q_α greater than 20.

The quality factor for S waves inferred by the coda method represents perhaps the most stable attenuation estimate. The simplest approach is that based on the hypothesis that coda waves are S-to-S singly scattered waves from uniformly distributed random heterogeneities. Three methods of coda Q_c calculation are most often used in practice. These are the methods derived from the backscattering model of Aki and Chouet (1975) and the single-scattering model of Sato (1977a) in the time domain, and the single-scattering model in the frequency domain (Phillips and Aki, 1986; Lee *et al.*, 1986). They are described in detail by Valdés and Novelo-Casanova (1989) and are briefly reported here.

The time dependence of rms (root-mean-square) displacement amplitudes $A(f, t)$ of coda waves, on a narrow-bandpass-filtered seismogram, can be written as (Aki and Chouet, 1975)

$$A(f,t) = C(f)t^{-1}\exp\left[-\pi ft/Q_c(f)\right], \qquad (8.145)$$

where $C(f)$ is the coda source factor, t is lapse time, and Q_c is the coda quality factor combining the effects of intrinsic absorption and scattering. Taking natural logarithms of equation (8.145) and rearranging terms, the following formula is obtained

$$\ln[A(f,t)t] = \ln[C(f)] - [\pi f/Q_c(f)]t, \tag{8.146}$$

representing a straight line as a function of time, where the slope is proportional to Q_c^{-1} and is found from a least-squares solution to equation (8.146). To find the dependence of Q_c on frequency, the procedure is applied to seismograms bandpass-filtered at different center frequencies f_m. The low- and high-frequency cutoffs of the filtered signal $A(f_m, t)$ are $(f_m - f_m/3)$ and $(f_m + f_m/3)$, respectively. The root-mean-square (rms) amplitudes of $A(f_m, t)$ are calculated by sliding windows

$$A(f_m, T_n) = \left[\sum_l A_l^2/(l+1) \right]^{1/2}, \tag{8.147}$$

where T_n is the central time of the nth window and l is the number of data points in window (Valdés and Novelo-Casanova, 1989).

When the source and the receiver are not coincident, the source–receiver distance r must be taken into account (Sato, 1977a). Then the rms coda wave amplitude can be described as

$$A(f, r|t) = C(f)K(r, x)\exp[-\pi ft/Q_c(f)], \tag{8.148}$$

where $x = t/t_s$, t_s is S-wave lapse time, $K(r, x) = (1/r)\{(1/x)\ln[(x+1)/(x-1)]\}^{1/2}$, and the other symbols have the same meaning as in equation (8.145). Taking natural logarithms of equation (8.148) and rearranging terms, we obtain

$$\ln[A(f, r|t)/K(r, x)] = \ln[C(f)] - [\pi f/Q_c(f)]t. \tag{8.149}$$

For narrow-bandpass-filtered seismograms $C(f)$ is a constant, and $Q_c(f)$ can be determined by applying exactly the same procedure as that described for the case of the collocated source and receiver.

The power spectrum $P(f, t)$ of the coda-wave displacement amplitude can be written as (Aki and Chouet, 1975)

$$P(f, t) = C(f)t^{-2}\exp[-2\pi ft/Q_c(f)]. \tag{8.150}$$

The factor t^{-2} is equivalent to the geometric spreading $K(x) = (1/x)[\ln(x+1)/(x-1)]$ of Sato (1977a). Thus relation (8.150) can be expressed as (Phillips and Aki, 1986; Lee et al., 1986)

$$P(f, t) = C(f)K(x)\exp[-2\pi ft/Q_c(f)]. \tag{8.151}$$

Taking the natural logarithms of this equation and rearranging terms, we obtain again

$$\ln[P(f,t)/K(x)] = \ln[C(f)] - [2\pi f/Q_c(f)]t. \qquad (8.152)$$

In this case coda waves can be measured immediately after the arrival of S waves. The power spectrum $P(f,t)$ in equation (8.151) can be evaluated by moving fast Fourier transforms (FFTs) in overlapping windows (Lee *et al.*, 1986). This involves a sliding window of selected length moving along the coda in steps of length usually half that of the window. At every step a direct FFT is evaluated from the data enclosed in a time window containing the number of samples equal to the window length divided by the sampling interval. This procedure provides a smoothed estimate of the spectrum as a function of the time corresponding to the center point of the sliding window. The Fourier spectrum is averaged for each center point time over selected frequency bands. The slope in relation (8.152) is readily evaluated by least-squares method for each frequency band, finally providing estimates of Q_c as a function of frequency.

The properties of coda waves permit to separate source, path, and site effects. The factor $C(f)$ in relation (8.150) incorporates in principle the source factor and the receiver site response factor as well. It can be assumed that $C(f)$ can be expressed as (Phillips and Aki, 1986; Chapman and Rogers, 1989)

$$C(f) = |S(f)|^2|T(f)|^2, \qquad (8.153)$$

where $S(f)$ is the source factor, common to all stations recording a particular earthquake, and $T(f)$ is the site response. Considering a set of n earthquakes recorded by a network of m stations, relation (8.153) can be rewritten in the following form (Phillips and Aki, 1986)

$$\tfrac{1}{2}\ln\left[P_{ij}(f,t)t^2\right] = a_i + b_j + ct, \qquad (8.154)$$

where $P_{ij}(f,t)$ is the coda power spectral density at frequency f and lapse time t, observed at station j from earthquake i, and $a_i = \ln|S_i(f)|$ for the ith source, $b_j = \ln|T_j(f)|$ for the jth site, and $c = -\pi f/Q_c$. When power spectral density measurements are available at k lapse times from all sources at all sites, then there are a maximum of nmk linear equations to solve for $n + m + 1$ unknowns. The source and site terms, however, cannot be determined uniquely, since an arbitrary constant can be added to each a_i and substracted from each b_j. The arbitrary condition, therefore, that the average of the network site terms is zero is usually imposed, and the problem is solved for the relative values of the coefficients a and b and the attenuation parameter c, using multiple linear regression technique (e.g., Chapman and Rogers, 1989).

Attenuation measurements are not as yet popular techniques in mines, and only a few results are published, although this is potentially a highly useful tool to monitor the rockmass heterogeneity changes in space and time in an active stope area. An impressive study of seismic coda properties and Q_c estimates from mine tremors in South Africa has been published by Cichowicz and Green (1989). A single-scattering model was used for the analysis of time changes in the mean energy density of scattered waves in a discrete random medium. The model of the mean energy density, originally proposed by Sato (1977a, 1982) for spherical radiation and isotropic scattering, was modified and applied to a medium in which scatterers are confined to a specific volume, and the mean free path can be shorter than the distance between the source and the receiver. The analysis of coda from microtremors occurring immediately in front of an advancing mine face provided an estimate of the size of the fracture zone induced by the stope. It was found that the rockmass contains a large proportion of fractured rock, which is the source of scattering, at a distance of about 15–20 m from the stope. The quality factors Q_α and Q_β were found to be about five times smaller within the stope fracture zone than those outside this zone. It was also estimated that the radii of scatterers are smaller than 3.5 m. The polarization properties of the coda waves were used by Cichowicz *et al.* (1988) to estimate the volume containing the near-source scatterers. The analysis of the *P*-coda waves suggested that the volume containing the scatterers can be approximated by a sphere with the radius of 35 m, encompassing the source.

Chapter 9 | Focal Mechanism of Mine Tremors

Earthquakes occur as a result of the sudden release of energy within some confined regions undergoing deformation. As the region is deformed, energy is stored in the rocks in the form of elastic strain until the accumulated strain exceeds the strength of the rock. Then fracture or faulting occurs, and the opposite sides of the fault rebound to a position of equilibrium, and the energy is released in the form of heat, in the crushing of rock, and in the generation of seismic waves. This is a statement of the elastic rebound theory of the immediate cause of earthquakes, formulated by Reid (1911) as a result of his studies of the geodetic measurements along the San Andreas fault before and after its rupture in the San Francisco earthquake of 1906. His theory is quite generally accepted.

If an earthquake is caused by faulting, then the direction of a fault should be related to the stresses acting within the Earth, which caused the fracture. It may be possible, on the other hand, to determine the orientation of the fault along which an earthquake occurred from analysis of records of seismic waves. This is the domain of the focal mechanism studies of earthquakes.

So far no systematic differences have been found between mine tremors and natural earthquakes, and most of what has been discovered about the mechanism of earthquakes can be applied to mine tremors (e.g., Gibowicz, 1984; McGarr, 1984). Studies of large mine tremors have confirmed that these events are caused by shear failures on fault planes in a rockmass (e.g., McGarr, 1971; Spottiswoode and McGarr, 1975; Gay and Ortlepp, 1979; McGarr et al., 1979). Small seismic events occurring at stope faces also tend to be shear failure events (e.g., Potgieter and Roering, 1984; Spottiswoode, 1984), but not always.

A common approach in the description of seismic sources is their approximation by a model of equivalent forces that correspond to the linear wave equations, neglecting nonlinear effects in the source area (Geller, 1976; Aki and Richards, 1980; Ben-Menahem and Singh, 1981; Kennett, 1983; Bullen and Bolt, 1985). Equivalent forces are defined as producing displacements at a given point that are identical to those from the real forces acting at the source.

The basic mechanical representation of a seismic source is a confined region with spatial dimension L and time duration L/c, where c is rupture velocity. When the source–receiver distance $r \gg L$ and the lengths l of observed seismic waves are relatively long ($l/L \gg 1$), the region can be considered as a point at which there is equilibrium of force and moment systems.

The theory of point sources is of fundamental importance in focal mechanism studies. Methods of determining the direction of motion at the source have been evolving for over 60 years. It was only after several decades of research that seismologists were able to establish that the earthquake source is best represented by a double couple. The general acceptance of the double-couple model is based on observational and theoretical evidence related to the radiation pattern of P and S waves. The analysis of seismic waves depends, in turn, on the solution of the elastic wave equation for the medium excited by a concentrated force. This problem was solved by Stokes in 1849 and extended by Love in 1903 to pairs of forces in opposite directions, or couples, and their combinations. These sources can be analyzed in a unified way by introducing the concept of seismic moment tensor (Gilbert, 1970), which completely describes, as a first-order approximation, the equivalent forces of general point sources (e.g., Aki and Richards, 1980; Ben-Menahem and Singh, 1981; Kennett, 1983; Jost and Herrmann, 1989).

In this chapter the basic theory of point sources in homogeneous media, from the single force to the general moment-tensor source, is presented following mostly the student's guide of Pujol and Herrmann (1990). The double couple and non-double-couple mechanisms of seismic sources induced by mining are considered, and the methods of focal mechanism estimation are described in some detail. The general properties of seismic moment tensors and their determination are also described, following to some extent the student's guide of Jost and Herrmann (1989).

9.1 Single Forces in Homogeneous Media

From equation (3.21) it follows that the x_i component of the displacement $u(\mathbf{x}, t)$ in a homogeneous, unbounded, isotropic, elastic medium generated by an external body force $f(\mathbf{x}, t)$ satisfies the following equation (Aki and Richards, 1980)

$$\rho \ddot{u}_i = f_i + (\lambda + \mu) u_{j,ji} + \mu u_{i,jj}, \tag{9.1}$$

where overdots are used to indicate time derivatives (e.g., $\ddot{u}_i = \partial^2 u_i / \partial t^2$), a comma followed by suffixes i and j indicates the partial derivative with respect to x_i and x_j (e.g., $u_{i,j} = \partial u_i / \partial x_j$), and the summation convention over repeated suffixes is used.

When the body force f is concentrated at the point ξ and acting in the x_i direction with time dependence $F(t)$, it can be represented as

$$f_i(\mathbf{x}, t) = F(t)\, \delta(\mathbf{x} - \xi)\, \delta_{il}, \qquad (9.2)$$

where $\delta(\mathbf{x} - \xi)$ is the three-dimensional delta function and the Kronecker delta $\delta_{ij} = 1$ for $i = j$ and $\delta_{ij} = 0$ for $i \neq j$. The force defined by relation (9.2) is known as a *single force* or a *concentrated force*.

In the theory of seismic disturbances, the modern procedure is to use the delta function $\delta(t)$ (also called the *Dirac function* or the *impulse*) and the related Heaviside step function $H(t)$. The delta function can be thought of as the limiting form of a rectangle as its width approaches zero and its height approaches infinity, in such a way that its area is not changed and is always equal to unity. The fundamental properties of the delta function are $\delta(t) = 0$ for $t \neq 0$ and

$$\int_{-\infty}^{\infty} \delta(t)\, dt = 1. \qquad (9.3)$$

The Fourier transform of the impulse is also equal to unity, and the impulse contains all frequencies in equal proportions. The so-called sifting property of the delta function states that

$$\int_{-\infty}^{\infty} \delta(t - t_0) g(t)\, dt = g(t_0), \qquad (9.4)$$

provided that $g(t)$ is continuous at point t_0. The property implies that the impulse takes on the value of the function $g(t)$ at the time the impulse is applied. An infinite sequence of impulses occurring at unit time intervals is often called an *infinite Dirac comb* or a *shah*. The Fourier transform in the limit of the shah in the time domain is another shah in the frequency domain, that is, the shah is its own Fourier transform. The three-dimensional delta function has the following two properties

$$\delta(x_1, x_2, x_3) = 0 \quad \text{for} \quad x_1 \neq 0 \quad \text{or} \quad x_2 \neq 0 \quad \text{or} \quad x_3 \neq 0$$

and

$$\int_{-\infty}^{\infty}\int_{-\infty}^{\infty}\int_{-\infty}^{\infty} \delta(x_1, x_2, x_3)\, dx_1\, dx_2\, dx_3 = 1, \qquad (9.5)$$

a natural generalization of the one-dimensional case. The Heaviside step function $H(t)$ is defined by the following equation:

$$\frac{d}{dt}H(t) = \delta(t).\tag{9.6}$$

The step function can be considered as a limit of ordinary functions with the limiting properties

$$H(t) = 0 \quad \text{for} \quad t < 0 \quad \text{and} \quad H(t) = 1 \quad \text{for} \quad t \geq 0.\tag{9.7}$$

For the body force described by relation (9.2), the displacement u_i is now given by (Aki and Richards, 1980)

$$u_i(\mathbf{x}, t) = \frac{1}{4\pi\rho}(3\gamma_i\gamma_j - \delta_{ij})\frac{1}{r^3}\int_{r/\alpha}^{r/\beta}\tau F(t - \tau)\,d\tau$$

$$+ \frac{1}{4\pi\rho\alpha^2}\gamma_i\gamma_j\frac{1}{r}F(t - r/\alpha) - \frac{1}{4\pi\rho\beta^2}(\gamma_i\gamma_j - \delta_{ij})\frac{1}{r}F(t - r/\beta),\tag{9.8}$$

where $r = |\mathbf{x} - \xi|$ and $\gamma_i = (x_i - \xi_i)/r$ are the direction cosines of the vector $(\mathbf{x} - \xi)$, forming the elements of a unit vector $\Gamma = (\gamma_1, \gamma_2, \gamma_3)$ directed along the source–receiver direction.

If $F(t)$ is an impulse, that is $F(t) = \delta(t)$, then the resulting displacement u_i is known as *Green's function* G_{ij} for equation (9.1). Green's function is simply the medium response to the delta function. From relation (9.8) it follows that the function G_{ij}, corresponding to the source located at the point ξ and acting at the time 0, can be written as follows (Hudson, 1980)

$$G_{ij}(\mathbf{x}, t) = \frac{1}{4\pi\rho}(3\gamma_i\gamma_j - \delta_{ij})\frac{t}{r^3}[H(t - r/\alpha)] - [H(t - r/\beta)]$$

$$+ \frac{1}{4\pi\rho\alpha^2}\gamma_i\gamma_j\frac{1}{r}\delta(t - r/\alpha) - \frac{1}{4\pi\rho\beta^2}(\gamma_i\gamma_j - \delta_{ij})\frac{1}{r}\delta(t - r/\beta),\tag{9.9}$$

where H is the Heaviside unit step function. The first term in this formula results from the fact that the integral in equation (9.8) is equal to zero unless $r/\alpha < \tau < r/\beta$.

For proceeding further, the convolution of signals should be briefly described. Nonperiodic functions $g(t)$ which decay sufficiently rapidly so that the integral

$$\int_{-\infty}^{\infty}|g(t)|^2\,dt\tag{9.10}$$

is finite, are called *energy signals* (Meskó, 1984). The convolution of two energy signals $g(t)$ and $h(t)$ is given by the integral

$$g(t)*h(t) = \int_{-\infty}^{\infty} g(\tau)h(t-\tau)\,d\tau, \tag{9.11}$$

where the asterisk between the functions to be convolved is a commonly used abbreviation for the integral. The convolution is commutative

$$g(t)*h(t) = h(t)*g(t), \tag{9.12}$$

associative

$$f(t)*[g(t)*h(t)] = [f(t)*g(t)]*h(t), \tag{9.13}$$

and distributive with respect to addition:

$$f(t)*[g(t) + h(t)] = f(t)*g(t) + f(t)*h(t). \tag{9.14}$$

If one of the functions to be convolved is one-sided, for example, $g(t) = 0$ for $t < 0$, the convolution integral is often rewritten as follows

$$g(t)*h(t) = \int_{0}^{\infty} g(\tau)h(t-\tau)\,d\tau. \tag{9.15}$$

If both functions are one-sided, that is, $g(t) = 0$ for $t < 0$ and $h(t) = 0$ for $t < 0$, the convolution integral becomes

$$\int_{0}^{t} g(\tau)h(t-\tau)\,d\tau = \int_{0}^{t} g(t-\tau)h(\tau)\,d\tau. \tag{9.16}$$

The delta function $\delta(t)$ does not belong to the class of energy functions. Its sifting property, however, allows the determination of the convolution with the impulse

$$\int_{-\infty}^{\infty} g(\tau)\,\delta(t-t_0-\tau)\,d\tau = g(t-t_0). \tag{9.17}$$

The Dirac delta function, convolved with another function, substitutes the original argument of that function with its own argument, performing a sifting operation. The original definition of the convolution, given by the integral (9.11) for energy signals, cannot be adapted for periodic functions as it would lead to unbounded variation. The time average of a modified convolution integral

$$\frac{1}{T} \int_{T}^{a+T} g(v)h(t-v)\,dv, \tag{9.18}$$

on the other hand, provides a function of bounded variation (Meskó, 1984). This formula resembles the original convolution integral (9.11), and the asterisk notation is usually retained. If both functions $g(t)$ and $h(t)$ are periodic, the notation $g(t)*h(t)$ abbreviates the integral (9.18). This convolution generates a periodic function with period T, and is called the *periodic* or *cyclic convolution* of periodic functions $g(t)$ and $h(t)$. The complex Fourier coefficients of the periodic convolution are equal to the products of the complex Fourier coefficients of the periodic functions. Thus the convolution of periodic functions in the time domain corresponds to simple multiplication in the frequency domain.

Going back to the displacement field generated by a single body force, the relation

$$F(t)*\delta(t - t_0) = F(t - t_0) \tag{9.19}$$

is used to show that from equations (9.8) and (9.9) it follows that the displacement field $u_i(\mathbf{x}, t)$ is equal to the convolution of a point force $F(t)$ by Green's function G_{ij}

$$u_i(\mathbf{x}, t) = F(t)*G_{ij}(\mathbf{x}, t). \tag{9.20}$$

If the body force is concentrated but its orientation is arbitrary $\mathbf{f}(\mathbf{x}, t) = \mathbf{F}(t)\delta(\mathbf{x} - \xi)$, where $\mathbf{F}(t) = (F_1, F_2, F_3)$, then the total displacement is equal to the sum of the displacements generated by the forces F_1, F_2, and F_3 directed along the x_1, x_2, and x_3 directions (e.g., Pujol and Herrmann, 1990):

$$u_i = F_1 * G_{i1} + F_2 * G_{i2} + F_3 * G_{i3} = F_j * G_{ij}. \tag{9.21}$$

The relative magnitude of different terms in the Green function in equation (9.8) depends on the source–receiver distance r. The first term behaves like r^{-2} for sources in which F is nonzero for times that are short as compared to $r/\beta - r/\alpha$. The remaining terms behave like r^{-1}, becoming dominant as $r \rightarrow \infty$. These terms, including $r^{-1}F(t - r/\alpha)$ and $r^{-1}F(t - r/\beta)$, are therefore called *far-field terms*. The first term, including $r^{-3}\int \tau F(t - \tau) d\tau$, is called a *near-field term*, since r^{-2} dominates over r^{-1} as $r \rightarrow 0$ (Aki and Richards, 1980).

Almost all seismic observations used in practice are collected in the far field, and the methods and techniques used in the focal mechanism studies are based on the far-field terms of displacement. There are important exceptions, however, such as observations of the final static offset caused by faulting, which is a near-field effect. The seismic data used in mines and in earthquake engineering are also occasionally collected in the near field. The near-field displacement u^N in equation (9.8) is defined by

$$u_i^N(\mathbf{x}, t) = \frac{1}{4\pi\rho} (3\gamma_i\gamma_j - \delta_{ij}) \frac{1}{r^3} \int_{r/\alpha}^{r/\beta} \tau F(t - \tau) \, d\tau. \qquad (9.22)$$

The displacement $\mathbf{u}^N$ is composed of both P-wave and S-wave motions, and it is not always possible to decompose it into its P-wave and S-wave components. The near-field displacement has both longitudinal and transverse motions. The longitudinal component is (Aki and Richards, 1980)

$$\mathbf{u}^N \cdot \Gamma = \gamma_j \frac{1}{4\pi\rho r^3} \int_{r/\alpha}^{r/\beta} \tau F(t - \tau) \, d\tau, \qquad (9.23)$$

where the vector Γ defines the source–receiver direction, and the transverse component is

$$\mathbf{u}^N \cdot \Gamma' = -\gamma_j' \frac{1}{4\pi\rho r^3} \int_{r/\alpha}^{r/\beta} \tau F(t - \tau) \, d\tau, \qquad (9.24)$$

where Γ' is the vector perpendicular to the Γ direction. It is possible to identify an arrival time and the duration of the near-field displacement $\mathbf{u}^N$ at a given receiver fixed at $\mathbf{x}$. If $t = 0$ is the chosen time when $F(t)$ first becomes nonzero and if $F(t)$ returns again to zero for all times $t > T$, then from equation (9.22) it follows that $\mathbf{u}^N$ is a motion that arrives at $\mathbf{x}$ at the P-wave arrival time r/α and that is active until the time $r/\beta + T$. Thus the motion has duration $(r/\beta - r/\alpha) + T$. If $F(t)$ never returns permanently to zero (T is not finite), then the near-field term persists indefinitely.

From equation (9.8) it follows that the far-field displacement can be written as

$$u_i = \frac{1}{4\pi\rho\alpha^2} \gamma_i\gamma_j \frac{1}{r} \dot{F}\left(t - \frac{r}{\alpha}\right) - \frac{1}{4\pi\rho\beta^2} (\gamma_i\gamma_j - \delta_{ij}) \frac{1}{r}\left(t - \frac{r}{\beta}\right). \quad (9.25)$$

The first term in this equation is proportional to γ_i and represents motion in the Γ direction, which corresponds to P-wave motion. The far-field P wave is therefore longitudinal (also called *radial*) in that its direction of particle motion is the same as the direction of wave propagation. The second term has the factor $(\gamma_i\gamma_j - \delta_{ij})$, which is the ith component of a vector perpendicular to the Γ direction, and represents S-wave motion. The far-field S wave is therefore a transverse wave, in that its direction of particle motion is normal to the direction of propagation.

The radiation pattern of the far-field displacement caused by a concentrated force depends on the source–receiver direction. It is convenient therefore to assume that $r = 1$ and to ignore all constant terms. Then from

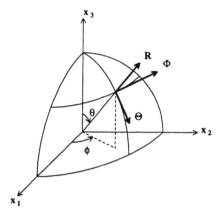

Figure 9.1 Spherical coordinates and unit vectors. [From Pujol and Herrmann (1990, Fig. 1).]

equation (9.25) it follows that for the force acting in the x_3 direction (Pujol and Herrmann, 1990)

$$u_i^P = \gamma_3 \gamma_i \qquad (9.26)$$

and

$$u_i^S = -\gamma_3 \gamma_i + \delta_{i3}, \qquad (9.27)$$

where the constant terms have been set to unity.

In spherical coordinates, shown in Fig. 9.1, the direction cosines are given by

$$\gamma_1 = \sin \theta \cos \phi, \qquad \gamma_2 = \sin \theta \sin \phi, \qquad \gamma_3 = \cos \theta. \qquad (9.28)$$

Now relations (9.26) and (9.27) can be expressed by the direction cosines as follows:

$$u_1^P = \tfrac{1}{2} \sin 2\theta \cos \phi, \qquad u_2^P = \tfrac{1}{2}\sin 2\theta \sin \phi, \qquad u_3^P = \cos^2 \theta, \quad (9.29)$$

$$u_1^S = -\tfrac{1}{2} \sin 2\theta \cos \phi, \quad u_2^S = -\tfrac{1}{2} \sin 2\theta \sin \phi, \quad u_3^S = \sin^2 \theta. \quad (9.30)$$

The amplitudes of the displacements are given by

$$|\mathbf{u}^P| = |\gamma_3| = |\cos \theta| \qquad (9.31)$$

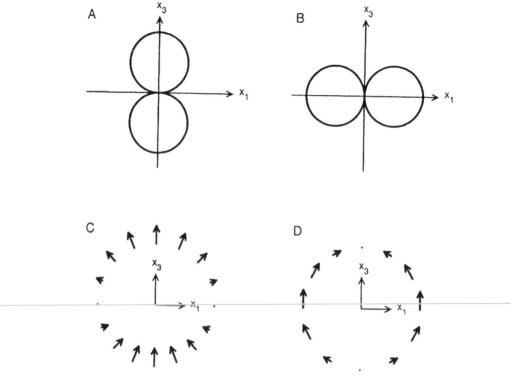

Figure 9.2 Radiation pattern of P wave (A) and S wave (B), and plot of the vector P motion (C) and S motion (D) in the (x_1, x_3) plane, generated by a single force acting in the x_3 direction. [From Pujol and Herrmann (1990, Fig. 2)]

and

$$|\mathbf{u}^S| = \left|\sqrt{1 - \gamma_3^2}\right| = |\sin \theta|. \tag{9.32}$$

Since the amplitudes are not dependent on ϕ, they are symmetric about the x_3 axis. The plot of $f(\theta) = \cos \theta$ corresponds to two circles with centers on the x_3 axis, shown in Fig. 9.2A. The radiation pattern for the P wave, therefore, has the shape of two spheres. In the (x_1, x_2) plane, for which $\theta = 90°$ (Fig. 9.1), motion is zero. The plot of $f(\theta) = \sin \theta$ corresponds to two circles with centers on the x_1 axis, shown in Fig. 9.2B, and the radiation pattern for the S wave resembles a doughnut without the central hole. There is no S-wave motion along the x_3 axis, for which $\theta = 0°$ (Pujol and Herrmann, 1990).

To determine the sense of motion (positive or negative direction), the sign of the vector components should be analyzed. For example, for positive x_1, $\phi = 0°$ and the P-wave displacement in the (x_1, x_3) plane is

$$u_1^P = \tfrac{1}{2} \sin 2\theta, \qquad u_2^P = 0, \qquad u_3^P = \cos^2 \theta, \qquad (9.33)$$

whereas for negative x_1, $\phi = 180°$ and

$$u_1^P = -\tfrac{1}{2} \sin 2\theta, \qquad u_2^P = 0, \qquad u_3^P = \cos^2 \theta. \qquad (9.34)$$

In both cases the sense of motion depends on the sign of u_1^P. For points with positive x_3 ($\theta < 90°$), motion is away from the source, and for points with negative x_3 ($\theta > 90°$), motion is towards the source (Fig. 9.2C). The corresponding plots for the S waves are shown in Fig. 9.2D.

9.2 Concentrated Force Couples

Although the single force is one of the simplest models of a seismic source, the supposed external application of a force is unlikely to occur in natural earthquakes. It is more probable that the force action is of self-balancing type, such as a pair of opposite forces acting simultaneously on two adjacent parts of the medium with the resultant force equal to zero.

Let us consider a pair of forces of equal magnitude acting along the positive and negative x_3 directions at a small distance ε apart in the x_2 direction. The two forces are $(0, 0, F_3)$ acting at point $(\xi + \varepsilon e_2/2)$ and $(0, 0, -F_3)$ acting at $(\xi - \varepsilon e_2/2)$, where e_2 is a unit vector in the x_2 direction. The total displacement u_i caused by the two forces is the sum of the displacements caused by each force (e.g., Pujol and Herrmann, 1990):

$$u_i = \varepsilon F_3 * \left[G_{i3}(\xi + \varepsilon e_2/2) - G_{i3}(\xi - \varepsilon e_2/2) \right] / \varepsilon. \qquad (9.35)$$

Taking the limit of u_i as F_3 tends to infinity and ε tends to zero, in such a way that the product εF_3 remains finite, the following relation is obtained:

$$u_i = M_{33} * \frac{\partial G_{i3}}{\partial \xi_2}, \qquad (9.36)$$

where $M_{32} = \varepsilon F_3$. This pair of forces is known in classic mechanics as a *couple* and the quantity M_{32} as the *moment of the couple*, which has dimension of force by length and may be a function of time.

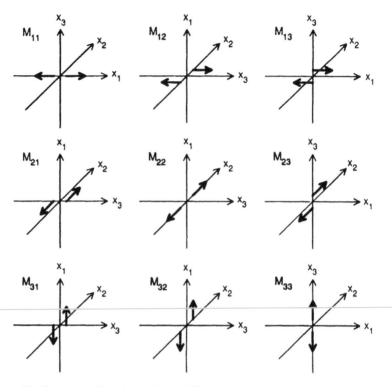

Figure 9.3 Representation of the nine possible couples M_{ij}. The subindices i and j denote the directions of the force and the arm of the couple, respectively. [From Aki and Richards (1980, Fig. 3.7).]

There are nine possible combinations of force and arm directions, shown in Fig. 9.3, represented by the moment M_{ij} of couple with forces in the x_i direction and arm in the x_j direction. When x_i and x_j are the same, the couple is known as a *vector dipole* or a *couple without moment*. All the other couples have nonzero moment, equivalent to torque. If a general body force representing a seismic source can be expressed as a linear combination of couples with moments M_{ij}, then the displacement caused by this force is the sum of the displacements caused by individual couples

$$u_k = M_{ij} * \frac{\partial G_{ki}}{\partial \xi_j} = M_{ij} * G_{ki,j}. \tag{9.37}$$

The set of nine terms M_{ij} is known as the *moment tensor* of the source, represented by a matrix **M** with elements M_{ij}. The full expression for u_k in

(9.37) is given by Aki and Richards (1980) as follows:

$$u_k = \left(\frac{15\gamma_k\gamma_i\gamma_j - 3\gamma_k\,\delta_{ij} - 3\gamma_i\,\delta_{kj} - 3\gamma_j\,\delta_{ki}}{4\pi\rho} \right) \frac{1}{r^4} \int_{r/\alpha}^{r/\beta} \tau M_{ij}(t - \tau)\, d\tau$$

$$+ \left(\frac{6\gamma_k\gamma_i\gamma_j - \gamma_k\,\partial_{ij} - \gamma_i\,\partial_{kj} - \gamma_j\,\partial_{ki}}{4\pi\rho\alpha^2} \right) \frac{1}{r^2} M_{ij}\left(t - \frac{r}{\alpha} \right)$$

$$- \left(\frac{6\gamma_k\gamma_i\gamma_j - \gamma_k\,\partial_{ij} - \gamma_i\,\partial_{kj} - 2\gamma_j\,\partial_{ki}}{4\pi\rho\beta^2} \right) \frac{1}{r^2} M_{ij}\left(t - \frac{r}{\beta} \right)$$

$$+ \frac{\gamma_k\gamma_i\gamma_j}{4\pi\rho\alpha^2} \frac{1}{r} \dot{M}_{ij}\left(t - \frac{r}{\alpha} \right) - \left(\frac{\gamma_k\gamma_i - \partial_{ki}}{4\pi\rho\beta^2} \right) \gamma_j \frac{1}{r} \dot{M}_{ij}\left(t - \frac{r}{\beta} \right). \qquad (9.38)$$

The near-field terms in this formula are proportional to $r^{-4}\int_{r/\alpha}^{r/\beta}\tau M_{ij}(t - \tau)\,d\tau$, and the far-field terms are proportional to $r^{-1}\dot{M}_{ij}(t - r/\alpha)$ for P waves and to $r^{-1}\dot{M}_{ij}(t - r/\beta)$ for S waves. It can be shown that components of the moment tensor M_{ij} are proportional to averaged particle displacements at the source (Aki and Richards, 1980). Then the time derivatives $\dot{M}_{ij}(t - r/\alpha)$ and $\dot{M}_{ij}(t - r/\beta)$, describing the pulse shape of displacement in the far field, are proportional to averaged particle velocities at the source. Some terms proportional to $r^{-2}M_{ij}(t - r/\alpha)$ and to $r^{-2}M_{ij}(t - r/\beta)$ are also present in formula (9.38). They are called the *intermediate-field terms*, since their asymptotic properties are intermediate to those of the near-field and far-field displacements. In practice, however, they are found to be small in the far field and are often comparable with the near-field displacements at short distances.

The far-field terms in equation (9.38) are composed of two terms, the first one in the direction of Γ, thus corresponding to the P-wave motion, and the second one in a direction perpendicular to Γ, thus corresponding to the S-wave motion. The total displacement vector $\mathbf{u}$, therefore, can be written as a sum of its P-wave and S-wave contributions:

$$\mathbf{u} = \mathbf{u}^P + \mathbf{u}^S. \qquad (9.39)$$

The vector $\mathbf{u}^S$, which lies in a plane perpendicular to Γ, can be decomposed into two vectors $\mathbf{u}^{SV}$ and $\mathbf{u}^{SH}$, one in a vertical plane that contains the source and the receiver, and the other in a horizontal plane, respectively. These vectors are shown in Fig. 9.4. The decomposition is readily made in spherical coordinates by introducing unit vectors $\mathbf{R}$, $\mathbf{\Phi}$, and $\mathbf{\theta}$ (Fig. 9.1). Vecot $\mathbf{R} = \Gamma$ is in the radial, source–receiver, direction, vector $\mathbf{\theta}$ is tangent to the meridian line, and vector $\mathbf{\Phi}$ is tangent to the parallel line with no component in the x_3

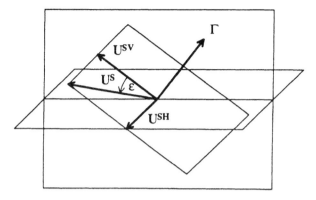

Figure 9.4   Decomposition of the S-wave displacement vector $\mathbf{u}^s$ into two vectors, $\mathbf{u}^{SV}$ and $\mathbf{u}^{SH}$. The three vectors lie in the plane perpendicular to Γ, and $\mathbf{u}^{SV}$ is in the vertical plane, which also contains Γ. The angle of polarization is marked by ε. [From Pujol and Herrmann (1990, Fig. 6).]

direction. They can be written as

$$\mathbf{R} = (\sin\theta\cos\phi, \sin\theta\sin\phi, \cos\theta),$$

$$\boldsymbol{\theta} = (\cos\theta\cos\phi, \cos\theta\sin\phi, -\sin\theta),$$

$$\Phi = (-\sin\phi, \cos\phi, 0). \tag{9.40}$$

Each vector is perpendicular to the other two, the absolute value of each of them is equal to 1, and the three vectors form a right-hand system.

Now the expressions for the displacement vectors $\mathbf{u}^P$, $\mathbf{u}^{SV}$, and $\mathbf{u}^{SH}$ can be derived. This can be found in the paper of Pujol and Herrmann (1990), who represented the far-field terms in equation (9.38) as a product of vectors and matrices and then derived the P, SV, and SH displacement in the form

$$\mathbf{u}^P = \frac{1}{4\pi\rho\alpha^3}\frac{1}{r}R^P\mathbf{R},$$

$$\mathbf{u}^{SV} = \frac{1}{4\pi\rho\beta^3}\frac{1}{r}R^{SV}\boldsymbol{\theta},$$

$$\mathbf{u}^{SH} = \frac{1}{4\pi\rho\beta^3}\frac{1}{r}R^{SH}\Phi, \tag{9.41}$$

where R^P, R^{SV}, and R^{SH} are the radiation patterns, which can be written in detail as

$$R^P = \gamma_i \dot{M}_{ij} \gamma_j = \gamma_1 v_1 + \gamma_2 v_2 + \gamma_3 v_3,$$

$$R^{SV} = \theta_i \dot{M}_{ij} \gamma_j = \theta_1 v_1 + \theta_2 v_2 + \theta_3 v_3,$$

$$R^{SH} = \phi_i \dot{M}_{ij} \gamma_j = \phi_1 v_1 + \phi_2 v_2 + \phi_3 v_3, \tag{9.42}$$

where θ_i and ϕ_i are the components of $\boldsymbol{\theta}$ and $\boldsymbol{\Phi}$, respectively, and

$$v_i = \dot{M}_{i1} \gamma_1 + \dot{M}_{i2} \gamma_2 + \dot{M}_{i3} \gamma_3. \tag{9.43}$$

These general formulas are very convenient for computations. It is often assumed in practice that all the components of the moment tensor have the same time dependence, and its time derivative can be ignored in radiation pattern calculations.

9.3   Double-Couple Sources

A double couple is obtained by combining two coplanar single couples of equal but opposite moments with mutually perpendicular force directions. Three fundamental double couples are shown in Fig. 9.5. Following Pujol and Herrmann (1990), the double couple formed by the M_{13} and M_{31} single couples, shown in Fig. 9.3, is considered first. The corresponding moment tensor $\mathbf{M}^{DC}$ is symmetric and is given by

$$M^{DC} = \begin{bmatrix} 0 & 0 & 1 \\ 0 & 0 & 0 \\ 1 & 0 & 0 \end{bmatrix}. \tag{9.44}$$

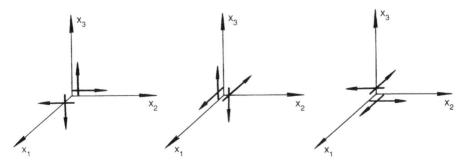

Figure 9.5   Three fundamental double couples.

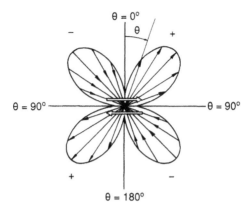

Figure 9.6   Radiation pattern of the P-wave displacement in a plane of constant azimuth, generated by a double-couple point source. The lobes are a locus of points having a distance from the origin that is proportional to $\sin 2\theta$. The pair of arrows at the center denotes the shear dislocation. [From Aki and Richards (1980, Fig. 4.5).]

Thus the far-field radiation patterns, determined from relations (9.42), are

$$R^P = 2\gamma_1\gamma_3 = \sin 2\theta \cos \phi,$$

$$R^{SV} = \gamma_1\theta_3 + \gamma_3\theta_1 = \cos 2\theta \cos \phi,$$

$$R^{SH} = \gamma_1\phi_3 + \gamma_3\phi_1 = -\cos \theta \sin \phi. \tag{9.45}$$

The pattern diagram for a plane of constant azimuth is shown in Fig. 9.6, taken from Aki and Richards (1980). The total amplitude of $\mathbf{u}^S$ is

$$|\mathbf{u}^S| = \left[(R^{SV})^2 + (R^{SH})^2 \right]^{1/2} = (\cos^2 2\theta \cos^2 \phi + \cos^2 \theta \sin^2 \phi)^{1/2}. \tag{9.46}$$

It has four lobes with maximum values along the x_1 and x_3 axes. Motion is zero along the lines for which R^P is maximum. The radiation pattern diagram of the transverse component in plane ($\phi = 0, \phi = \pi$) is shown in Fig. 9.7, taken from Aki and Richards (1980).

It can be shown (e.g., Pujol and Herrmann, 1990) that the described double couple generates the same radiation pattern as a pair of tensional and compressional dipoles at right angles with respect to each other and at 45° with respect to the x_1 and x_3 axes. The double couple and the corresponding double dipole system are shown in Fig. 9.8, taken from Pujol and Herrmann (1990). The equivalence of the double couple and the pair of tensional and compressional dipoles is very important. This equivalence allows to introduce two vectors $\mathbf{t} = (1/\sqrt{2})(1,0,1)$ and $\mathbf{p} = (1/\sqrt{2})(1,0,-1)$ which are directed along the axes that contain the dipoles. A third vector $\mathbf{b}$ perpendicular to the

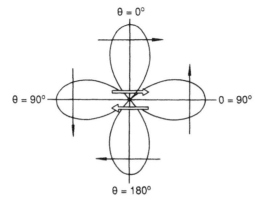

Figure 9.7 Radiation pattern of the *S*-wave displacement in plane ($\phi = 0$, $\phi = \pi$), generated by a double couple. The central pair of arrows shows the sense of shear dislocation, and arrows imposed on each lobe show the direction of particle displacement associated with the lobe. [From Aki and Richards (1980, Fig. 4.6).]

other two is obtained from the vector product of **p** and **t**, which is **b** $= (0, 1, 0)$, a vector in the x_2 direction. The axis corresponding to the vector **t** is known as the *T* or tensional axis, and is situated in one of two quadrants in the (x_1, x_3) plane generating compressions at the receiver (motion away from the source). The axis corresponding to **p**, known as the *P* or compressional axis, lies in a dilatational (motion toward the source) quadrant. The third axis corresponds to **b** and is called *B* or null axis, since the radiation of *P* waves

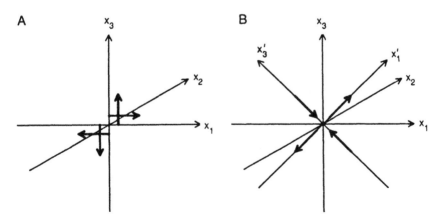

Figure 9.8 The double couple (A) and the corresponding double-dipole system (B) that gives the same radiation pattern as the double couple. [From Pujol and Herrmann (1990, Fig. 9).]

is zero along this axis. These three axes are determined routinely in focal mechanism studies as an equivalent representation of double-couple sources.

The radiation patterns described by relations (9.45) for the particular double couple with the corresponding moment tensor $\mathbf{M}^{DC}$ is related to the far field of displacement only. The near-field and intermediate-field radiation patterns might be of interest in mining situations as well. Earthquakes are assumed to be generated in most cases by shear failure or slip on a fault plane, which means that seismic forces act on a surface. The question then arises as to what distribution of equivalent body forces causes exactly the same displacement as that produced by surface forces. The answer to this question was given by Burridge and Knopoff (1964). In the particular case of slip on the (x_1, x_2) plane and in the x_1 direction, the equivalent force system when the fault is finite but much smaller than the wavelength of the seismic waves is given by $M_0 \mathbf{M}^{DC} H(t)$ (Pujol and Herrmann, 1990). M_0 is known as the scalar moment and is given by $M_0 = \mu \bar{u} A$, where μ is the shear modulus, $\bar{u}$ is the average slip, and A is the fault area (Aki and Richards, 1980).

The total displacement caused by this particular double couple is (Aki and Richards, 1980)

$$\mathbf{u}(\mathbf{x}, t) = \frac{1}{4\pi\rho} R^{N} \frac{1}{r^4} \int_{r/\alpha}^{r/\beta} \tau M_0(t - \tau)\, d\tau$$

$$+ \frac{1}{4\pi\rho\alpha^2} R^{1P} \frac{1}{r^2} M_0(t - r/\alpha) + \frac{1}{4\pi\rho\beta^2} R^{1S} \frac{1}{r^2} M_0(t - r/\beta)$$

$$+ \frac{1}{4\pi\rho\alpha^3} R^{FP} \frac{1}{r} \dot{M}_0(t - r/\alpha) + \frac{1}{4\pi\rho\beta^3} R^{FS} \frac{1}{r} \dot{M}_0(t - r/\beta), \quad (9.47)$$

where the near-field, intermediate-field, and far-field P and S have radiation patterns given, respectively, by

$$R^{N} = 9 \sin 2\theta \cos \phi \mathbf{R} - 6(\cos 2\theta \cos \phi \boldsymbol{\theta} - \cos \theta \sin \phi \boldsymbol{\Phi}),$$

$$R^{1P} = 4 \sin 2\theta \cos \phi \mathbf{R} - 2(\cos 2\theta \cos \phi \boldsymbol{\theta} - \cos \theta \sin \phi \boldsymbol{\Phi}),$$

$$R^{1S} = -3 \sin 2\theta \cos \phi \mathbf{R} + 3(\cos 2\theta \cos \phi \boldsymbol{\theta} - \cos \theta \sin \phi \boldsymbol{\Phi}),$$

$$R^{FP} = \sin 2\theta \cos \phi \mathbf{R},$$

$$R^{FS} = \cos 2\theta \cos \phi \boldsymbol{\theta} - \cos \theta \sin \phi \boldsymbol{\Phi}. \quad (9.48)$$

The far-field radiation patterns in (9.48) agree with those given in relations (9.45). It should be noted that while only the radial component is present in the far-field P wave and only the transverse component is present in the far-field S wave, the near-field and the intermediate-field displacements of both P and S waves involve both radial and transverse components.

There are two commonly used descriptions of double couple sources. The first is in terms of an angular description of the nodal planes in the seismic radiation from a pure slip motion, or a shear dislocation, on a fault. The second, which is of increasing importance, is the description of the source by the six independent components of the symmetric moment tensor, assumed to have a common dependence on time, most often in the form of a step function:

$$M_{ij}(t) = M_{ij} H(t). \tag{9.49}$$

This assumption means that the far field radiation is a delta function in time for a uniform medium. The moment tensor components in an isotropic medium for a double couple of equivalent forces of arbitrary orientation are given by (Aki and Richards, 1980; Ben-Menahem and Singh, 1981)

$$M_{ij} = M_0(s_i n_j + s_j n_i), \tag{9.50}$$

where s is a unit slip vector lying in a fault plane and n is a unit vector normal to the fault. It should be noted that the contributions of the slip vector s and the vector n normal to the fault are symmetric in formula (9.50). Since the tensor M is also symmetric, the vectors s and n could be interchanged without affecting the displacement field, which means that the vector normal to the fault could equivalently be the slip vector and vice versa. Thus two nodal planes, called the fault plane and the auxiliary plane, cannot be distinguished from the seismic radiation of a point source alone. They are known therefore as *conjugate planes*. Studies of locations of aftershocks, surface faulting, or static final displacements are needed to resolve this ambiguity. The mass of the rock below an inclined fault plane is known as the *footwall*, and the mass of the rock above it as the *hanging wall*. The slip vector s indicates the relative motion of the hanging wall with respect to the footwall.

The orientation of the fault is usually described in a geographic coordinate system with the x_1 axis pointing to the north, the x_2 axis pointing to the east, and the x_3 axis pointing downward (Fig. 9.9). In this coordinate system fault orientation is specified by two angles, and the third angle is used to specify the direction of slip. The line of intersection of the fault plane with the surface of the Earth is known as the *strike direction* of the fault. Its orientation is expressed in terms of an angle, known as the *strike angle*, the *strike azimuth*, or the *strike of the fault*. The strike angle ϕ_s is measured clockwise from north to the strike direction when looking down into the Earth; $0 \leq \phi_s < 2\pi$. The strike direction is chosen in such a way that the fault dips downward to the right when looking in the direction of strike. A line on the surface of the Earth perpendicular to the strike direction, drawn in the direction in which the fault plane dips, is known as the *dip direction*. A line in

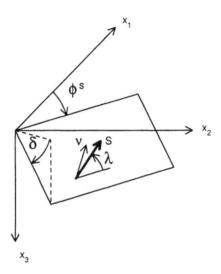

Figure 9.9   Definition of the fault orientation in a geographic coordinate system with the x_1 axis pointing to the north, the x_2 axis pointing to the east, and the x_3 axis pointing downward. The strike angle ϕ_s is measured clockwise from north, the dip angle δ is measured down from the horizontal plane, and the rake angle λ is the angle between the stike direction and the slip vector **S**, which is taken as the direction of the hanging-wall motion with respect to the footwall.

the fault plane perpendicular to the strike direction, drawn downward, is the dip of the fault. An angle δ between the dip direction and the dip of the fault is known as the *dip angle*, and is measured down from the horizontal plane; $0 \le \delta \le \pi/2$. The third angle λ is known as the *rake* or the *rake angle* or the *slip angle* and is the angle between the strike direction and the slip vector, which is taken as the direction of the hanging wall, relative to the footwall. The rake is measured in the fault plane and is positive when measured counterclockwise from the horizontal strike direction; the convention $-\pi < \lambda \le \pi$ is widely adopted in practice (the convention $0 \le \lambda < 2\pi$ can also be used). To clarify the adopted convention for the rake, it should be noted that rake angles between $-\pi$ and $0°$ correspond to dilatations of P waves along the vertical axis and rake angles between $0°$ and π have compressions along the vertical axis (e.g., Herrmann, 1975).

These are the angles used in the routine publications of the National Earthquake Information Service of the U.S. Geological Survey and the International Seismological Centre. Several other conventions exist to describe the direction of the slip relative to the strike. The angle in the vertical plane, for example, is known as the *plunge p* ($\sin p = \sin \lambda \sin \delta$) and the angle in the horizontal plane as the *trend t* ($\sin t = \sin \lambda \cos \delta / (1 - \sin^2 \lambda \sin^2 \delta)^{1/2}$).

The six independent components of the moment tensor for a shear dislocation are expressed in the geographic coordinate system as follows (Aki and Richards, 1980; Kennett, 1988; Pujol and Herrmann, 1990)

$$M_{11} = -M_0\left(\sin\delta\cos\lambda\sin2\phi_s + \sin2\delta\sin\lambda\sin^2\phi_s\right),$$

$$M_{12} = M_{21} = M_0\left(\sin\delta\cos\lambda\cos2\phi_s + \left(\tfrac{1}{2}\right)\sin2\delta\sin\lambda\sin2\phi_s\right),$$

$$M_{13} = M_{31} = -M_0\left(\cos\delta\cos\lambda\cos\phi_s + \cos2\delta\sin\lambda\sin\phi_s\right),$$

$$M_{22} = M_0\left(\sin\delta\cos\lambda\sin2\phi_s - \sin2\delta\sin\lambda\cos^2\phi_s\right),$$

$$M_{23} = M_{32} = -M_0\left(\cos\delta\cos\lambda\sin\phi_s - \cos2\delta\sin\lambda\cos\phi_s\right),$$

$$M_{33} = M_0\sin2\delta\sin\lambda. \tag{9.51}$$

Equations (9.51) and (9.42) can be combined to give the relations for the P, SV, and SH radiation patterns in the far field in terms of the fault and receiver parameters. The distribution of radiation is considered on a unit sphere surrounding the source point in a uniform isotropic medium. For distant observation points, the directly arriving seismic waves leave the source downward, and the lower hemisphere of this focal sphere is usually taken into consideration. At local distances, such as in mines, if the seismic energy that reaches far-field stations leaves the source upward, the upper hemisphere of the focal sphere would be of interest. The points on this focal sphere can be parametrized by looking at the orientation of a radial unit vector Γ and two vectors θ and Φ orthogonal to Γ, described by the takeoff angle i measured from the vertical and the azimuth ϕ measured clockwise from the north (Fig. 9.10).

Thus from relations (9.42) it follows that the far-field radiation pattern of P waves from a double couple source, when all the moment tensor components have the same time dependence, can be expressed as (e.g., Kennett, 1988)

$$R^P = \gamma_i\gamma_j M_{ij}$$
$$= \sin^2 i\left(\cos^2\phi M_{11} + \sin2\phi M_{12} + \sin^2\phi M_{22} - M_{33}\right)$$
$$+ 2\sin i\cos i\left(\cos\phi M_{13} + \sin\phi M_{23}\right) + M_{33}. \tag{9.52}$$

Similarly, the SV radiation pattern is given by

$$R^{SV} = \theta_i\gamma_j M_{ij}$$
$$= \sin i\cos i\left(\cos^2\phi M_{11} + \sin2\phi M_{12} + \sin^2\phi M_{22} - M_{33}\right)$$
$$+ \cos2i\left(\cos\phi M_{13} + \sin\phi M_{23}\right), \tag{9.53}$$

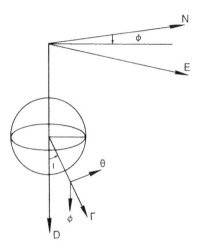

Figure 9.10   The takeoff angle i and azimuth ϕ in relation to the focal sphere surrounding the source. The orientations of the unit vectors Γ in the P-wave direction, θ in the SV direction, and Φ in the SH direction are also indicated.

and the SH radiation pattern is

$$R^{SH} = \phi_i \gamma_j M_{ij}$$
$$= \sin i \left[\sin \phi \cos \phi (M_{22} - M_{11}) + \cos 2\phi M_{12} \right]$$
$$+ \cos i (\cos \phi M_{23} - \sin \phi M_{13}). \tag{9.54}$$

Introducing the moment tensor components from relations (9.51) into equations (9.52)–(9.54), the radiation patterns R^P, R^{SV}, and R^{SH} are directly expressed in terms of the fault parameters ϕ_s, δ, and λ and the receiver parameters ϕ and i (e.g., Aki and Richards, 1980)

$$R^P = \cos \lambda \sin \delta \sin^2 i \sin 2(\phi - \phi_s) - \cos \lambda \cos \delta \sin 2i \cos(\phi - \phi_s)$$
$$+ \sin \lambda \sin 2\delta \left[\cos^2 i - \sin^2 i \sin^2 (\phi - \phi_s) \right]$$
$$+ \sin \lambda \cos 2\delta \sin 2i \sin(\phi - \phi_s), \tag{9.55}$$

$$R^{SV} = \sin \lambda \cos 2\delta \cos 2i \sin(\phi - \phi_s) - \cos \lambda \cos \delta \cos 2i \cos(\phi - \phi_s)$$
$$+ (1/2)\cos \lambda \sin \delta \sin 2i \sin 2(\phi - \phi_s)$$
$$- (1/2)\sin \lambda \sin 2\delta \sin 2i \left[1 + \sin^2(\phi - \phi_s) \right], \tag{9.56}$$

$$R^{SH} = \cos \lambda \cos \delta \cos i \sin(\phi - \phi_s) + \cos \lambda \sin \delta \sin i \cos 2(\phi - \phi_s)$$
$$+ \sin \lambda \cos 2\delta \cos i \cos(\phi - \phi_s)$$
$$- (1/2)\sin \lambda \sin 2\delta \sin i \sin 2(\phi - \phi_s). \tag{9.57}$$

Vectors **p**, **t**, and **b**, defining the compressional, tensional, and null axes, respectively, can also be written in terms of the fault parameters. It can be shown (Pujol and Herrmann, 1990) that the components of the slip vector **s**, the vector **b**, and the normal vector **n** are given by the elements of a rotation matrix **R**, transforming the original coordinates **x** to the new system **x′** in which **n** is in the positive x_3' direction and **s** is in the x_1' direction:

$$\mathbf{s} = (R_{11}, R_{12}, R_{13}), \tag{9.58}$$

$$\mathbf{b} = (R_{21}, R_{22}, R_{23}), \tag{9.59}$$

$$\mathbf{n} = (R_{31}, R_{32}, R_{33}). \tag{9.60}$$

Vectors **p** and **t** are defined by the vectors **s** and **n**

$$\mathbf{p} = (1/\sqrt{2})(\mathbf{s} - \mathbf{n}) \tag{9.61}$$

and

$$\mathbf{t} = (1/\sqrt{2})(\mathbf{s} + \mathbf{n}). \tag{9.62}$$

The elements of the rotation matrix **R** are expressed in terms of the fault parameters as follows (Pujol and Herrmann, 1990)

$$R_{11} = \cos \lambda \cos \phi_s + \sin \lambda \cos \delta \sin \phi_s,$$

$$R_{12} = \cos \lambda \sin \phi_s - \sin \lambda \cos \delta \cos \phi_s,$$

$$R_{13} = -\sin \lambda \sin \delta,$$

$$R_{21} = -\sin \lambda \cos \phi_s + \cos \lambda \cos \delta \sin \phi_s,$$

$$R_{22} = -\sin \lambda \sin \phi_s - \cos \lambda \cos \delta \cos \phi_s,$$

$$R_{23} = -\cos \lambda \sin \delta,$$

$$R_{31} = -\sin \delta \sin \phi_s,$$

$$R_{32} = \sin \delta \cos \phi_s,$$

$$R_{33} = -\cos \delta. \tag{9.63}$$

The components of vectors **p**, **t**, and **b** can now be obtained from relations (9.58)–(9.63).

Determination of the orientation of the fault and auxiliary planes or of the P and T axes for the simple shear dislocation model of the source is known as a *fault-plane solution*.

9.4   Determination of Fault-Plane Solutions

Determination of the orientation of the focal mechanism is greatly simplified by the use of a focal sphere. Seismic stations are projected on points on the surface of the focal sphere by ray tracing back to the source. Their positions on the focal sphere are given by i, the takeoff angle at the source measured from the vertical, and ϕ, the azimuth measured from the north (Fig. 9.10). For stations at teleseismic distances, values of i are obtained from travel-time tables, assuming a certain velocity in the focal region. For local stations at short distances, precise knowledge of the depth and velocity structure is needed to determine i.

In the case of a symmetrical focal mechanism, only half of the focal sphere is needed. When seismic rays arriving at a seismic network are downgoing rays from the source, it is convenient to use the lower focal hemisphere. In underground mines the situation might be different and upgoing rays would be recorded most often. Then the upper focal hemisphere can be used. If the station position on the lower hemisphere is specified by (i, ϕ), then its position on the upper hemisphere is given by $(180° - i, \phi + 180°)$, provided that the angle i in both cases is measured from the same vertical direction.

Fault-plane solutions based on first-motion polarities of P waves are still most widely used. In many cases they provide the only method available to obtain the focal mechanism. The use of only vertical-component seismometers is such a case, often encountered in seismic networks in underground mines. Deriving a fault-plane solution from P waves amounts to finding the two orthogonal nodal planes separating the first motions of P waves into compressional and dilatational quadrants on the focal sphere. An interesting account of the development of fault-plane studies for the earthquake mechanism was recently reviewed by Udias (1989).

It is not convenient to plot data on a spherical surface, and a projection of a three-dimensional sphere onto a plane is needed for displaying the focal sphere and observational data on a two-dimensional diagram. Various projection techniques have been used to plot the P-wave polarity data. The most common are stereographic and equal-area projections. The stereographic projection, also called the *Wulff net*, extensively used in structural geology and rock mechanics, preserves angles and is an equal-angle projection. The equal-area projection, also called the *Schmidt net*, is very similar to the stereographic projection and generally is preferred for fault-plane solutions since area on the focal sphere is preserved on the diagram.

In equal-area projection, the coordinates (i, ϕ) specifying a point on the focal sphere are transformed to plane polar coordinates (R, ϕ) by the formula (e.g., Lee and Stewart, 1981)

$$R = \sqrt{2} \sin(i/2), \tag{9.64}$$

where the maximum value of R is conveniently taken as unity. The set of n observations of P-wave first-motion polarities (either compression or dilatation) may then be represented by plotting the proper symbols at (R_n, ϕ_n) on a Schmidt net. First arrival times of P waves and their corresponding directions of motion are often read from vertical-component seismograms. Compression corresponds to upward motions (away from the source) and is usually marked by the symbol C or $+$, and dilatation corresponds to downward motions (toward the source) and is marked either by D or $-$. The azimuth ϕ is measured clockwise from the north from the earthquake epicenter to a given station, and is computed from the corresponding coordinates of these two points. The takeoff angle i of the seismic ray from the earthquake hypocenter to the given station is usually determined in the course of computing the travel-time derivatives during the location procedure when the ray path of P waves does not follow a straight line as a result of velocity gradients or refraction. In a simple case of the ray path following a straight line, the angle i is readily computed from the coordinates of the hypocenter and station.

As an example, a fault-plane solution of the seismic event of September 30, 1980 in the Szombierki coal mine in Poland is shown in Fig. 9.1. Arc *FBCG* represents the projection of a nodal plane; its strike is 140° and its dip is 50° to the southwest (50°SW). The pole of this nodal plane is point A, which is 90° from the great-circle arc *FBCG*. The second nodal plane is perpendicular to the first and the great-circle arc representing its projection must pass through point A. This arc is *WBAE* in Fig. 9.11; its strike is 270° and its dip is 50°N, as measured from the equal-area net. The pole of the second nodal plane is point C, which is 90° from arc *WBAE* and is situated on the first nodal plane. The two nodal planes do not separate the compressional and dilatational quadrants perfectly, which is exactly what often happens. The quality of resolution depends on the number and distribution of reliable observations, the adequacy of the velocity model used to trace the rays, and proper polarity of seismic recorders. In fact, one of the most difficult tasks in operating an underground seismic network in mines is to ensure that the direction of motion recorded on the seismograms corresponds to the true direction of ground motion.

The intersection of the two nodal planes is a point B in Fig. 9.11, which is the null axis. The plane normal to the null axis is represented by the great-circle arc *APCT*, containing four axes. These are the A and C axes normal to the two nodal planes, and the P and T axes, which are compression (or pressure) and tension axes. The P axis is 45° from the A and C axes and is situated in the dilatational quadrant, and the T axis is 90° from the P axis and lies in the compressional quadrant. In the case here the P axis is determined by the trend of 115° and the plunge of 65°, and the T axis by its trend of 205° and its plunge of 3°. It is often assumed that the P axis

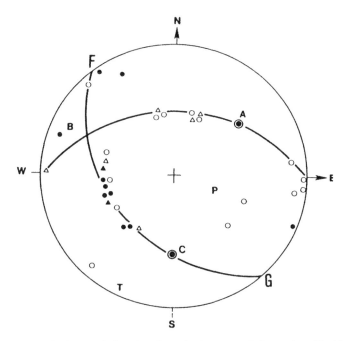

Figure 9.11   Fault-plane solution for the seismic event of September 30, 1980 in the Szombierki coal mine, Poland; a lower-hemisphere equal-area projection is used. Solid circles and triangles mark compressional arrivals, and open circles and triangles represent dilatational arrivals of *P* waves. *P* and *T* are the axes of compression and tension, and *B* is the null axis. *A* and *C* are the poles of two nodal planes (the fault and the auxiliary planes).

represents the direction of maximum compressive stress and the *T* axis corresponds to the direction of maximum tensile stress. If the first nodal plane (arc *FBCG* in Fig. 9.11) is selected as the fault plane, then point *C* (the axis normal to the auxiliary plane) is the slip vector with the rake −52°. If the second nodal plane (arc *WBAE*) is the fault plane, then point *A* is the slip vector with the rake −128°. The slip vector is commonly assumed to be parallel to the resolved shearing stress in the fault plane. The *P* and *T* axes do not necessarily correspond exactly to the stresses that caused the earthquake, and the relations between fault-plane solutions and directions of principal stresses in real heterogeneous media are rather complex. There are some advantages, however, in using the *P* and *T* axes instead of nodal planes alone to describe fault-plane solutions. The *P* and *T* axes are unambiguously defined by the quadrants of compression and dilatation, which is not the case for the fault plane. They indicate the direction of the maxima of the *P*-wave amplitudes, and it is sometimes possible to estimate an approximate orientation of these axes, even from a few observations, by picking the average orientation of compressions for the T axis and dilatation for the P axis.

The amplitudes of P-wave first motion should be used, whenever possible, together with polarity observations for precise determination of fault-plane solutions. Two approaches in this respect are possible. In the first approach, which can be applied to polarity data alone as well, the observed P-wave first motions on the focal sphere are matched with those that are expected from a pair of orthogonal planes at the source, and which are defined by the radiation pattern (9.55). The second approach is similar to that used to determine the nodal planes from a P-wave polarity plot. The relative amplitudes, normalized to the largest one, are indicated on a focal sphere projection by the scaled symbols, corresponding to compressions and dilatations, at observation points defined by the azimuth ϕ and the takeoff angle i. The fault-plane solution is then determined exactly in the same manner as that found from polarity observations alone.

A fault-plane solution indicates the type of faulting taking place in the source area. If the rake $\lambda = 0$ or π, the fault is known as a *strike-slip, lateral, or transcurrent fault*. In contrast, if $\lambda = \pi/2$ or $-\pi/2$, the fault is known as *dip-slip fault*. Elementary types of fault-plane solutions for strike-slip and dip-slip faults are shown in Fig. 9.12. A more general fault corresponds to the superposition of a strike-slip fault and a dip-slip fault, and is known as *oblique-slip fault*. Our example of fault-plane solution in Fig. 9.11 corresponds to such a fault. There are two kinds of strike-slip faults and two kinds of dip-slip faults. In a right-lateral fault, also known as *dextral strike-slip fault*, the direction of the relative displacement of the side of the fault opposite the observer who is facing the fault is to the right (Fig. 9.13A). In a left-lateral fault, or a *sinistral strike-slip fault*, the direction of the relative displacement of the opposite side is to the left (Fig. 9.13B). For a vertical strike-slip fault ($\delta = \pi/2$) the choice of hanging wall and footwall is arbitrary,

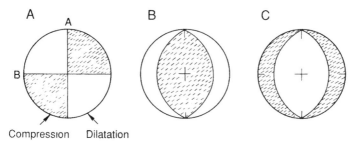

Figure 9.12 Elementary types of fault-plane solutions: (A) vertical strike-slip fault; (B) dip-slip reverse fault; and (C) dip-slip normal fault.

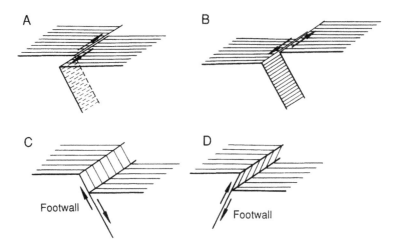

Figure 9.13 Two types of strike-slip faults: right-lateral fault (A) and left-lateral fault (B), and two types of dip-slip faults: normal fault (C) and reverse fault (D). [From Ben-Menahem and Singh (1981, Figs. 4.22 and 4.23).]

and there are two possible choices for the strike direction. The convention is to select either one of the two possible strike directions and to label the right-hand block (looking along the strike) as the hanging wall. Then the rake $\lambda = 0$ defines a left-lateral strike-slip fault, and $\lambda = \pi$ is for a right-lateral fault (Aki and Richards, 1980). Similarly, left-lateral faults with the dip angle $\delta < \pi/2$ are also characterized by the rake $\lambda = 0$ and right-lateral faults by $\lambda = \pi$. The left-lateral strike-slip components of an oblique-slip fault have λ either in the range of 0 to $\pi/2$ or in the range of $-\pi/2$ to 0, whereas the right-lateral strike-slip components correspond to λ values either between $\pi/2$ and π or between $-\pi$ and $-\pi/2$.

Two kinds of dip-slip faults are distinguished. A *normal fault* is an inclined fracture along which the hanging wall moves down in relation to the footwall (Fig. 9.13C), and a *reverse fault* is an inclined fracture along which the hanging wall moves upward relative to the footwall (Fig. 9.13D). A reverse fault with the dip angle $\delta < 45°$ is also known as a *thrust fault*, and with $\delta < 10°$, an *overthrust fault*. A pure dip-slip reverse fault has the rake $\lambda = \pi/2$, and a pure dip-slip normal fault is characterized by $\lambda = -\pi/2$. In a more general sense, the faults with λ within the range from 0 to π are termed reverse faults, and the faults with λ in the range of $-\pi$ to 0 are called normal faults.

Thus, the sign and the value of the rake define the type of faulting occurring at the source. If the first nodal plane (arc *FBCG*) in Fig. 9.11 is chosen as the fault plane, then the diagram represents normal faulting with a minor left-lateral component. If, on the other hand, the second nodal plane (arc *WBAE*) is selected as the fault plane, then the diagram indicates normal faulting with a minor right-lateral component. It should be noted that fault-plane solutions found from first-motion data describe the focal mechanism corresponding to the hypocenter, where the rupture is initiated. The mechanism may change during rupture propagation, especially during large seismic events, and then the determination of its final state must be based on an inversion of the whole waveform data.

To restrain the range of possible fault-plane solutions for a given set of data from *P* waves, observations from *S* waves are occasionally used, although this approach is severely restricted by the well-known difficulty in identification of the initial motion of *S* waves. To determine the fault-plane solution from the observations of *S* waves, three-component seismograms are needed. Similarly as for *P* waves, either the sign only of *SV* and *SH* waves or their signs and amplitudes are used. The signs are defined following the generally accepted convention; *SV* is positive when directed upward from the ray in the plane of incidence, and *SH* is positive when directed to the right of the ray when viewed from the source toward the receiver. The plane of incidence is a vertical plane that contains the source and the receiver (the vector Γ in Fig. 9.4).

The technique most often used in focal mechanism studies from *S* waves is based on their polarization. The direction of motion of *S* waves in the plane transverse to the seismic ray has a preferred orientation determined by the character of the source, which means that the *S* waves are plane-polarized. The polarization is best described in terms of the polarization angle ε, shown in Fig. 9.4, which is the angle between the *S* motion and the plane of incidence and is defined by the relation

$$\tan \varepsilon = \frac{u^{SH}}{u^{SV}} = \frac{R^{SH}}{R^{SV}}, \qquad (9.65)$$

where u^{SH} and u^{SV} are the *SH* and *SV* components of the *S* motion, and R^{SH} and R^{SV} are the radiation patterns of the *SH* and *SV* waves, given by equations (9.56) and (9.57). It is accepted that ε remains constant along a given ray, from the source to the receiver.

To determine the polarization angle of *S* waves, the amplitudes of the three components should be read at the same instant of time, as close as possible to the predicted *S* arrival time, and then corrected to a common instrument magnification. The use of the polarization of *S* waves has one big advantage over the other methods of focal mechanism determination. It does not depend essentially on reading the very first motion or first half-amplitude of the *S* wave. The polarization is described by the orientation of a line in space, and it can be determined even when a small initial motion of *S* wave is missed and the reading is taken a half-period too late. Then the sense of the motion will be in error (180° out of phase), but its orientation would remain correct. If the azimuth from the receiver back to the epicenter is ϕ_r, the displacements u^N and u^E of the N and E components can be rotated to form the radial u^R and tangential u^T components of the S-wave motion by the transformation (e.g., Herrmann, 1975)

$$u^R = -u^N \cos \phi_r - u^E \sin \phi_r, \qquad (9.66)$$

$$u^T = u^N \sin \phi_r - u^E \cos \phi_r, \qquad (9.67)$$

where u^N and u^E are taken to be positive in the north and east directions, respectively. The u^R component is positive in a direction away from the epicenter, and the u^T component is positive in a direction directed to the right of the ray when viewed from the epicenter toward the receiver. Since u^{SV} was defined as being positive when directed up from the ray in the plane of incidence, the radial component u^R must be of opposite sign to u^{SV}. Furthermore, the radial component u^R and the vertical component u^Z must be of opposite sign as well, since u^Z is taken to be positive upward at the receiver. This provides a useful check if a particular set of S-wave observations is correct.

If the free-surface interaction is not present, the polarization angel ε can be determined directly from the values of the radial and tangential components

$$\tan \varepsilon = -\frac{u^T}{u^R}. \qquad (9.68)$$

The polarization angles can be plotted directly on the focal sphere projection at the point corresponding to each seismic station. This point should be placed at the center of the projection and a small line drawn making an angle ε with respect to the azimuth of the station; either to the right from the azimuth for positive values of ε or to the left for negative values of ε. The S-wave polarization tics should converge at the T and P axes and should be perpendicular to the nodal planes of P waves. An example of such a pattern is shown in Fig. 9.14, reproduced from Herrmann (1975). The polarization of S waves is usually interpreted jointly with the observations from P waves. The interpretation consists of determining two perpendicular nodal planes that satisfy the P-wave polarity data and the S-wave polarization data.

Other amplitude ratios as well were used extensively by Russian seismologists in the 1950s, to restrain the fault-plane solutions by fewer observations. The method was modified in the early 1980s and introduced in the form of S to P amplitude ratios for computing the double-couple focal mechanism from regional and local observations (Kisslinger, 1980, 1982; Kisslinger et al.,

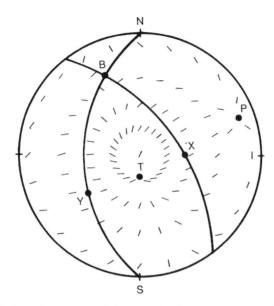

Figure 9.14   Polarization pattern of S waves showing convergence at P and T axes and perpendicularity with respect to nodal planes. [From Herrmann (1975, Fig. 2).]

1981). This approach is essentially based on the radiation patterns given by equations (9.55)–(9.57), adopted for the ratio u^{SV}/u^P to the case when only vertical components are recorded by a seismic network. The appropriate formulas for local observations are given by Kisslinger *et al.* (1981) and corrected by Kisslinger (1982). A combination of both *P* polarity and *S*/*P* ratio data provides an effective means of constraining fault-plane solutions from fewer observations.

If the number of *P*-wave first-motion observations, or any other focal mechanism-related data, is small, it is a common practice to determine a composite fault-plane solution. The composite solution is obtained from joint processing of *P*-wave polarity observations or *S*-wave polarization data from a number of seismic events, occurring in a given area and within a limited time interval, for a common focal mechanism. This technique is used extensively in studies of natural microearthquakes and occasionally in studies of small mine tremors (e.g., Gibowicz and Cichowicz, 1986; Williams and Arabasz, 1989; Wong *et al.*, 1989). Such a practice should be used with caution because the assumption that the focal mechanisms for all involved seismic events are identical may not be true, especially under rapidly changing stress conditions in underground mines.

Many attempts have been made to determine fault-plane solutions by computer. Essentially, most computer programs for fault-plane solutions try to match the observed P-wave first motions on the focal sphere with those that are expected theoretically from a pair of orthogonal planes at the source. To find the best match, a pair of orthogonal planes is led to assume a sequence of positions, sweeping systematically through the complete solid angle at the focus. In any position, a score may be computed to see how well the data are matched, and the pair of orthogonal planes with the best score may be chosen as the nodal planes.

An excellent review of numerical methods and computer programs used for determining fault-plane solutions is given by Udias (1989). Knopoff (1961a, 1961b) was the first to formulate the problem in terms of a probability function, describing the polarity of first motion in the presence of noise.

Udias and Baumann (1969) developed a mixed method that combined signs of first motion of P waves and polarization angles of S waves. The program searched for a minimum combined error E of the two types of data

$$E = \left[\frac{1}{N} \sum_{i=1}^{N} (\varepsilon_i - \varepsilon_{ci})^2 \right]^{1/2} + 100 \sum_{i=1}^{N} |\operatorname{sgn} u_{pi} - \operatorname{sgn} u_{pci}|, \qquad (9.69)$$

where ε_i and ε_{ci} are the observed and calculated polarization angles of the S wave, and $\operatorname{sgn} u_{pi}$ and $\operatorname{sgn} u_{pci}$ are the observed and calculated signs of first motion of the P wave; E is a sum of the standard error of the residuals of the polarization angles of S waves and the number of inconsistencies of P-wave data multiplied by 100, which is equivalent to using the P-wave observations as a constraint. The program searches for the orientation of the source that best satisfies the S-wave data within the region of those solutions with a minimum error in P-wave data. The solution is obtained by a systematic change of the X and Y axes, defining the orientation of the source, in small increments of the polar angles. This method was improved by Chandra (1971), who introduced a weighting procedure to give P and S wave data similar weight.

The problem of combining polarities of P waves and polarization angles of S waves was considered again by Pope (1972) and Dillinger et al. (1972) with a more rigorous statistical approach. A likelihood function for the combined solution was proposed as the product of the P and S likelihood functions, and the program applied discrete incremental rotations of the X and Y axes of the solution to find the values where the likelihood function is the greatest. Recently, a joint inversion of S polarizations and P polarities for fault mechanisms of local earthquakes was described by Zollo and Bernard (1991). The method is based on the joint probability density function of observations and fault parameters. Another method to find focal mechanisms of local earthquakes, proposed by De Natale et al. (1991), is based on the nonlinear estimation of the probability function for strike, dip, and rake of the source fault from P polarities, S polarizations, and S/P amplitude ratios. The method was shown to be effective for small earthquakes recorded at local distances, with S/P amplitude ratios providing strong constraints on fault parameters.

A probabilistic formulation of the problem was proposed by Keilis-Borok et al. (1972), based on a maximum-likelihood method and using P-wave polarity data. If a_k are the observed signs of P-wave motion and α_k are the theoretical signs, the maximum-likelihood function L is given by

$$L = \sum_{k=1}^{N} \pi_k^{(a_k+1)/2}(1-\pi_k)^{(a_k-1)/2}, \qquad (9.70)$$

where π_k is the probability of a correct reading of P-wave polarities at station k with respect to the model and is given by

$$\pi_k = p\frac{1+\alpha_k}{2} + (1-p)\frac{1-\alpha_k}{2}, \qquad (9.71)$$

in which p is the probability of reading a compression and α_k is a function of the angles of the X and Y axes defining the model. The maximum-likelihood estimates of these parameters are found by maximization of the function L by a search procedure. The appropriate confidence regions for these parameters can also be obtained in a similar manner.

An extension of the problem to consider fault-plane solutions for a group of earthquakes in a given region was presented by Brillinger et al. (1980). The method simultaneously handles data from several earthquakes, with individual solutions considered as particular cases of the general problem. Several probability distributions for the observed first motion are compared and for simplicity only P-wave first motions are considered, although an extension to include S-wave polarization angles is not difficult. For a single earthquake, the model assumes that the probability of observing a compression or dilatation at a given station is a function of theoretical amplitude expected at that station. The probability of reading a compression at station j may then be written as

$$\pi_j = \gamma + (1-2\gamma)\Phi[\rho A_j(\theta)], \qquad (9.72)$$

where γ is a small constant ($0 \leq \gamma \leq 1/2$) that accounts for reading errors; Φ is the normal cumulative distribution function; $\rho = \alpha/\sigma$, in which α is a proportionality constant and σ is the standard deviation of the noise; and A_j are the normalized expected amplitudes corresponding to a particular orientation of the source defined by any combination of the three independent angles.

The solution of the problem consists in finding the orientation of the source (T and P or X and Y axes) that maximize the probability that this orientation corresponds to the set of observations Y_j. This is done by maximizing the likelihood function L given by

$$L = \sum_{j=1}^{N} \log\left\{\tfrac{1}{2}\left[1 + (2\pi_j - 1)Y_j\right]\right\}.$$ (9.73)

Since L is a differentiable function of the parameters involved, available efficient routines can be used for its maximization. The program uses an iterative process from a given initial solution. Standard errors of the estimates are determined, and methods for testing hypotheses concerning the parameter values are applied. To measure the goodness of a solution the score p or the proportion of correct polarities was introduced:

$$p = \frac{1}{N} \sum_{j=1}^{N} \left(1 + Y_j \operatorname{sgn} A_j\right).$$ (9.74)

This value, together with the number N of data used and the standard deviations of the focal parameters, provides the necessary indication of the solution quality.

For a group of earthquakes, the probability of reading a compression at station j from event i is now given by an expression similar to (9.72), in the form (Brillinger *et al.*, 1980)

$$\pi_{ij} = \gamma + (1 - 2\gamma)\Phi\left[\rho_i A_{ij}(\boldsymbol{\theta})\right],$$ (9.75)

where ρ_i is a set of parameters assigned to all observations of each event as a weight. These parameters enter as variables during the maximization process of the likelihood function. The likelihood function corresponding to relation (9.73) is

$$L = \sum_{i=1}^{M} \sum_{j=1}^{N_i} \log\left\{\tfrac{1}{2}\left[1 + (2\pi_{ij} - 1)Y_{ij}\right]\right\},$$ (9.76)

where M is the number of earthquakes, N_i is the number of observations for each earthquake, and Y_{ij} are the observations of polarities. The number of variables in the maximization process of L is $M + 3$; the orientation of the joint solution described by three angles and the values of ρ_i for $i = 1, \ldots, M$. To start the iterative process, a set of initial equal values of ρ_i must be given, together with an initial solution. During the process, negative values may be assigned to ρ_i for events whose observations are predicted to have opposite signs by the joint solution. To know to what extent events agree with the joint solution, the scores p_i [relation (9.74)] for each event and the plots of p_i versus ρ_i are examined. For perfect agreement $p_i = 1$ and ρ_i tends to infinity. To separate events, a threshold value of p is used, below which events are considered as not having the same focal mechanism as that found for the joint solution. The procedure can be continued by looking for successive joint solutions. This statistical rigorous approach to composite fault-plane solutions might be highly useful for focal mechanism studies of seismic events induced by mining.

Although many other methods have been developed to study focal mechanisms, such as moment-tensor inversion and waveform analyses, the fault-plane method is still widely used because of its simplicity. Fault-plane solutions today provide standard information in seismotectonic studies, and many national and international seismological agencies calculate and publish them routinely. The double-couple source, however, is not necessarily the only possible mechanism for a wide range of seismic events, especially for those associated with volcanic phenomena and those associated with human activity. Inversion of the seismic moment tensor is the most general approach for determining a point source mechanism and could possibly answer the growing question as to whether certain seismic events in mines are nonshearing events.

9.5 Moment-Tensor Inversion

Moment tensors, already introduced [equations (9.36) and (9.37)], describe completely in a first-order approximation the equivalent forces of general seismic point sources; the double-couple source is just one of them. The seismic moment tensor was introduced by Gilbert (1970) to calculate the displacement that can be expressed as a sum of the moment-tensor elements times the corresponding Green function. The linearity between the moment tensor and Green function elements was first used by Gilbert (1973) to calculate moment-tensor elements from observations, known as *moment-tensor inversion*. The concept of seismic moment tensors was further extended by Backus and Mulcahy (1976) and Backus (1977a, 1977b).

A convenient description of seismic sources is provided by considering them as arising from deviations from the elastically predicted stress state. The stress tensor is divided into a purely elastic part and an inelastic part, which Backus and Mulcahy (1976) called the "stress glut." The physical source region is characterized by the existence of the equivalent forces, arising as the result of differences between the model stress and the actual physical stress. Outside the source region, the stress glut and the equivalent forces vanish.

The displacement u_k generated at a point $\mathbf{x}$ at the time t by a distribution of equivalent body-force densities f_i within a source volume V is given by (e.g., Aki and Richards, 1980; Kennett, 1988; Jost and Herrmann, 1989;

Vasco and Johnson, 1989)

$$u_k(\mathbf{x}, t) = \int_{-\infty}^{\infty} \int_V G_{ki}(\mathbf{x}, t; \mathbf{r}, t') f_i(\mathbf{r}, t') \, dV dt', \tag{9.77}$$

where $G_{ki}(\mathbf{x}, t; \mathbf{r}, t')$ are the Green functions containing the propagation effects between the source $(\mathbf{r}, t')$ and the receiver $(\mathbf{x}, t)$. The Green functions may be expanded in a Taylor series (Stump and Johnson, 1977) about the point $\mathbf{r} = \xi$ as

$$G_{ki}(\mathbf{x}, t; \mathbf{r}, t') = \sum_{n=0}^{\infty} \frac{1}{n!} (r_{J_1} - \xi_{J_1}) \cdots (r_{J_n} - \xi_{J_n}) G_{ki, J_1 \cdots J_n}(\mathbf{x}, t; \xi, t'), \tag{9.78}$$

where the comma between indices describes partial derivatives with respect to the coordinates after the comma. The location of the reference point would normally be the hypocenter for small sources, whereas for extended faulting an improved representation is obtained by considering the centroid of the source. Defining the time-dependent force moment tensor as

$$M_{iJ_1 \cdots J_n}(\xi, t') = \int_V (r_{J_1} - \xi_{J_1}) \cdots (r_{J_n} - \xi_{J_n}) f_i(\mathbf{r}, t') \, dV, \tag{9.79}$$

the displacement can be written as a sum of terms which resolve additional details of the source, known as multiple expansion (Backus and Mulcahy, 1976; Stump and Johnson, 1977; Aki and Richards, 1980; Kennett, 1983; Jost and Herrmann, 1989; Vasco and Johnson, 1989)

$$u_k(\mathbf{x}, t) = \sum_{n=0}^{\infty} \frac{1}{n!} G_{ki, J_1 \cdots J_n}(\mathbf{x}, t; \xi, 0) * M_{iJ_1 \cdots J_n}(\xi, t), \tag{9.80}$$

where $*$ denotes the temporal convolution.

If the source dimensions are small in comparison to the observed wavelengths of seismic waves, only the first term in relation (9.80) needs to be considered, and then the displacement can be written as

$$u_k(\mathbf{x}, t) = G_{ki, j}(\mathbf{x}, t; \mathbf{0}, 0) * M_{ij}(\mathbf{0}, t) \tag{9.81}$$

for $\xi = \mathbf{0}$. Assuming that all components of the time-dependent seismic moment tensor have the same time dependence $s(t)$, the case known as *synchronous source* (Silver and Jordan, 1982), the displacement can be expressed as

$$u_k(\mathbf{x}, t) = M_{ij} [G_{ki, j} * s(t)], \tag{9.82}$$

where $s(t)$ is often called the *source time function*. Thus the displacement u_k is a linear function of the moment-tensor elements and the terms in the square brackets. If the source time function is a delta function, the only term left in the square brackets is $G_{kl,j}$ describing nine generalized couples (e.g., Jost and Herrmann, 1989), shown in Fig. 9.3.

In general, the source moment tensor **M** of second order, describing a general dipolar source, has nine components M_{ij} and is represented as a 3×3 matrix in a given reference frame. It can be written as (Ben-Menahem and Singh, 1981)

$$\begin{aligned}
\mathbf{M} = M_{ij}e_ie_j \\
= \tfrac{1}{3}(M_{11} + M_{22} + M_{33})(e_1e_1 + e_2e_2 + e_3e_3) + \tfrac{1}{3}(2M_{11} - M_{22} - M_{33})e_1e_1 \\
+ \tfrac{1}{3}(2M_{22} - M_{33} - M_{11})e_2e_2 + \tfrac{1}{3}(2M_{33} - M_{11} - M_{22})e_3e_3 \\
+ \tfrac{1}{2}(M_{32} + M_{23})(e_3e_2 + e_2e_3) + \tfrac{1}{2}(M_{32} - M_{23})(e_3e_2 - e_2e_3) \\
+ \tfrac{1}{2}(M_{13} + M_{31})(e_1e_3 + e_3e_1) + \tfrac{1}{2}(M_{13} - M_{31})(e_1e_3 - e_3e_1) \\
+ \tfrac{1}{2}(M_{21} + M_{12})(e_2e_1 + e_1e_2) + \tfrac{1}{2}(M_{21} - M_{12})(e_2e_1 - e_1e_2),
\end{aligned} \quad (9.83)$$

where e_i and e_j are the unit vectors along the x_i and x_j directions. The first term on the right-hand side of this equation describes a center of compression, and the successive terms describe three dipoles along the coordinate axes, three double couples, and three torques about the coordinate axes, respectively. This is known as the *decomposition theorem*. The center of compression comes from the isotropic part of the moment tensor, corresponding to a volume change in the source. The remaining nine sources form the deviatoric part of the moment tensor. This deviatoric part can be further decomposed; a multitude of different decompositions are possible.

The conservation of angular momentum for the equivalent forces leads to the symmetry of the seismic moment tensor (Gilbert, 1970). If the moment tensor is symmetric, then $M_{ij} = M_{ji}$ and the torques in equation (9.83) vanish. The eigenvalues m_1, m_2, and m_3 of a symmetric second-order tensor are all real and its eigenvectors $\mathbf{a}_1$, $\mathbf{a}_2$, and $\mathbf{a}_3$ are mutually orthogonal. Then from equation (9.83) it follows that a moment tensor can be decomposed into an isotropic part and three vector dipoles (Ben-Menahem and Singh, 1981; Jost and Herrmann, 1989)

$$\begin{aligned}
\mathbf{M} = \tfrac{1}{3}(m_1 + m_2 + m_3)\mathbf{I} + \tfrac{1}{3}(2m_1 - m_2 - m_3)\mathbf{a}_1\mathbf{a}_1 \\
+ \tfrac{1}{3}(2m_2 - m_3 - m_1)\mathbf{a}_2\mathbf{a}_2 + \tfrac{1}{3}(2m_3 - m_1 - m_2)\mathbf{a}_3\mathbf{a}_3,
\end{aligned} \quad (9.84)$$

where $\mathbf{I} = \delta_{ij}$ is the identity matrix. The isotropic component of the source mechanism is often thought to be associated with a phase transition expected to occur in the source of deep-focus earthquakes. Most recent results show, however, that a sudden implosive phase change can rather be ruled out as the primary physical mechanism for deep-focus earthquakes (Kawakatsu,

1991a). The source process of a shallow-focus earthquake, on the other hand, which occurred on May 14, 1985 off the northern Mozambique coast, has been considered as a combination of a normal fault and a subsequent isotropic source (Honda and Yomogida, 1991).

Equation (9.84) may also be written in the form

$$\mathbf{M} = \tfrac{1}{3}(m_1 + m_2 + m_3)\mathbf{I} + \tfrac{1}{3}m_1(2\mathbf{a}_1\mathbf{a}_1 - \mathbf{a}_2\mathbf{a}_2 - \mathbf{a}_3\mathbf{a}_3)$$
$$+ \tfrac{1}{3}m_2(2\mathbf{a}_2\mathbf{a}_2 - \mathbf{a}_3\mathbf{a}_3 - \mathbf{a}_1\mathbf{a}_1) + \tfrac{1}{3}m_3(2\mathbf{a}_3\mathbf{a}_3 - \mathbf{a}_1\mathbf{a}_1 - \mathbf{a}_2\mathbf{a}_2), \quad (9.85)$$

where $2\mathbf{a}_1\mathbf{a}_1 - \mathbf{a}_2\mathbf{a}_2 - \mathbf{a}_3\mathbf{a}_3$ represents a compressional dipole of strength 2 in the direction of the eigenvector $\mathbf{a}_1$ and two dilatational dipoles each of unit strength along the $\mathbf{a}_2$ and $\mathbf{a}_3$ axes. This type of source is known as a *compensated linear vector dipole* (CLVD). A general dipolar source with a symmetric moment tensor, therefore, is equivalent to a center of compression and three mutually orthogonal CLVDs.

Alternatively, a symmetric moment tensor can be decomposed into an isotropic component and a major and minor double couple, introduced by Kanamori and Given (1981). The major couple seems to be the best approximation of a general seismic source by a double couple, since the direction of the principal axes of the moment tensor remain unchanged (Jost and Herrmann, 1989). To construct the major double couple, the eigenvector of the smallest eigenvalue (in the absolute sense) is taken as the null axis, and it is assumed that the purely deviatoric eigenvalues m_i^{d} of the moment tensor

$$m_i^{\mathrm{d}} = m_i - \frac{m_1 + m_2 + m_3}{3} \qquad (9.86)$$

are such that $|m_3^{\mathrm{d}}| \geq |m_2^{\mathrm{d}}| \geq |m_1^{\mathrm{d}}|$. Then the complete decomposition can be written as (Jost and Herrmann, 1989)

$$\mathbf{M} = \tfrac{1}{3}(m_1 + m_2 + m_3)\mathbf{I} + m_3^{\mathrm{d}}(\mathbf{a}_3\mathbf{a}_3 - \mathbf{a}_2\mathbf{a}_2) + m_1^{\mathrm{d}}(\mathbf{a}_1\mathbf{a}_1 - \mathbf{a}_2\mathbf{a}_2), \quad (9.87)$$

in which the second term represents the major double couple and the third term represents the minor double couple. A best double couple can be constructed in a similar way by replacing m_3^{d} by the average of the largest (in the absolute sense) two eigenvalues (Giardini, 1984).

Following Knopoff and Randall (1970) and Fitch *et al.* (1980), a moment tensor can be decomposed into an isotropic part, a compensated linear vector dipole, and a double couple. Assuming again that $|m_3^{\mathrm{d}}| \geq |m_2^{\mathrm{d}}| \geq |m_1^{\mathrm{d}}|$ in relation (9.86) and that the same principal stresses produce the CLVD and the double couple radiation, the following decomposition is obtained (Jost and Herrmann, 1989):

$$\mathbf{M} = \tfrac{1}{3}(m_1 + m_2 + m_3)\mathbf{I} + m_3^{\mathrm{d}}F(2\mathbf{a}_3\mathbf{a}_3 - \mathbf{a}_2\mathbf{a}_2 - \mathbf{a}_1\mathbf{a}_1)$$
$$+ m_3^{\mathrm{d}}(1 - 2F)(\mathbf{a}_3\mathbf{a}_3 - \mathbf{a}_2\mathbf{a}_2), \qquad (9.88)$$

where $F = -m_1^{\mathrm{d}}/m_3^{\mathrm{d}}$. Such a decomposition seems to be the most interesting one for source studies of seismic events induced by mining. The CLVD source was considered as a model for sudden phase transitions in deep-focus earthquakes (Knopoff and Randall, 1970), tensile failure of rock in the presence of high-pressure fluids (Julian and Sipkin, 1985; Foulger, 1988), and source complexity (e.g., Frohlich et al., 1989). The CLVD source corresponding to a sort of uniaxial compression could possibly explain one of the mechanisms of pillar-associated seismic events, observed in situ in deep hard-rock mines in South Africa and reported by Lenhardt and Hagan (1990).

There are various methods of inversion for moment-tensor elements. The inversion can be done in the time or frequency domain, and different data can be used separately or in combination. The moment tensor inversion in the time domain can be based on the formulation described by equation (9.82) (e.g., Gilbert, 1970; Stump and Johnson, 1977; Strelitz, 1978; Fitch et al., 1980; Langston, 1981). If the source time function is not known or cannot be assessed or the assumption of a synchronous source is not upheld, the frequency-domain approach is chosen (e.g., Gilbert, 1973; Gilbert and Buland, 1976; Stump and Johnson 1977; Patton and Aki, 1979; Kanamori and Given, 1981). The displacement in the frequency domain, corresponding to the formulation in (9.82), can be written as

$$u_k(\mathbf{x}, f) = M_{ij}(f)G_{ki,j}(f) \qquad (9.89)$$

for each frequency f. Both approaches (9.82) and (9.89) lead to linear inversions in the time or frequency domain, respectively, for which a number of fast computational algorithms are available (e.g., Lawson and Hanson, 1974; Press et al., 1990).

Both equations either (9.82) or (9.89) can be written in a matrix form:

$$\mathbf{u} = \mathbf{G}\mathbf{m}. \qquad (9.90)$$

In the time domain, the vector $\mathbf{u}$ consists of n sampled values of the observed ground displacement at various stations, $\mathbf{G}$ is a $n \times 6$ matrix containing the Green's functions calculated using an appropriate algorithm and Earth model, and $\mathbf{m} = (M_{11}, M_{12}, M_{22}, M_{13}, M_{23}, M_{33})$ is a vector containing the six moment-tensor elements to be determined. In the frequency domain, equations (9.90) are written separately for each frequency. The vector $\mathbf{u}$ consists of real and imaginary parts of the displacement spectra, the matrix $\mathbf{G}$ and the vector **m** contain real and imaginary parts as well, and **m** contains also the transform of the source time function of each moment-tensor element. The details of solving equations (9.90) for **m** are given by Aki and Richards (1980). A detailed description of the procedure for regional and local seismograms is given by Oncescu (1986). The application of moment-tensor inversion to microseismic events is described by O'Connell and Johnson (1988).

Following Jost and Herrmann (1989), an outline of the processing steps in a moment tensor inversion is briefly described here. The first step is the data acquisition and preprocessing. Data with good signal to noise ratio that have a good coverage of the focal sphere (see Satake, 1985) are needed. The horizontal components should be rotated into radial and transverse components. Linear trends have to be identified and removed. The instrumental effect should be considered. The nominal instrument response can be used or the calibration of the instrument can be checked by using the calibration pulse on the record, and the polarity of the instruments should be verified. High-frequency noise in the data is removed by lowband-pass filtering.

In the second step synthetic Green's functions are calculated, taking into account geological structure between the source and the receiver, the location of the point source, and the position of the receiver. The source time function in equations (9.82) is often assumed to be a step function (Knopoff and Gilbert, 1959) or a ramp function (Haskell, 1964) and used in the inversion (e.g., Gilbert, 1970; Stump and Johnson, 1977; Dziewonski *et al.*, 1981; Kanamori and Given, 1981). Powerful waveform inversion methods are available when source time functions need to be recovered (e.g., Burdick and Mellman, 1976; Wallace *et al.*, 1981).

The third step is the proper inversion—the solution of equations (9.90). The inversion is usually formulated in terms of least-squares problems (norm L_2) (Gilbert and Buland, 1976; Mendiguren, 1977; Stump and Johnson, 1977), although other norms (e.g., norm L_1) can have advantages when less sensitivity to gross errors is required.

The main difficulty in the moment-tensor inversion is a proper calculation of Green's functions for geologically complex media. The Green function is in general different for different displacement components and takes different values for particular stations. The simplest approach in the time domain is to use directly the source radiation formulation for P, SV, or SH waves, described by formulas (9.41), (9.42), and (9.43). This approach was used by Fitch *et al.* (1980) and De Natale *et al.* (1987) and others. A more rigorous approach is based on the method of matrix propagators of Haskell–Thomson (Haskell, 1953), modified by Knopoff (1964) and Dunkin (1965), and used in the frequency domain. Another method of evaluating Green's functions, especially valuable for highly complex structures including possible lateral inhomogeneities, is an empirical one using Green's functions determined from observations and a known source. This relative moment tensor determination was first proposed by Strelitz (1980) in a study of subevents of complex deep-focus earthquakes. The method was extended by Oncescu (1986) to individual small events recorded at a few stations.

If the focal depth is not known with sufficient precision, then a linear inversion can be done for a number of trial depths. The most probable depth will minimize the corresponding error between observed and theoretical waveforms (e.g., Mendiguren, 1977; Patton and Aki, 1979; Sipkin, 1982).

It is convenient to characterize the moment tensor by its eigenvalues. This can be done by a rotation of the moment tensor from geographic coordinates into its principal axes. Then the moment tensor can be written in the diagonal form

$$M_{ij} = m_i \, \delta_{ij}, \tag{9.91}$$

where m_i are the eigenvalues of $\mathbf{M}$. For a general moment tensor all eigenvalues m_i are different. Seismic sources with no volume change can be obtained by constraining the moment tensor to have zero trace

$$\text{tr} \, \mathbf{M} = m_1 + m_2 + m_3 = 0, \tag{9.92}$$

or in a more general form

$$\text{tr} \, \mathbf{M} = M_{11} + M_{22} + M_{33} = 0. \tag{9.93}$$

The sum of the diagonal elements of the moment tensor divided by 3 is a measure of the volume change associated with the source. It can be readily shown that for the moment tensor of the double-couple source, one principal value of $\mathbf{M}$ must vanish, which means that the determinant $\det \mathbf{M}$ must also vanish

$$\det \mathbf{M} = m_1 m_2 m_3 = 0, \tag{9.94}$$

or in a general form

$$\det \mathbf{M} = M_{11} M_{22} M_{33} + 2 M_{12} M_{23} M_{13} - M_{11} M_{23}^2 - M_{22} M_{13}^2 - M_{33} M_{12}^2 = 0. \tag{9.95}$$

The vanishing of $\det \mathbf{M}$ and $\text{tr} \, \mathbf{M}$ are therefore necessary and sufficient conditions for a double-couple source. In the more general case, the eigenvalues of the moment tensor $\mathbf{M}$ can be readily found, following, for example, Kennett (1988).

The nonisotropic constraint of zero trace on the moment tensor is linear, whereas for double-couple sources the constraint of zero determinant on the moment tensor is nonlinear. To solve the linear system of equations (9.90) under these constraints, the method of Lagrange multipliers is used (Strelitz,

1980; Oncescu, 1986). The system must be solved iteratively until the determinant det $\mathbf{M}$ and the trace tr $\mathbf{M}$ converge to zero. The scalar seismic moment M_0 can be determined from a given moment tensor by

$$M_0 = \tfrac{1}{2}(|m_1| + |m_2|), \tag{9.96}$$

where m_1 and m_2 are the largest eigenvalues in the absolute sense. The seismic moment can equivalently be estimated by the following relations (Silver and Jordan, 1982)

$$M_0 = \left(\frac{\Sigma M_{ij}^2}{2} \right)^{1/2} = \left(\frac{\Sigma m_i^2}{2} \right)^{1/2}. \tag{9.97}$$

After the recovery of moment tensor, the deviation of the solution from the pure double-couple model can be evaluated from the ratio (Dziewonski et al., 1981)

$$\varepsilon = \frac{|m^{\mathrm{d}}|_{\mathrm{min}}}{|m^{\mathrm{d}}|_{\mathrm{max}}}, \tag{9.98}$$

where $|m^{\mathrm{d}}|_{\mathrm{min}}$ is the smallest and $|m^{\mathrm{d}}|_{\mathrm{max}}$ is the largest deviatoric eigenvalue in the absolute sense. The values of this ratio can range from 0 for a pure double-couple source to 0.5 for a pure compensated linear vector dipole. Alternatively, the ratio ε can be expressed in percentages of CLVD by multiplying ε by 200. The percentage of a pure double couple is $100(1-2\varepsilon)$. The variation of ε against seismic moment and earthquake space distribution was studied by Dziewonski and Woodhouse (1983) and Giardini (1984). Silver and Jordan (1982) have developed a method for the estimation of the isotropic and deviatoric components of the moment tensor, introducing the isotropic, deviatoric, and total scalar seismic moments. Graphical methods have been recently suggested for identifying non-double-couple moment-tensor components (Pearce et al., 1988; Hudson et al., 1989; Riedesal and Jordan, 1989). A method for the exact mapping of error bounds on seismic waveforms into bounds on certain moment-tensor properties was presented by Vasco (1990). The properties are the three invariants of the moment tensor: the trace, the determinant, and the sum of the determinants of the diagonal minors. Finding upper and lower bounds on these unique coordinate-free invariants allows determination of whether significant volume change is associated with the source or whether a non-double-couple mechanism is needed to satisfy the data. Furthermore, the range of models that fit the data is an indication of how well constrained the source properties are.

In general, moment-tensor inversions involve two major assumptions: (1) it is assumed that the earthquake may be treated as a point source for a given frequency of seismic waves; and (2) that the effect of the Earth's structure on the seismic waves is properly modeled. If the earthquake cannot be represented as a point source or the assumed model of structure is incorrect, the apparent moment tensor may contain a large non-double-couple component, even if the source mechanism is a double couple (Strelitz, 1978; Barker and Langston, 1982). Increasing the complexity of the source structure model to improve the fit of the Green functions to the seismograms, improving the azimuth coverage, and leaving the time function free to compensate for the deficiencies of the Green functions decrease the size of the non-double-couple component (Johnston and Langston, 1984).

Moment-tensor inversions have been routinely performed for several years by the U.S. Geological Survey. Centroid-moment-tensor solutions (simultaneous inversion of the waveform data for the hypocentral parameters of the best point source and for the six independent elements of the moment tensor) are regularly published by the Harvard University group for all larger earthquakes recorded at teleseismic distances.

The application of a moment-tensor inversion technique, however, to local events is a relatively recent innovation. Saikia and Herrmann (1985, 1986) used this technique for the interpretation of the observed body-wave amplitudes at local distances for two aftershocks of the 1982 Miramichi, Canada, earthquake and three 1982 Arkansas (U.S.A.) swarm earthquakes. Oncescu (1986) used a simple and efficient method for relative moment-tensor determination of 95 intermediate-depth small earthquakes from the Vrancea region, Romania, recorded by a local seismic network. A moment tensor inversion was performed by De Natale *et al.* (1987) for 10 small volcanic events from the Campi Flegrei in Italy. O'Connell and Johnson (1988) made moment-tensor inversions in the frequency domain for three microearthquakes from The Geysers geothermal field in California. Most recently, Ebel and Bonjer (1990) performed moment-tensor inversion of small earthquakes, with magnitudes from 0.5 to 2.2, in southwestern Germany and confirmed that direct *P*- and *S*-wave amplitudes can be inverted for the source focal mechanism. Koch (1991) examined two methods for moment-tensor inversion of waveform data for applicability to high-frequency near-source data. An algorithm was developed for near-source data, in which a stabilization procedure was introduced. Both methods, one in the time domain and one in the frequency domain, allow the retrieval of the complete time-dependent moment tensor. Ohtsu (1991) applied moment-tensor analysis to acoustic emission recorded during an in situ hydrofracturing test. He used moment-tensor components to classify crack types and to determine crack orientations.

A few works only have been published that are related to the use of moment-tensor inversion in studies of the source mechanism of seismic events induced by mining.

Sato and Fujii (1989) have studied the source mechanism of a large-scale gas outburst at Sunagawa coal in Japan, which occurred in January 1986. They used a new method to evaluate the moment tensor in the frequency domain and applied it to 15 seismic events recorded by the mine underground seismic network. Out of 15 studied tremors associated with the outburst, 12 seismic events could be interpreted in terms of a double-couple focal mechanism. In contrast to these results, the moment-tensor inversion performed on the observations from two small seismic events at Horonai coal mine in Japan has shown that they are non-double-couple events (Fujii and Sato, 1990). The tremors were associated with longwall mining and were located in the vicinity of the longwall face.

The Integrated Seismic System (ISS) recently introduced into the Welkom seismological network in South Africa (Mendecki, 1990; Mendecki et al., 1990) includes a software package that calculates all components of the moment tensor. Moment-tensor inversions for three seismic events with volume sources, which occurred within a dike in a mine in the Orange Free State mining district on April 9, 1991, provide good examples to illustrate the kind of information (elaborated by G. van Aswegen from the ISS International Ltd.) becoming available from moment-tensor analyses. The main event with moment magnitude $M = 2.1$ occurred at 07^h09^m, and two aftershocks with the same moment magnitude $M = 1.8$ occurred at 09^h24^m and 12^h27^m, respectively. The dike was vertical and striked EW (east–west). The reef being mined was planar and diped $16°$ easterly. The main event created intense but highly localized damage, and the dike "exploded." The routine moment-tensor inversion provided the following results (in a diagonalized form in a coordinate system defined by eigenvectors):

$$
\underset{\text{Moment tensor}}{\begin{bmatrix} 0.14 & 0.00 & 0.00 \\ 0.00 & -0.06 & 0.00 \\ 0.00 & 0.00 & 1.41 \end{bmatrix}} = 0.494 \underset{\text{Isotropic}}{\begin{bmatrix} 1 & 0 & 0 \\ 0 & 1 & 0 \\ 0 & 0 & 1 \end{bmatrix}} + 0.355 \underset{\text{CLVD}}{\begin{bmatrix} -1 & 0 & 0 \\ 0 & -1 & 0 \\ 0 & 0 & 2 \end{bmatrix}}
$$

$$
+ 0.202 \underset{\text{DC}}{\begin{bmatrix} 0 & 0 & 0 \\ 0 & -1 & 0 \\ 0 & 0 & 1 \end{bmatrix}}
$$

with the largest isotropic component corresponding to extension and the smallest double-couple (DC) component.

The two aftershocks occurred also in the dike. Damage during the first aftershock was more widespread than that during the main event, but it was less intense. The moment-tensor inversion showed more shear than volume change corresponding this time to contraction. The third event caused no

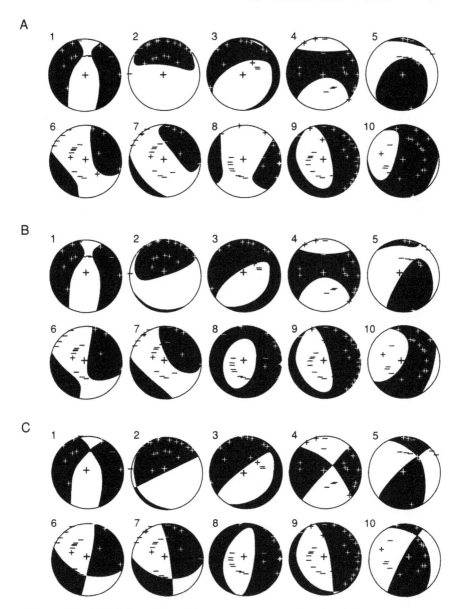

Figure 9.15 Nodal lines deduced from moment tensor inversion for 10 selected seismic events that occurred in 1991 at Rudna copper mine, Poland; a lower-hemisphere equal-area projection is used. Shaded areas represent the regions of upward motion for P waves (individual observations marked by $+$), and unfilled areas represent the regions of downward motion (observations marked by $-$). Three solutions are shown for each event: (A) for a general six-free-component moment tensor, (B) for a constrained moment tensor corresponding to sources with no volume changes, and (C) for a constrained moment tensor corresponding to double-couple sources. [From Wiejacz (1991, Fig. 40).]

apparent damage and was characterized by even larger double-couple component and smaller volume change corresponding to extension.

Wiejacz (1991) has studied the source mechanism of 60 small seismic events (in the seismic moment range 10^{11}–10^{12} N · m) at Rudna copper mine in the Lubin mining district in Poland, which occurred in 1990 and 1991. He performed the moment tensor inversion in the time domain using the first-motion amplitudes and signs of P and the amplitudes of SV waves recorded by the mine underground network composed of over 20 vertical seismometers. The time-independent solutions were obtained for a general six-free-component moment tensor, constrained solutions corresponding to sources without volume changes, and constrained solutions corresponding to double-couple sources. Examples of such solutions for 10 selected events are shown in Fig. 9.15. In general, the solutions that are well constrained by observations (good coverage of the focal sphere) have dominant shear components, although occasionally the isotropic component (showing both extension and contraction) could be as large as 25 percent of the mechanism. The CLVD component corresponds to uniaxial compression in all cases and is usually larger than the isotropic component.

9.6   Non-Double-Couple Seismic Events

Recent results from focal mechanism studies indicate growing evidence that alternative mechanisms other than shear failure are possible. In a number of cases of large shallow earthquakes, the deviation of moment tensor solutions from the double-couple mechanism may be considerable (e.g., Dziewonski and Woodhouse, 1983). There are several possible explanations for finite values of the deviation. One is the curvature of the plane of failure or an improper selection of the focal depth; another explanation could be the presence of subevents occurring on fault planes of different geometries and separated in time (e.g., Frohlich, 1990; Kawakatsu, 1991b). Waveform inversions at shorter wavelengths have also provided some indications of non-double-couple events. The 1980 Mammoth Lakes, California, earthquake sequence, believed to be associated with renewed magmatic activity at depth, is a recent example of the controversy over its mechanism since the waveform data could be satisfied by a number of alternative models, such as a double-couple with a complex rupture, lateral heterogeneity, and tensile failure (e.g., Julian and Sipkin, 1985; Wallace, 1985).

There is considerable similarity between volcanic seismicity and mine-induced seismicity in the sense that both types of seismic activity are responses to fast-changing stresses; in the former case, induced by magma movements and in the latter, by mining operations. In most cases volcanic events are well interpreted in terms of the double-couple mechanism (e.g.,

De Natale *et al.*, 1987). Even earthquake swarms associated with intrusive events at Kilauea volcano in Hawaii, have been found to be represented by strike-slip mechanisms, well correlated with inferred magma propagation directions and seismicity patterns (Karpin and Thurber, 1987). On the other hand, the focal mechanisms of microearthquakes associated with the cooling lava lake of Kilauea Iki, Hawaii, have been hypothesized to be of non-double-couple nature (Chouet, 1979).

The most prominent cases of what appear to be anomalous focal mechanisms are reported from mine seismicity studies. The results from moment-tensor inversions are described in the previous section. Here the results from studies mostly based on first-motion polarity and radiation patterns, reviewed by Gibowicz (1990b), are briefly described.

The results of first-motion data analyses from seismic events occurring close to the stope face in South African gold mines fall into two groups. In one group the dominance of dilatational first arrivals of *P* waves has been reported, suggesting the convergence of the surrounding rockmass on a volume of rock that failed suddenly near an excavation (Joughin and Jager, 1983). First-motion radiation patterns consistent with shear motion along a fracture plane have been reported in the other group (Potgieter and Roering, 1984; Spottiswoode, 1984).

First-motion analysis for seismic events in the North Staffordshire coal field in Great Britain shows that two different focal mechanisms exist: one of the shear type and the other implosional. Larger events have a shear-type mechanism and are believed to be generated by pillar failures, whereas smaller events have an implosion mechanism and are generated by waste collapses (Westbrook *et al.*, 1980; Kusznir *et al.*, 1980, 1984).

The results of a large seismic monitoring experiment undertaken during July–August 1984 in the eastern Wasatch Plateau, Utah, an area of active underground coal mining and intense microseismicity, were reported by Williams and Arabasz (1989) for the Eastern Mountain area and by Wong *et al.* (1989) for the Gentry Mountain area. Many seismic events in the Eastern Mountain area were located within 1 km below mine level, and some appear to extend down to at least 2 km below mine level. In the Gentry Mountain mining area the majority of seismicity was also located below mine level, and some events extended as deep as 3 km.

Several focal mechanisms have been determined for seismic events recorded with both compressional and dilatational first motion and located at or directly below the level of mine workings in the Eastern Mountain area (Williams and Arabasz, 1989). The focal mechanisms imply reverse faulting, as does the composite focal mechanism for three events located between the Eastern Mountain and Gentry Mountain areas (Wong *et al.*, 1989). An unexpected result of both these studies was the observation of numerous seismic events recorded with dilatational first motions at all stations. Williams and Arabasz (1989) have noted that the majority of the dilatational events can be fitted with a double-couple normal faulting mechanism if they occurred above mine level. In the Gentry Mountain area the vast majority of the largest events located by Wong *et al.* (1989) were characterized by a non-double-couple mechanism and were determined to be predominantly below mine level.

A combined shear-tensional and shear-implosional mechanism has been proposed by Teisseyre (1980), who introduced a qualitative source model with consecutive propagation of both dislocations to describe the non-double-couple behavior observed from mine tremors. From the new earthquake rebound theory proposed by Teisseyre (1985a, 1985b), describing the relation between creep processes and rapid energy release, it follows that shear and extensional (implosional) processes take place in time one after the other or are originated simultaneously but are shifted in space. For small events, such as mine tremors, a model of simultaneously radiating shear and implosive displacements can be accepted for radiation pattern computations (Rudajev and Šileny, 1985; Rudajev *et al.*, 1986). A shear-tensional mechanism is well simulated by simple laboratory experiments on frictional sliding under uniaxial compression with a precut slit, acting as a stress concentrator (e.g., Nemat–Nasser and Horii, 1982).

The radiation pattern of *P* waves was computed for a focal model with shear and implosive components and applied to the observations from seismic events in the Kladno coal district, Czechoslovakia (Rudajev and Šileny, 1985; Rudajev *et al.*, 1986) and in Upper Silesia, Poland (Rudajev *et al.*, 1986). The first-motion *P*-wave amplitudes were calculated and compared with those observed from tremors in the Kladno district (Rudajev *et al.*, 1986; Šileny, 1986), and the calculated maximum *S*-wave amplitudes were compared with those observed from tremors in the Karvina coal mining area, Czechoslovakia (Šileny, 1989). In the Kladno district the combined focal model has provided an adequate description of the focal mechanism for about one third of the considered tremors. The implosive component was rather small, in general not exceeding 10 percent of the shear component value. In the Karvina area the combined model seems to be adequate for about half of the selected tremors, but only *S*-wave amplitudes from five stations were used. Thus these results, like most of the others, are rather inconclusive because of the highly limited available data.

One more seismic event with a possible non-double-couple source has been reported recently from Alabama (Long and Copeland, 1989). The event occurred on May 7, 1986 and was felt in Tuscaloosa. At the same time a roof collapse occurred in a coal mine some 10–15 km from Tuscaloosa, at a depth of 610 m. The event was recorded by seismic stations at regional distances only, and the data were not sufficient to confirm its location in the mine. The stations recorded only dilatational first motions, but the focal mechanism is weakly constrained allowing either an implosional mechanism or reverse or strike-slipping faulting (Long and Copeland, 1989).

A recent study of the source parameters of seismic events at the Heinrich Robert coal mine in the Ruhr basin, Germany, provides more evidence for non-double-couple events (Gibowicz *et al.*, 1990). It has been found that the ratio of *S*- to *P*-wave energy ranges from 1.5 to 30 for selected tremors occurring in a cluster. The high *P*-wave energy and low apparent stress events are thought to be the most likely candidates for non-double-couple events. A similar result was also found from a study of the source parameters of seismic events induced by the excavation of a shaft in granite at the Underground Research Laboratory in Manitoba, Canada (Gibowicz *et al.*, 1991). The observed high *P*-wave energy events were most probably the events with non-double-couple focal mechanisms, implying that tensile failures, or at least shear failures with tensile components, were generated by the excavation of the shaft.

Chapter 10 | Seismic Source Modeling

The body-force equivalent used to describe the point seismic source, considered in Chapter 9, is a formal concept and it is necessary to relate its characteristics to some physical concepts of the real earthquake source. The elastic rebound theory of earthquakes, that earthquakes are the result of fracture of the Earth's material caused by tectonic stresses, formulated by Reid (1911), is such a concept, which led to the dominance of the faulting theory of earthquakes. Only after a long delay, the fact that the vast majority of shallow tectonic earthquakes arise from faulting instabilities was proved by seismological observations. The modern era of earthquake source studies began in the early 1960s and it was only then that dynamic faulting was widely accepted as the origin of the majority of seismic events. Fracture mechanics introduced concepts and methods to consider and analyze the phenomena of fracture nucleation, propagation, and arrest at the seismic source.

In this chapter several topics, such as recent developments in fracture mechanics, dislocation and crack models of seismic sources, complex source models, earthquake sequences, and faults and fractals, are briefly presented following to some extent a review of Gibowicz (1986). For a more detailed treatment see, for example, Aki and Richards (1980), Kasahara (1981), Das et al. (1986), Tullis (1986), Atkinson (1987a), Kostrov and Das (1988), Scholz and Mandelbrot (1989), and Scholz (1990).

10.1 Recent Developments in Fracture Mechanics

Laboratory studies of fracture and frictional behavior of rocks provide the physical basis for understanding the phenomena of earthquake rupture. Following the suggestion of Brace and Byerlee (1966), that stick-slip occurring in laboratory sliding experiments may be analogous to the mechanism of crustal earthquakes, many other laboratory observations have been correlated with various characteristics of earthquakes. It is now generally accepted that crustal earthquakes are caused by a sudden drop in shear stress accompanied by unstable slip on a fault.

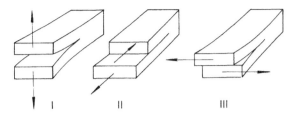

Figure 10.1 The three crack propagation modes· (I) tensile opening: (II) in-plane shear, (III) antiplane shear.

The mechanism of rupture in unstable slip can be described either as a brittle fracture or as a stick-slip friction instability. These two approaches are mathematically equivalent in relating motion in the medium to a drop in shear stress on the fault surface, but the rupture process is considered differently (e.g., Scholz, 1990). In fracture mechanics it is assumed that a characteristic fracture energy per unit area, which is a material property, is required for the propagation of the crack, whereas in the stick-slip model rupture is assumed to occur when the stress on the fault reaches the value of static friction and the condition for dynamic instability exists.

By an elastic–brittle-crack model, a model is understood in which material outside the crack remains ideally elastic and in which there is an abrupt drop in stress on the plane of the crack. For an ideal flat, perfectly sharp crack of zero thickness, there are three basic modes of crack tip displacement (Fig. 10.1). Mode I is the tensile, or opening, mode in which the crack wall displacements are normal to the crack. Mode II is the in-plane shear mode in which the displacements are in the plane of the crack and normal to the crack edge, and mode III is the antiplane shear mode in which the displacements are also in the plane of the crack but parallel to the edge.

In general, the analysis of shear and tensile cracks is not much different, except that, unlike tensile cracks, shear crack faces are not stress-free and that the fracture energy release rate in shear cracks is often two orders of magnitude higher than that for tensile cracks, at least in laboratory experiments. Two well known approaches are used in elastic–brittle-crack studies. In the first approach the stress intensity factor K_n ($n = $ I, II, III denoting the three modes of deformation), introduced by Irwin (1960), is used to characterize the intensity of the crack-tip stress field. It is supposed that when K_n reaches a critical value K_{nc}, crack extension occurs. This fracture criterion $K_n = K_{nc}$ represents a balance of the crack driving stress intensity with a critical stress intensity that the material can sustain. The critical stress intensity factor K_{nc} is also sometimes called the *fracture toughness*. The second approach is an extension of the work of Griffith (1920) by characterizing fracture as a balance between the available energy G to drive the crack and the energy G_c absorbed by the inelastic breakdown processes of the material at the crack tip; G_c is also called the *fracture energy*. It can be shown that the fracture criterion $G = G_c$ is consistent with the criterion $K_n = K_{nc}$ in a linear elastic body. This fracture characterization by a single parameter is similar to the classical concept of strength that relates the shear stress to the shear strength at failure of a specimen undergoing uniform deformation (e.g., Li, 1987). The strength concept, however, cannot in general characterize the crack driving force and fails to predict the load level that the material with flaws can sustain. The strength concept and brittle elastic crack mechanics may be regarded as two opposite limiting conditions of a more general slip-weakening model described later.

If the crack is assumed to be planar and perfectly sharp, then the near-field approximations to the crack-tip stress and displacement are given by completely general expressions (Lawn and Wilshaw, 1975)

$$\sigma_{ij} = K_n (2\pi r)^{-1/2} f_{ij}(\theta) \tag{10.1}$$

and

$$u_i = (K_n / 2E)(r/2\pi)^{1/2} f_i(\theta), \tag{10.2}$$

where r is the distance from the crack tip, θ is the angle measured from the crack plane, E is Young's modulus, and $f_{ij}(\theta)$ and $f_i(\theta)$ are well-defined functions of θ depending on the loading mode, which can be found in Lawn and Wilshaw (1975). Equation (10.1) indicates that there is a stress singularity at the crack tip, resulting from the assumption of perfect sharpness of the slit. There must be a region of nonlinear deformation near the crack tip that relaxes this singularity, and various models have been proposed to describe this nonlinear area (e.g., Dugdale, 1960; Barenblatt, 1962).

The stress intensity factors depend on the geometry and magnitudes of the applied loads. They are tabulated in Tada *et al.* (1973). For a two-dimensional crack in any mode the stress intensity factor is given by

$$K = C\sigma_r(\pi c)^{1/2}, \tag{10.3}$$

where σ_r is the remote applied stress; C is a numerical factor to account for crack geometry, loading conditions, and edge effects; and c is half the crack length for penny-shaped internal cracks.

An alternative approach to crack extension is to consider the strain energy release rate G, or crack extension force. The energy release rate (with respect to crack length and not with respect to time) is the loss of energy per unit of new crack separation area formed during an increment of crack extension. For plane strain and for fracture in each of the three crack fundamental modes it is given by

$$G_I = K_I^2(1 - \nu^2)/E, \qquad (10.4)$$

$$G_{II} = K_{II}^2(1 - \nu^2)/E, \qquad (10.5)$$

$$G_{III} = K_{III}^2/2\mu = K_{III}^2(1 + \nu)/E, \qquad (10.6)$$

where ν is Poisson's ratio and μ is the shear modulus. For plane stress the factor $(1 - \nu^2)$ in equations (10.4) and (10.5) is replaced by unity. The energy release rates for different crack modes are additive. Since the near-tip stress has the universal form described by equation (10.1) for cracks in a linear elastic solid, the energy release rate G can be expressed in terms of the stress intensity factors as

$$G = \left(K_I^2 + K_{II}^2\right)\left[(1 - \nu^2)/E\right] + K_{III}^2(1 + \nu)/E \qquad (10.7)$$

for plane strain, or

$$G = \left(K_I^2 + K_{II}^2\right)/E + K_{III}^2(1 + \nu)/E \qquad (10.8)$$

for plane stress (e.g., Atkinson, 1987b).

Thus the condition for crack propagation is

$$G_c = K_c^2/E = 2\gamma \qquad (10.9)$$

for plane stress, with a similar expression for plane strain. To account for additional contributions to the crack extension force, resulting from distributed cracking, plastic flow, and other dissipative processes within the nonlinear zone at the crack tip, equation (10.9) can be rewritten as (e.g., Scholz, 1990)

$$G_c = 2\Gamma, \qquad (10.10)$$

where Γ is a lumped parameter that includes all dissipation within the crack-tip area.

The nonphysical effect of a stress singularity at the crack tip may be avoided by assuming that the crack breakdown occurs over some finite distance, which has the effect of smearing out the stress drop and its associated stress concentration. In brittle fracture this may correspond to the development of the process region and in friction approach to the breakdown between the static and dynamic friction values, which requires a critical slip distance (e.g., Scholz, 1990). Laboratory observations from triaxial experiments in rocks indicate a complex breakdown process in the localized shear band in the postpeak stage. On a larger scale, direct shear testing of rock joints indicates shearing off and crushing of asperities in jointed rockmass. Both triaxial rock specimens and direct-shear jointed rock specimens in the laboratory show a decreasing shear load-carrying capacity as a function of the amount of sliding (e.g., Li, 1987).

The conceptual model of shear fracture is a strength degradation process in which the frictional resistance is overcome to initiate slip, with the strength decaying to a reduced level. The simplest model for such a process is the slip-weakening model in which the shear strength along the fault is a decreasing function of the slip. The model was motivated by the well-known "cohesive zone" models of tensile fracture developed by Dugdale (1960) and Barenblatt (1962) for metal, and it was adapted to shear faulting by Ida (1972) and Palmer and Rice (1973).

A simple general form of the slip-weakening constitutive model may be written as (Li, 1987)

$$\tau = f(\delta, \bar{\sigma}_n, T), \tag{10.11}$$

where $\bar{\sigma}_n \equiv \sigma_n - p$ is the effective normal compressive stress, that is, normal stress σ_n reduced by pore pressure p, acting across the slip surface and T is temperature. The model is shown in Fig. 10.2, reproduced from Rice (1983), and in Fig. 10.3, reproduced from Wong (1986). In the plot stress τ against slip δ (Fig. 2A), strength degrades in the slip δ^* from a peak resistance τ^p to initiate slip down to a fixed residual friction level τ^t, sustained for larger amounts of slip. The plot is for constant normal stress σ_n and pore pressure p. If they are changed then τ^p and τ^f are also altered as in Fig. 10.2B, where they are considered to depend on the effective normal stress.

Triaxial tests in rocks suggest that both τ^p and τ^t increase in such a manner that the stress drop $\tau^p - \tau^f$ first increases and then decreases with increasing σ_n, indicating a transition from brittle to ductile deformation (Wong, 1986). Laboratory experiments at constant normal stress also show that $\tau^p - \tau^f$ decreases with temperature (e.g., Li, 1987). These considerations of normal stress and temperature dependence are important when the

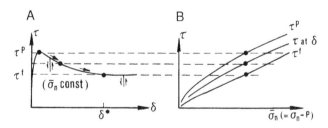

Figure 10.2 (A) Slip-weakening stress versus slip relations for constant effective normal stress, (B) dependence of peak, residual, and intermediate strengths on effective normal stress. [From Rice (1983, Fig. 2)]

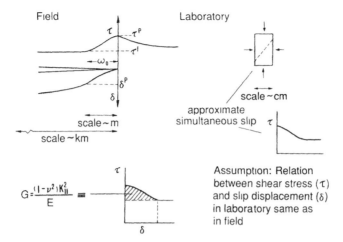

Figure 10.3 Slip-weakening model as applied in the field and laboratory. The distribution of shear stress and slip with respect to a fault in the field is shown [From Wong (1986, Fig. 1).]

slip-weakening constitutive law is applied to the Earth's crust; such applications can be found in a number of works (Stuart and Mavko, 1979; Li and Rice, 1983; Li and Fares, 1986; Li, 1987).

The slip-weakening constitutive model describes a single fault slip sequence and is not appropriate to characterize repeated seismic slip along a given fault because no provision is made for restrengthening (Dieterich, 1979a, 1979b; Rice, 1983). Thus, seismicity would be expected to cease or to be replaced by quasistatic fault slip to accommodate fault motion. Starting with laboratory observations of time- and slip rate-dependent rock frictional sliding along simulated fault surfaces, rate- and state-dependent fault constitutive relations, or state variable friction laws, have been developed, and it has been shown that they predict a wide range of frictional sliding behavior and describe laboratory observations of quasistatic sliding (Dieterich, 1978, 1979a, 1979b; Ruina, 1983; Gu *et al.*, 1984; Tullis and Weeks, 1986).

A frictional constitutive law is a relation between frictional resistance (shear stress) and the various factors that may affect it. The simplest is Amonton's law, which states that the frictional shear stress is proportional to the normal stress, the constant of proportionality being the coefficient of friction. A more complex description is needed to take into account the common observation that the coefficient of sliding friction is smaller than the coefficient of static friction.

Rate- and state-dependent rock friction models were developed empirically to describe laboratory experiments for stable sliding along simulated fault surfaces. The slip rate dependence of the coefficient of friction is of

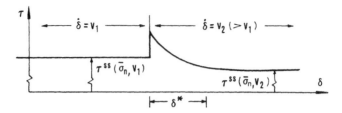

Figure 10.4   Rate- and state-dependent response to sudden increase of slip rate at constant normal stress. [From Rich (1983, Fig. 3).]

particular importance to models of the mechanism of instability (Dieterich, 1979a; Ruina, 1983). The essential features of the state variable rock friction model are (Rice, 1983; Okubo and Dieterich, 1986): direct velocity effect (when a sudden change in slip rate is introduced, the frictional resistance changes in the same sense), steady-state effect (for sliding at constant slip rate and under constant normal load, friction evolves toward a steady-state value dependent on that slip rate), and characteristic slip distance (the evolution of friction toward the steady-state value at a given slip rate is controlled by an exponential decay over a characteristic amount of slip δ^*). These features are shown in Fig. 10.4, reproduced from Rice (1983), where a schematic record of stress τ at the slip surface against displacement δ is presented for slip at constant normal stress, just before and after a suddenly imposed change in slip rate.

Two specific constitutive friction models are widely used to describe stress changes in laboratory experiments. The formulation for the coefficient k of friction proposed by Dieterich (1979a), which can be written in the form

$$k = k_0 + A \log(Bt + 1),   (10.12)$$

where k_0, A, and B are constants and t is the time of contact of the saw-cut surfaces, accounts for the direct velocity effect and the steady-state effect. The evolution of the state, or the slip history, is contained in the formulation for a single state variable B, determined empirically, and related to an effective time of contact between asperities. The state variables may also be identified phenomenologically as parameters describing experimental results. This approach was taken by Ruina (1983), who elaborated on the interpretation of state variables and proposed a simpler form for the coefficient of friction.

A state variable constitutive law employs one or more internal variables that characterize the state of the sliding surface and together with the sliding velocity determine the frictional resistance. The state variable constitutive law in the form given by Rice and Gu (1983), closely following Ruina (1983), in terms of sliding velocity V and state variables ψ_i, can be expressed as

$$\tau = \sigma_n \left[k_0 + b_1\psi_1 + \cdots + b_t\psi_t + a \ln(V/V_0) \right], \tag{10.13a}$$

$$d\psi_1/dt = -\left[(V/L_1)(\psi_1 + \ln(V/V_0)), \right.$$

$$\vdots$$

$$d\psi_t/dt = -\left[(V/L_t)(\psi_t + \ln(V/V_0)), \right. \tag{10.13b}$$

where τ is shear stress, σ_n is normal stress, k_0 is a constant that may be thought of as the Amonton's friction, b_t and L_t are the empirically determined constants and critical slip distances, a is an empirical constant, and V_0 is an arbitrary reference velocity.

For the low slip rates, characteristic for the stable sliding experiments, the reduction of Dieterich's friction law to Ruina's law is straightforward; but these two formulations would predict different behavior at high slip rates, which are achieved during stick-slip processes. For stick-slip frictional instabilities Okubo and Dieterich (1984) have noted the occurrence of slip-weakening-like behavior at the onset of stick-slip sliding and that this behavior is similar to that observed for stable sliding when slip rate along the simulated fault is suddenly increased. Thus it might be expected that similar constitutive relations could be used for both stable and stick-slip sliding. This possibility was explored by Okubo and Dieterich (1986), who applied model calculations based on the two formulations of the state variable friction law to stick-slip data. They found that the stick-slip observations can be fitted by model calculations using the frictional model proposed by Dieterich (1979a) as long as high-speed cutoffs to the velocity related effects are included in the model.

There is some uncertainty as to whether the constitutive description developed from the laboratory results by Dieterich and Ruina is also applicable at the normal stresses relevant to crustal earthquakes. The steady-state frictional resistance proportional to the negative logarithm of the sliding velocity is called *velocity weakening* (shown in Fig. 10.4). It is important to evaluate what experimental conditions may cause negative slip velocity dependence of the shear stress or friction coefficient. The experiments conducted by Tullis and Weeks (1986) to define the constitutive behavior for frictional sliding of granite at room temperature and normal stresses up to 84 MPa consistently found velocity weakening at normal stresses from 20 to 84 MPa and other constitutive parameters similar to those reported by Dieterich and Ruina at normal stresses from 2 to 10 MPa.

Shimamoto and Logan (1984) have argued that empirical laws, fitting laboratory friction data considerably well, predict entirely different results when extrapolated to slow natural deformations. Long-term experiments, continuing upward of a few years, would be necessary to test the validity of friction laws. The solution to this problem is indicated by the laboratory experiments on simulated fault gouge of halite (Shimamoto and Logan, 1986). Halite is selected as an analog for natural fault gouge material because it provides a wide range of mechanical behavior under accessible laboratory conditions. Shimamoto and Logan (1986) have conducted triaxial experiments on dry specimens of Tennessee sandstone with 0.3–1.0-mm-thick layer of halite along a 35° precut, with slip rates along the precut ranging from 300 down to 0.003 μm/s (about 10 cm/year) and confining pressure from 10 to 250 MPa, corresponding to the behavior from dominantly brittle to clearly ductile. They found that a negative velocity dependence, present at low to moderate normal stress, changes to positive dependence at low velocities and that this negative-velocity dependence also changes with increasing velocity to a region of no dependence and subsequently to one of positive dependence.

The described rate and state dependences of the stress constitute an extremely small part of the total stress required to slip a fault, at least for variations in velocity by factors of 10^3 or less (Rice, 1983). These dependences can be regarded as modest variations of a classically described critical stress for slip. Although the rate- and state-dependent constitutive framework is much more comprehensive than the rate-independent slip-weakening concepts, it is much more complex and relatively little progress has been made toward the understanding of instability with realistic fault models. Application of experimental results to natural faults requires theoretical extension to a system in which the amount of slip varies along the sliding surface by elastic accommodation of the surroundings. Theoretical analysis of this type is reported by Tse and Rice (1986).

10.2 Dislocation and Crack Source Models

It is generally assumed that crustal earthquake ground motion results from unstable slip accompanying a sudden drop in shear stress on a geologic fault. Even seismic events induced by mining, at least the larger of them, display a double-couple radiation pattern, implying that shearing processes are at the roots of their generation. Seismic radiation is the main source of information on the faulting processes of earthquake phenomena. The radiation can be computed by a space–time convolution of a slip function with a Green function. The slip function describes the fault displacement during an earthquake as a function of time and position on the fault plane, while Green's function represents the Earth's response to the slip. The slip function and Green's function therefore express quantitatively the source and propagation effects on seismic motion.

Thus, from seismic radiation together with field observations, the rupture history of a specific event could be reconstructed, and in an ideal case seismic radiation would be used to perform a tomographic study of the fault (e.g., Ruff, 1984, 1987; Frankel *et al.*, 1986; Beroza, 1991). This task, however, is very difficult because of the limited azimuthal coverage by seismic stations and limited frequency band of available instruments. As a result, highly idealized models of rupture with a highly restricted number of source parameters are in general use.

Seismic radiation can be calculated using either dislocation or a crack approach. A dislocation is considered to be a defect in an ideally elastic or viscoelastic medium formed by a cut along a given surface and a finite relative displacement of the two faces of the cut, which means that the dislocation is represented by a discontinuity in the displacement (e.g., Rybicki, 1986). Dislocations for which the relative displacement is an arbitrarily defined function of the position vector on the dislocation surface are called *Somigliana dislocations*. Their particular case for which the relative displacement is of the rigid-body motion type is called *Volterra dislocation*. Thus, in the dislocation approach the earthquake is represented in terms of the slip function on the fault plane, and its form is generally chosen intuitively, without rigorous analysis of the time-dependent stress acting in the area. Frequently used models of this type are the propagating dislocation model of Haskell (1964) and the model of Brune (1970, 1971). The last one, although assuming an infinite rupture velocity, is rationalized in terms of the dynamic properties of the source.

In contrast to dislocation models, in crack models an explicit account of the driving and resisting stresses in the source region is taken, and the resulting slip is derived by solving the equations of motion. Thus, to describe the fracture at an earthquake source as a crack, it is necessary to know the initial distribution of stress on the fracture surface before the earthquake and the laws governing the fracture propagation and interaction of the fault faces. The distribution of the displacement on the fault becomes then one of the unknowns. When fracture is described as a dislocation, the model is called *kinematic*, and when it is described as a crack, the model is called *dynamic* (e.g., Kostrov and Das, 1988). Although the dynamic crack modeling is physically more proper, the kinematic dislocation models are much simpler and involve more efficient computation, preserving at the same time enough degree of freedom to represent some realistic models.

The various dislocation solutions can be classified according to the model dimensions (two- and three-dimensional models), the characteristics of Earth structure (uniform and layered models), the various methods of solution (Cagniard-de Hoop, synthesis of Green's functions in the time or frequency domains, discrete wavenumber representation), and according to the range of

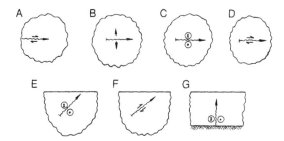

Figure 10.5 Two-dimensional dislocation models: (A) uniformly moving line shear dislocation in an unbounded medium with slip perpendicular to the rupture front; (B) suddenly starting
extensional line dislocation propagating with constant velocity in an unbounded medium; (C)
suddenly starting line shear dislocation with slip parallel to the rupture front; (D) line shear
dislocation with slip normal to the rupture front; (E) infinitely long strike-slip fault in a uniform
elastic half-space; (F) infinitely long thrust fault in an elastic half-space; (G) vertical strike–slip
fault on a layered elastic slab supported on a rigid half-space. [From Luco and Anderson (1983,
Fig. 1).]

considered rupture velocities (sub-, trans-, and supercritical) and the type of
slip function used. A review of dislocation models, including extensive
references, has been given by Luco and Anderson (1983). Figure 10.5,
showing various types of two-dimensional dislocation models, and Fig. 10.6,
presenting three-dimensional models, are reproduced from their paper.

As an example, the simplest slip functions used for source modeling are
briefly introduced here. Considering a moving dislocation model with the
geometry shown in Fig. 10.7, the displacement $u(t)$ on the fault surface can
be expressed in the simplest form as

$$u(t) = u_0 H(t - x/V_r), \qquad (10.14)$$

where u_0 is the final displacement, $H(t)$ is the Heaviside unit step function,
and V_r is the rupture velocity. This is the propagating Heaviside dislocation
model proposed by Knopoff and Gilbert (1959). Haskell (1964) replaced
relation (10.14) by the expression

$$u(t) = u_0 G(t - x/V_r), \qquad (10.15)$$

to construct a ramp dislocation model. Here $G(t)$ represents a ramp function
that is zero at $t < 0$ and increases linearly with time until it reaches 1 at
$t = T$, which is called the *rise time* or the *characteristic time*. The function
describes a process in which slips occur progressively along a fault with the
velocity V_r, resulting in a uniform slip over the whole fault surface. The
source time functions of these two models are shown in Fig. 10.8, reproduced
from Kasahara (1981).

A B C

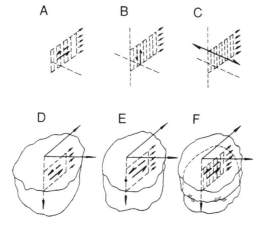

D E F

Figure 10.6 Three-dimensional dislocation models: (A) uniform shear dislocation moving with constant rupture over a rectangular fault buried in an unbounded medium with slip normal to the rupture front; (B) shear dislocation with slip parallel to the rupture front; (C) tensile dislocation; (D), (E) rectangular faults embedded in a uniform elastic half-space; (F) rectangular vertical fault in a layered half-space. [From Luco and Anderson (1983, Fig. 2).]

On the basis of the Haskell model, Savage (1972) constructed the seismic far-field radiation solution in terms of particle velocity

$$\dot{\mathbf{u}} = R_\alpha(\theta, \phi, R)\mu h I_\alpha + R_\beta(\theta, \phi, R)\mu h I_\beta \qquad (10.16)$$

with

$$I_c = D \int_0^L \left[\frac{d^2}{dt^2} G\left(t - \frac{R}{c} - \frac{\xi}{c} - \left(\frac{c}{V_r} \cos\theta \right) \right) \right] d\xi, \qquad (10.17)$$

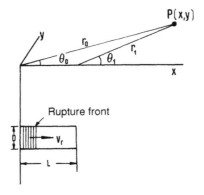

Figure 10.7 Geometry of a moving dislocation source model with a receiver at point P.

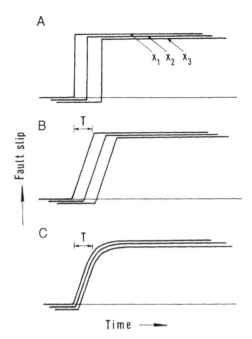

Figure 10.8   Source time functions in three source models due to (A) Knopoff and Gilbert (1959); (B) Haskell (1964), and (C) Brune (1970). [From Kasahara (1981, Fig. 5.12).]

where θ, ϕ, and R are the spherical polar coordinates; ξ is the x coordinate of the point of integration in the fault; R_c is the radiation pattern from an appropriate double-couple source; h is the fault width; L is the fault length; D is the offset; and c stands for either α or β. The first term in relation (10.16) represents the P-wave radiation, and the second term represents the S-wave radiation. Thus equations (10.16) and (10.17) describe the seismic radiation generated by the moving rupture. Integration of formula (10.17) and then a relation (10.16) provides the far-field displacement $\mathbf{u}_c$ for each wave as

$$\mathbf{u}_c = R_c(\theta, \phi, R)(M_0/\tau_0)\{G[t - (R/c)] - G[t - (R/c) - \tau_0]\}, \quad (10.18)$$

where $M_0 = \mu hLD$ is the seismic moment and τ_0 is the width of the pulse.

Another popular source model is the model of Brune (1970). It is assumed that a tangential stress step is applied to the interior of a dislocation surface resulting in the fault blocks movement in the opposite directions. For simplicity, the fault propagation effects are neglected and the stress step is assumed to apply instantaneously over the fault surface. It is also assumed that the fault surface reflects elastic waves during rupture. The stress step sends a pure shear stress wave propagating perpendicularly to the fault surface and the initial time function of the pulse is given by

$$\sigma_0(y,t) = \sigma_{\text{eff}} H(t - y/\beta),\qquad(10.19)$$

where σ_{eff} is the effective stress and β is the shear wave velocity. If $\mu(\partial u/\partial y) = -\sigma_0(y,t)$ and $u = 0$ for $t = y/\beta$, where μ is the rigidly of the elastic medium surrounding the fault, then the displacement is

$$u = (\sigma_{\text{eff}}/\mu)\beta(t - y/\beta)H(t - y/\beta)\qquad(10.20)$$

and its time derivative is

$$\dot{u} = (\sigma_{\text{eff}}/\mu)\beta H(t - y/\beta).\qquad(10.21)$$

The initial particle velocity at $y = 0$ and $t = 0$ is given by

$$\dot{u}_0 = (\sigma_{\text{eff}}/\mu)\beta.\qquad(10.22)$$

Thus, at a point near the fault displacement increases linearly with time as the stress pulse propagates away from the fault, and then it levels off as the finiteness of the fault is felt at the observation point. The initial particle velocity begins to decrease to zero when the effects of fault finiteness reach the observation point. Brune (1970, 1971) introduced a time constant τ, equivalent to the travel time of the signal $\tau \propto r_0/\beta$, where r_0 is the radius of a circular dislocation representing the fault surface, and replaced relations (10.20) and (10.21) by the following expressions

$$u(y = 0, t) = (\sigma_{\text{eff}}/\mu)\beta\tau(1 - e^{-t/\tau}),\qquad(10.23)$$

and

$$\dot{u}(y = 0, t) = (\sigma_{\text{eff}}/\mu)\beta e^{-t/\tau}.\qquad(10.24)$$

The source time function of Brune's model is also shown in Fig. 10.8.

If the slip on the fault is calculated from the stress drop and strength of the fault, the dislocation model is identical to a crack model. The solution to static slip of a circular shear crack is very well known (e.g., Madariaga, 1983b). Assuming that the crack is loaded by an initial stress σ_{xz}^0 and that the stress drop

$$\Delta\sigma = \sigma_{xz}^0 - \sigma_{xz}^f,\qquad(10.25)$$

where σ_{xz}^f is a final stress on the fault, is uniform on the fault surface, the slip is given by

$$\Delta u_x(r) = \frac{24}{7\pi}\frac{\Delta\sigma}{\mu}(r_0^2 - r^2)^{1/2},\qquad(10.26)$$

where r is radial distance from the center of the crack on the (x, y) plane. This slip is quite different from the uniform slip assumed in kinematic models.

Construction of a dynamic model of faulting requires a description of the dynamic stress drop, the rupture velocity and the rupture complexity over the rupture region. These dynamic characteristics may be considered as high-frequency analogs to the more familiar static characteristics of the slip and static stress drop. The spectral model proposed by Boatwright (1982) illustrates this analogy between two descriptions of the faulting process. The low-frequency level of the displacement spectrum is related to the product of the slip and the rupture area (see Chapter 11), whereas the high-frequency level of the acceleration spectrum is related to the product of the dynamic stress drop, the peak rupture velocity, and the square root of the rupture area. The different dependence on the rupture area is a result of the high-frequency acceleration being incoherent, whereas the low-frequency displacements are coherent. The high- or low-frequency seismic radiation can be calculated from the dynamic or static description of the fault. In this sense the models should be considered as frequency-dependent descriptions of faulting.

Analytic solutions of crack propagation for any given initial and boundary conditions is extremely difficult, and the existing solutions involve several simplifying assumptions. For this reason only a few simple geometric models have been solved analytically; examples are the elliptical self-similar crack studied by Kostrov (1964), Burridge and Willis (1969), and Richards (1976), and the antiplane dynamic shear crack studied by Kostrov (1966). Practically all the other available solutions have been obtained with the use of numerical methods.

Although there are no simple analytic solutions equivalent to solution (10.26) for a dynamic circular shear crack of finite radius, a highly useful analytic solution is known for a self-similar circular shear crack. This is a crack that starts from a point and then grows symmetrically with a constant rupture velocity and never stops. The slip is driven by a stress drop and the solution of this problem is very simple (Kostrov, 1964). The slip is everywhere parallel to the x axis ($\Delta u_y = 0$) and is given by

$$\Delta u_x(r, t) = \frac{\Delta\sigma}{\mu} C(V_r)\left(V_r^2 t^2 - r^2\right)^{1/2}, \qquad (10.27)$$

where $C = C(V_r)$ is an almost constant function of the rupture velocity V_r and is close to 1 for the whole subsonic range of this velocity.

The next step in approximation of realistic seismic sources is to assume that the self-similar shear crack stops suddenly once a final radius r_0 is reached. The solution of this problem was obtained numerically by Madariaga

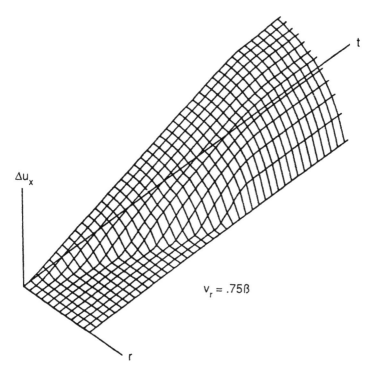

Figure 10.9 Three-dimensional plot of the source slip function $\Delta u_x(x, t)$ for a circular fault with the rupture velocity $V_r = 0.75\beta$. This is a composite plot of the slip history as a function of the radius r on the fault. [From Madariaga (1976, Fig. 10).]

(1976), and an example of one of his numerical simulations is shown in Fig. 10.9. Initially, until the sudden stop of the rupture at the final radius, the slip function is a numerical approximation to equation (10.27). The stopping of rupture generates very strong P, S, and Rayleigh waves that propagate inward from the edge of the fault. After the arrival of a Rayleigh wave the slip velocity is reduced to zero. Madariaga (1976) called this process "healing." At that time it is assumed that friction is sufficiently high to freeze the slip at that point on the fault. The slip functions shown in Fig. 10.9 are very different from the simple ramplike functions used in dislocation models. Slip here is clearly a function of position on the fault plane.

Around the edges of a circular crack there is a mixed mode of rupture with a variable stress intensity. As a consequence, the apparent simplicity of the solutions should disappear as soon as a rupture criterion is introduced. The crack will rapidly deform into more or less elliptical shapes (e.g., Madariaga, 1983b). The only possible way to study spontaneous rupture in three dimensions is by numerical modeling. It should be noted, however, that there are considerable differences between analytic and numerical solutions (Dmowska and Rice, 1986). The analytic solutions are usually obtained with the use of Griffith's theory and for a single crack development describing the fracture phenomenon. In numerical solutions cracks are usually smaller in comparison to grid size than might be desirable. All quantities are discretized, and some discrepancies between various solutions are associated with discretization. Stress concentrations at crack tips are finite, and the numerical fracture criteria of different type correspond only approximately to those used in fracture mechanics.

In numerical modeling three calculation techniques are in general use. These are finite-element (e.g., Archuleta and Day, 1980), finite-difference (e.g., Virieux and Madariaga, 1982), and boundary integral equations (e.g., Das, 1980, 1981) methods. The boundary integral method seems to be more accurate than the other two, since it does not suffer from dispersion of wavelengths on the order of the element size, which is the case with the finite-element and finite-difference methods (Andrews, 1985).

A simpler class of numerical solutions is that for which crack motion is specified a priori rather than being derived from a failure criterion. These fixed rupture velocity fault models have been studied for faulting represented by a circular area (Madariaga, 1976; Das, 1980), a semicircular area (Archuleta and Frazier, 1978), and rectangular regions (Madariaga, 1977; Archuleta and Day, 1980; Day, 1982a). These numerical solutions satisfactorily quantify some important three-dimensional geometric effects such as the influence of fault width on the slip function, controlling static slip and slip rise time (Day, 1982a, 1982b).

Changes in rupture velocity of a propagating fault are the predominant source of high-frequency radiation (Madariaga, 1977, 1983a). To understand an unsteady rupture propagation requires a spontaneous rupture dynamic model. That is, a definite fracture criterion must be imposed for crack advance and crack motion is not prescribed a priori. A boundary integral method was developed by Hamano (1974) and Das and Aki (1977) for two-dimensional problems, who prescribed the critical stress level of material along the fault plane. Such a procedure can be considered as an attempt to simulate numerically the critical stress intensity factor criterion from fracture mechanics.

Andrews (1976) combined a finite difference method with the slip-weakening fracture criterion to study the rupture propagation of a finite two-dimensional shear crack in an unbounded medium. For the case of in-plane shear crack, he found that the terminal rupture velocity could be smaller than the Rayleigh velocity or higher than the shear-wave velocity. This was confirmed by three-dimensional finite-difference implementation of the slip-weakening criterion (Day, 1982b). More recently Andrews (1985) adapted the boundary integral method to a slip-dependent friction law on the crack plane. He found that spontaneous plane–strain shear ruptures can make a transition from sub-Rayleigh to near-P-wave propagation velocity. Results from the boundary integral method agree with his earlier results from a finite-difference method on the location of this transition in space but the methods differ in their prediction of rupture velocity following the transition. Numerical crack modeling with a slip-weakening friction law shown in Fig. 10.10 is reproduced from his paper.

Similar features appear in numerical modeling of in-plane shear rupture using the critical stress criterion. Such features, however, including transonic rupture velocities, are intrinsic to numerical methods only and would not appear in analytic solutions based on the critical stress intensity factor criterion (Virieux and Madariaga, 1982). For a finite crack the maximum stress criterion depends on the number of grid points inside the crack. Thus, for a given maximum stress intensity, the finer the numerical mesh, the higher the maximum stress that has to be adopted.

Three-dimensional solutions for spontaneous shear cracks are highly limited in number. Numerical solutions have been obtained by Day (1982b), using the slip-weakening failure criterion, and by Das (1981) and Virieux and Madariaga (1982), using the critical stress level criterion. Boatwright and Quin (1986), using a rapid hybrid solution of the three-dimensional integral equation for computing source time functions developed by Das (1980), have generated a class of rupture models with spontaneous rupture growth, slip weakening, and spatially variable loading and strength. They found that the average rupture velocity is controlled by the normalized strength, the complexity of the rupture process is proportional to the variation of the loading and strength, and the radiation efficiency depends on the average rupture velocity but does not depend, to first order, on the complexity of the rupture process. As the rupture phases become more complex, the amplitudes of stopping phases decrease, while the radiation from the stress release inside the rupture increases.

Numerical three-dimensional modeling of spontaneous rupture propagation is probably the best approximation to real earthquake source processes, in which heterogeneity of fault planes is a dominant factor in their development.

10.3 Complex Source Models: Asperities and Barriers

Seismological and geologic observations suggest that the mechanical properties of a fault zone are not homogeneous. The heterogeneity of faults and faulting is present on all scales, and dynamic faulting is also heterogeneous at all scales. Heterogeneity is, in fact, a fundamental part of the earthquake

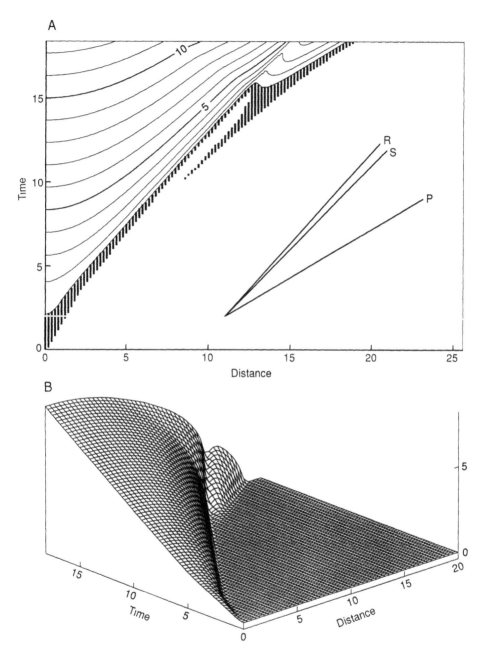

Figure 10.10 Numerical crack modeling with a slip-weakening friction law: (A) slip contours plotted in dimensionless distance and time, with cohesive zone indicated by shading; (B) perspective view of half the bilaterally symmetric solution for dimensionless slip. [From Andrews (1985, Fig. 4).]

process. Surface rupture observed during large earthquakes is usually complex, exhibiting irregular variations in displacement along the fault (e.g., Das and Aki, 1977; Aki, 1979; King and Yielding, 1984; Steidl *et al.*, 1991; Beroza, 1991). Direct evidence from fractures in mines shows that faults are usually highly complex with side steps and unbroken but strongly deformed ligaments (McGarr *et al.*, 1979). Seismic body waves from large earthquakes are also complex, and their interpretation is often based on multiple event source models (e.g., Kanamori and Stewart, 1978; Deschamps *et al.*, 1982). Modeling of short-period displacement waveforms implies that much of the higher frequency energy arrives from small, high-stress-drop zones on the fault plane (e.g., Wallace *et al.*, 1981; Mori and Shimazaki, 1984). The highly random nature of strong motion accelerograms suggests large variations in effective stress during fault rupture. High-frequency waves dominating accelerograms are too complex for their simulation by deterministic models. They are affected by numerous small-scale heterogeneities of the fault plane that require too many parameters for their description. The usual approach to simulation of acceleration-time histories is to introduce a composite model of rupture containing deterministic and stochastic models, in which gross features of rupture propagation are specified deterministically, while the details of the process are described stochastically with a limited number of parameters (e.g., Boore and Joyner, 1978; Hanks, 1979; Hanks and McGuire, 1981; Boatwright, 1982; Papageorgiou and Aki, 1983a, 1983b; Gusev, 1983).

Complexity of rupture results from the heterogeneity of physical properties of the fault zone. Two major factors control the spread of rupture: the distribution of stress drop, that is the difference between prestress and sliding friction, and the distribution of fracture energy or strength on the fault. Since faults cut highly different rocks and have complex geometry, it is expected that fracture energy is a highly variable function of position on the fault. Once the earthquake starts, the growth and arrest of fracture is controlled in a very complex manner by the distribution of stress and strength.

Two types of heterogeneity, generally called *asperities* and *barriers*, are recognized. Asperities are patches on the fault surface with strong cohesion resisting the break; in other words, they have stronger resistance to slip movements. They are therefore focal points of stress accumulation. Laboratory experiments on rock friction demonstrate that for two rough surfaces in contact, the real contact area could be as small as a few percent of the total surface area (Teufel and Logan, 1978). If the fault therefore is considered as two such rough surfaces held together by compressive normal stresses, it can be expected to consist of separate asperities, and the fracture process resulting in an earthquake may be considered as the rupturing of these asperities. Such a concept has been applied to earthquake studies (e.g., Kanamori and Stewart, 1978; Madariaga, 1979; Rudnicki and Kanamori, 1981; Das and Kostrov, 1983) and has been termed the *asperity model*.

In contrast to asperities, Aki (1979, 1984) defines barriers as areas that a seismic event skips over. On an idealized rectangular fault containing small circles representing cracks, a slip occurs on cracks during the fault rupture, but the region between cracks remains unbroken after the rupture. The possibility of such segmented ruptures was shown by Das and Aki (1977) by numerical experiment. A rupture front may be stopped by a barrier, but in the case of shear crack, elastic waves generated by the slip can break the fault plane ahead of the barrier, and then the rupture can propagate over the whole fault plane, leaving unbroken barriers behind. This model is termed the *barrier model*.

It is important to distinguish the barrier model from the asperity model, as explained by Aki (1984). In the barrier model the initial state of stress over the fault is uniform. After the earthquake the rupture has slipped irregularly because of the presence of barriers, where the stress concentration exists. In the asperity model initial stress concentrations are present at asperities that lock the fault. Local failures at these sites correspond to foreshocks. The asperities are broken by the earthquake and the stress is uniform over the fault. The barrier and asperity models for the aftershock and foreshock processes are illustrated in Fig. 10.11, reproduced from Aki (1984). The breaking of asperities may be seen as a smoothing process, while the existence of barriers may be seen as a roughening process of the fault surface. By virtue of these concepts, the preseismic processes are controlled by asperities, while the seismic processes are governed by barriers.

In his discussion of barriers Aki (1979) distinguished two types of barriers: one related to strength and the other geometric. King and Yielding (1984) subdivide geometric barriers into two types: conservative and nonconservative. Both are associated with the transfer of rupture between two fault planes of different orientation. The difference is in the orientation of the slip vectors on the faults. In a conservative barrier the slip vector is common to

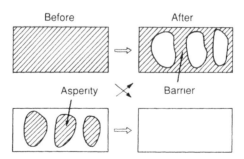

Figure 10.11 The barrier and asperity source models for the aftershock and foreshock processes. [From Aki (1984, Fig. 1).]

both fault planes, while in a nonconservative barrier the slip vectors of the two planes are different. The first kind of barrier conserves the slip vector when motion transfers from one plane to the other, whereas the second does not conserve the slip vector, and new faulting must be created.

A thorough theoretical study of the asperity problem has been done by Das and Kostrov (1983). They considered the problem of spontaneous rupture of a single, circular asperity on an infinite fault plane and the resulting far-field seismic radiation. The numerical three-dimensional boundary integral equation method and a critical stress level fracture criterion were used to describe the rupture process. The rupture was initiated at one or more points along the crack edge. It was found that for asperities of constant strength, the rupture first propagates around the edge of the asperity and then inward.

A dynamic model of the failure of a single asperity on a fault plane of Das and Kostrov (1983) has been used by Das and Boatwright (1985) to interpret accelerograms of a small aftershock of the 1975 Oroville earthquake and to demonstrate the expected characteristics of the acceleration radiated by the failure of an asperity. Other works involving waveform modeling have also demonstrated the existence of highly nonuniform stress distributions on fault planes of moderate and large earthquakes (e.g., Hartzell and Helmberger, 1982; Beck and Ruff, 1984; Liu and Helmberger, 1985).

Michael and Eberhart-Phillips (1991) have developed three-dimensional P-wave velocity models for the regions surrounding five large earthquakes in California, leading to the recognition of relations between fault behavior and the material properties of the rocks that contact the fault at depth. They found that regions of the high release of seismic moment appear to correlate with high seismic velocities, whereas rupture initiation or termination may be associated with lower seismic velocities. These relations lead to an understanding why faults are divided into segments that can fail independently.

Application of the isochrone method (Bernard and Madariaga, 1984; Spudich and Frazer, 1984), based on the ray theory, is very promising for calculations of high-frequency radiation from earthquake sources with spatially variable rupture and slip velocity, enhancing the ability to calculate ground motions in laterally varying media. It was demonstrated that at any instant of time the high-frequency waves reaching an observer come from a line on the fault plane, called an *isochrone*, and it was shown that wavefront discontinuities, such as critical or stopping phases, are radiated every time an isochrone becomes tangent to a barrier (Bernard and Madariaga, 1984). By definition, an isochrone velocity depends on rupture velocity and resembles the usual directivity functions, which is a consequence of the spacing of the isochrones on the fault. Observed ground motions are directly dependent on this isochrone velocity. Ground acceleration may also be related to spatial variations of slip velocity on the fault (Spudich and Frazer, 1984). A compari-

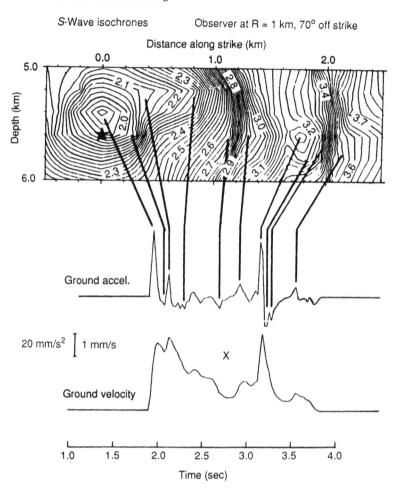

Figure 10.12 Unfiltered ground velocity and acceleration from rupture compared with isochrones, which are contours on the fault plane of the time of an *S* wave, radiated from any point on the fault, that would arrive at an observer. Since the isochrones show exactly which parts of the fault contribute to each time in the seismograms, individual pulses can be assigned to origin locales on the fault. [From Spudich and Frazer (1984, Fig. 11).]

son of the observed unfiltered ground velocity and acceleration from rupture with the *S*-wave isochrones on the fault plane is shown in Fig. 10.12; the figure is reproduced from Spudich and Frazer (1984). They found simple relations showing that ground motions can be caused by either the temporal or spatial derivatives of slip velocity and isochrone velocity. Using the isochrone approach, the forward problem of calculating earthquake ground motions is similar to that of calculating travel times, and a new approach to the inverse problem is possible.

Deployment of a dense seismograph array is another technique providing a means of understanding in detail the ground motion at a given site. In such a case the ground motion can be resolved into constituent waves. The so-called frequency–wavenumber analysis decomposes the wavefield into monochromatic plane waves, each with its own propagation slowness across the array and its own direction of travel. When the wavefield is decomposed into plane waves, the waves can be identified as body or surface waves incident from different directions. Although dense arrays have been used for decades in teleseismic observations and in seismic reflection and refraction profiling, their use to study high-frequency ground motions from local earthquakes is rather recent (e.g., McLaughlin *et al.*, 1983; Spudich and Cranswick, 1984; McMechan *et al.*, 1985). Successful dense array observations of rupture dynamics were made by Spudich and Cranswick (1984), who used records of the 1979 Imperial Valley earthquake from the five-element El Centro differential array. The array was located only 5.6 km from the surface trace of the fault, and the 30-km rupture length caused the propagation direction of waves impinging on the array to change by 150° during the earthquake. Spudich and Cranswick (1984) were able to track the progress of the rupture front along the entire fault length and to observe changes in rupture velocity. An example of their interpretation is shown in Fig. 10.13, reproduced from their paper.

A simple inhomogeneous fault model has been introduced by McGarr (1981) for the analysis of ground-motion parameters of tremors recorded in a deep gold mine in South Africa. The simple source model of Brune (1970, 1971), which assumes a constant stress drop over the circular fault area, could not account for source processes playing a role in determining the observed values of peak accelerations and ground velocities. The model of McGarr (1981) has many features in common with Brune's model, such as a circular geometry and simple assumptions regarding fault dynamics. The model involves the failure of a circular asperity within an annular faulted region. The asperity fails with a high stress drop and on a time scale corresponding to its small dimensions. Following the small-scale failure, the large fault zone deforms under the influence of the ambient state of stress, and the large-scale deformation results in a lower average stress drop over the total area of faulting. Although the proposed inhomogeneous faulting model is very simple in concept, it has been demonstrated by McGarr (1981) to be quite useful for the analysis of ground-motion observations.

Simple barrier and asperity models were proposed by Kuhnt *et al.* (1989) to interpret the source parameters of low-stress-drop seismic events induced by room and pillar mining in a deep mine and directly associated with a stope

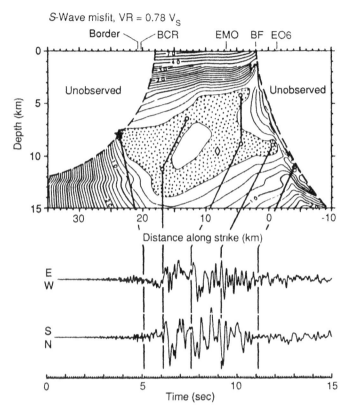

S-Wave misfit, VR = 0.78 V$_S$

Figure 10.13 Absolute values of the difference between S-wave source times and the time the fault would rupture, based on a 0.78 ratio of rupture velocity to shear velocity. on the west side of the Imperial fault in California. Regions with differences smaller than 0.2 s are dotted Regions marked "unobserved" would radiate slownesses higher or lower than those observed on the horizontal accelerograms The DA 1 horizontal accelerograms of the 1979 Imperial Valley earthquake are shown with lines indicating potential sources of various parts of the accelerogram. [From Spudich and Cranswick (1984. Fig. 22).]

area. In the barrier model the crack begins at the stope and propagates toward the caving zone. The pillars act as barriers with higher strength, and the crack is stopped without breaking the pillars. In the asperity model pillars act as stress concentrators, and with increasing stress level the pillar asperities are broken. For simplicity, the asperities, barriers, and cavity zones are modeled by circular or annular areas. The source is represented by a circular shear crack containing a single asperity or barrier of annular form. For the barrier model, constant stress drop is assumed outside of the annular barrier; and for the asperity model, inside the annular asperity.

10.4 Earthquake Sequences

Shallow earthquakes seldom occur as isolated events; they usually form well-defined sequences in space and time. Mogi (1963) has divided these sequences into three types: main shock–aftershocks, foreshocks–main shock–aftershocks, and swarm, which he considered as indicating increasing heterogeneity of the source area, although this idea seems to be too general (Scholz, 1990). Foreshock and aftershock sequences are associated with a larger event called the *main shock*. Earthquake sequences not associated with a dominant event are called *swarms*. Main characteristics of the three types of earthquake sequences are illustrated in Fig. 10.14, where the rates of earthquake occurrence as a function of time are shown schematically. Occasionally, two or more main shocks may be closely related in space and time; they are called *doublets* and *multiplets*.

Foreshocks are smaller events that precede the main shock. They usually occur closely to the main shock hypocenter and are probably a part of the nucleation process (e.g., Scholz, 1990). Their occurrence is rather irregular. Many earthquakes have no foreshocks, and foreshock sequences range from one or two events to small swarms. Jones and Molnar (1979), who made a global survey of foreshocks, found that about 60 percent of large earthquakes with magnitude $M \geq 7$ were preceded by foreshocks. Their activity typically becomes evident 5–10 days before the main shock and rapidly accelerates up to its occurrence. No relation was found between the size of the largest foreshock and the magnitude of the main shock. Jones and Molnar (1979) found that the time sequence of foreshocks can be expressed by an empirical relation

$$n = at^{-c}, \tag{10.28}$$

where n is the rate of occurrence of foreshocks, t is the time before the origin time of the main shock, and a and c are constants. The exponent c is close to one (Papazachos, 1975; Kagan and Knopoff, 1978).

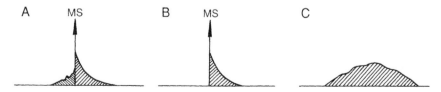

Figure 10.14 Schematic illustration of main characteristics (the rate of occurrence as a function of time) of the three types of earthquake sequences: (A) main shock (MS) with foreshocks and aftershocks, (B) main shock with aftershocks; (C) earthquake swarm.

The asperity model has been used to model precursory seismic patterns (Jones and Molnar, 1979; Dmowska and Li, 1982). Jones and Molnar (1979) proposed that foreshocks represent accelerating failure of asperities resulting from concentration of stress on the unbroken asperities. Mikumo and Miyatake (1983) were able to explain with simple asperity models a number of the spatial and temporal seismicity patterns, including the occurrence of foreshocks. The basic assumptions of asperity models relating small earthquakes to large seismic strain accumulation and release are as follows: fault surfaces are held together by a number of asperities; weaker asperities fail during small earthquakes with increasing tectonic stress, transferring more stress to the remaining asperities; and the fault becomes unstable when most of the asperities have broken. These assumptions lead to two characteristics of foreshocks: stress drops of foreshocks should be higher than those of previous events from the region, if stress drop is proportional to tectonic stress, and the foreshocks should be clustered along strong asperities and should occur as groups of shocks with similar location, focal mechanism, and waveforms (Pechmann and Kanamori, 1982).

It is very difficult, however, to detect temporal changes in stress drop since its determination depends on many factors, and efforts in this direction produced differing results. From an analysis of four events of an extended foreshock sequence of the 1978 Miyagi-Oki, Japan, earthquake, carried out by Choy and Boatwright (1982), it follows that the dynamic stress drop of the main shock is apparently bounded by the dynamic stress drop of the foreshocks.

Waveforms of foreshocks are studied only occasionally. Ishida and Kanamori (1978) found that seismograms of five events preceding the 1971 San Fernando, California, earthquake were remarkably similar. Frankel (1981) observed that several preshocks to a magnitude 4.8 earthquake in the Virgin Islands had very similar waveforms. But waveforms of foreshocks to the 1952 Kern County, California, earthquake differed rather significantly from event to event (Ishida and Kanamori, 1980). The seismograms of two foreshocks to the 1966 Parkfield, California, earthquake were also different, similarly as those of two foreshocks to the 1975 Oroville earthquake (Bakun and McEvilly, 1979). It seems evident that the number of foreshocks and the similarity of their waveforms will depend on the number, strength, and distribution of asperities in the area.

Pechmann and Kanamori (1982) have studied in detail waveforms and spectra of preshocks and aftershocks of the 1979 Imperial Valley earthquake. They found much stronger evidence to support the prediction of the asperity model related to waveforms rather than to spectra. In particular, the waveform data imply that the preshocks originated from a relatively small number of highly localized sources in comparison to aftershocks. The preshocks occurred in groups of events with strikingly similar waveforms over the entire length of records. The close match in waveforms implies similar source mechanisms and close clustering of hypocenters. Aftershock waveforms are more variable from one event to the next, although groups of similar events were found as well. These observations can be explained by the asperity model, which predicts localization of failure on strong unbroken asperities along the fault during the period preceding large earthquakes. A second prediction of the asperity model about higher stress drop for preshocks was not supported by data. It appears that the waveforms of small shocks are a more sensitive indicator of seismic potential than are the spectra.

Dmowska and Li (1982) suggested that the end areas of asperities are often locations of clustering of foreshocks. In their model a slip front progresses from the bottom of the lithosphere upward. When the front crosses an asperity, the edge of the asperity shows seismic activity while the center remains aseismic. When the asperity is elongated, foreshocks would occur at the two ends of the asperity and the precursory pattern will be very similar to the foreshock pattern of the 1979 Petatlán, Mexico, earthquake. Its foreshocks and aftershocks were studied by Hsu et al. (1984) who proposed a simple two-asperity model to account for the observed foreshocks and after-shocks of the Petatlán earthquake. It has been unclear whether an asperity would cause quiescence due to high strength or cause seismic activity due to high stress before the main shock (Habermann, 1983). Both cases seem to be possible depending on the stages of the stress environment.

Inward migration of foreshocks toward the epicenter of the main shock has been observed in some cases (e.g., Kagan and Knopoff, 1976; Rikitake, 1982). Jones et al. (1982) has studied the foreshock sequence of the 1975 Haicheng, China, earthquake: the first major event to be accurately pre-dicted. The Haicheng earthquake was preceded by over 500 foreshocks very near the epicenter of the main shock in the 4 days prior to the main event; 60 foreshocks were located. Jones et al. (1982) have examined the spatial distribution and radiation patterns of the located events and found that the main shock did not occur on the same plane as the foreshocks. The spatial distribution of the foreshock cluster changed with time, expanding from a very small spherical volume to a larger elongated volume. Two different faulting mechanisms were activated during the foreshock sequence, which can be correlated with different parts of the zone. These features may result from the process of transferring strain across an echelon step in the fault. The large distance between the foreshocks and the main event, compared with the dimensions of the rupture area of foreshocks, implies that only small changes in shear stress could have been induced on the fault of the main shock by foreshocks.

Aftershocks follow almost all shallow earthquakes and form the most frequently observed earthquake sequences. The globally observed character-istics of aftershocks is the hyperbolical time dependence of their frequency, which is known as the *Omori law* and can be expressed as

$$n = \frac{c}{(1+t)^p},$$ (10.29)

where n is the rate of occurrence of aftershocks, t is the time from the origin time of the main shock, and c and p are constants. The exponent p is usually found to be very close to one. The largest aftershock in the sequence is typically at least one magnitude smaller than the main shock (Utsu, 1971). The total seismic moment of aftershocks usually amounts to only about 5 percent of the moment of the main shock (Scholz, 1972). This rule was found to be also valid for the 1977 Lubin, Poland, tremor, which occurred in a deep copper mine (Gibowicz *et al.*, 1979). Aftershocks begin immediately after the main shock over the entire rupture area and its surroundings, although they are usually concentrated in places where large stress concentrations produced by the main rupture are expected (e.g., Mendoza and Hartzell, 1988).

All aftershock models attempt to reproduce their time dependence described by the Omori law [equation (10.29)]. In a detailed review of the aftershock mechanics, Hull (1983) divides the aftershock models into two categories: time-dependent reloading of the rupture surface and time-dependent strength loss of the surrounding material after an earthquake. These theories can be grouped into three general classes: fractures caused by (1) reloading through rock creep, (2) the weakening of rock by stress corrosion or pore fluid flow, and (3) increased static shear stress or local stress concentrations resulting from the main shock.

In the barrier model of Das and Aki (1977) and Aki (1979) aftershocks are the result of stress concentrations. A barrier is a site of high stress concentration on the earthquake fault plane. The barrier fails later as a result of strength loss, causing an aftershock. The aftershocks, however, are restricted to the plane of the main shock. It is well documented that aftershocks tend to cluster at the ends of a rupture, and that aftershock clusters often occur away from the rupture plane.

The change in the stress field around a main shock has been calculated by many people. Niewiadomski and Rybicki (1984), for example, computed the stress field from two interacting antiplane shear cracks in a laterally inhomogeneous half-space. They found a decrease in stress in the regions adjacent to a crack except at the crack tips where an increase in stress is observed. These results explain aftershock clusters at the ends of a rupture, and they are similar to those found in the antiplane case of a two-dimensional Volterra dislocation (Rybicki, 1971).

The stress off the crack plane in the normal direction shows entirely different patterns for in-plane and antiplane cracks. The increase in the shear stress at distance normal to the fault could be the cause of off-fault aftershocks, which also sometimes occur in locations distinctly distant from the rupture plane of the main shock (e.g., Das and Scholz, 1981b). This stress pattern, induced by an in-plane shear crack, has been analyzed and used to explain the off-fault aftershock activity in several papers (Das and Scholz, 1981a, 1981b; Kostrov and Das, 1982). A similar effect has been found using a Volterra three-dimensional rectangular dislocation model (Yamashina, 1978; Stein and Lisowski, 1983). The increase in the shear stress off the fault is only about 10 percent of the stress drop. If this stress increase is responsible for the aftershock clustering, either the rocks must be near failure or the stress drop must be large. Stein and Lisowski (1983) have speculated that the rocks surrounding a fault are weakened or exist in a high-stress environment. The coseismic stress redistribution itself, however, cannot generate aftershocks. Some time-dependent process must be responsible for the time delay (Hull, 1983). Das and Scholz (1981a) have separated the cause of aftershocks into two mechanisms: one loading of the rocks causing them to fail, and the other providing the time delay. They presented a static stress field that exists immediately following a main shock and provided a time-delay mechanism with the theory of stress corrosion.

The 1979 Homestead Valley, California, earthquake sequence exhibited aftershock clusters off the earthquake trend. A plot of the fault model and the location of earthquakes which occurred from March 20, 1979 to February 10, 1981 are shown in Fig. 10.15, reproduced from Stein and Lisowski (1983). There were four main shocks, and the cluster of aftershocks started to form within one day of the first main shock. Aftershocks tended to occur where the static stress increased by a few bars; few aftershocks occurred where the stress decreased. An important conclusion of the study by Stein and Lisowski (1983) on the Homestead Valley earthquakes is that the stress that causes aftershocks depends on the combination of the normal and shear stresses altered by the earthquake. This explains the diagonal cross distribution of the aftershocks.

Earthquake swarms are sequences of earthquakes that often start and end gradually, without a single earthquake dominating in size. They are most often associated with volcanic regions, but this is not a universal rule (Sykes, 1970). There is a tendency for earthquake swarms to occur in and around the focal region long before the large earthquake (Evison, 1977). This is the seismicity pattern used as an intermediate-term precursor to the occurrence of earthquakes.

Little is known about tremor sequences in mines. McGarr and Green (1978) have studied foreshock–aftershock sequences of two mine tremors of magnitudes 1.5 and 1.2, which occurred in May 1973 in a region of active mining in the East Rand Propiety Mines near Johannesburg, South Africa. A highly sensitive seismic array and a tiltmeter in the source area were used for the study. The seismic coverage of the sequences was complete down to

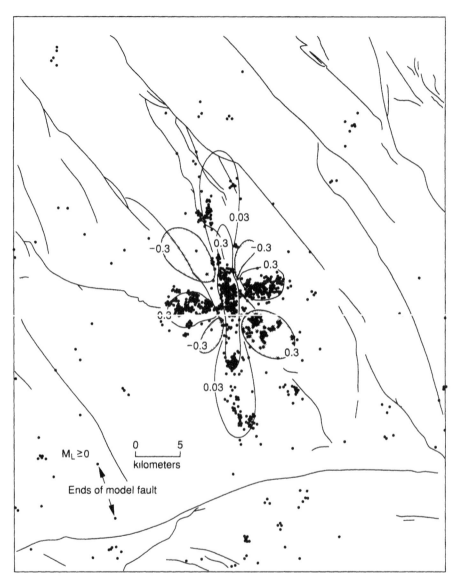

Figure 10.15 Plot of the fault model and the locations of earthquakes with magnitude $M_L > 0$ which followed, from March 20, 1979 to February 10, 1981, the 1979 Homestead Valley, California earthquake. Calculated failure stress increases (solid perimeter) and decreases (stippled areas) are contoured, and quaternary faults are indicated. [From Stein and Lisowski (1983, Fig. 9a)]

magnitudes of -3.5. In the first hour after each event some 140 microaftershocks were recorded with a rate of occurrence decreasing with time in a regular manner. Although the two events occurred within about 50 m of one another and within 12 days, the seismicity before the tremor of magnitude 1.5 was at its normal ambient level of about 4 events per hour, whereas the seismicity before the event of magnitude 1.2 was unusually high, at about 30 events per hour. No indications were found suggesting any evidence of an instability developing before either of the tremors. McGarr and Green (1978) have shown that a main shock–aftershock sequence and sequence similar to a swarm can occur in the same volume of rocks under similar strain conditions.

The 1977 tremor of magnitude 4.5, which occurred at the Lubin copper mine in Poland, was followed by a regular aftershock sequence with the largest aftershock of magnitude 3.4 (Gibowicz *et al.*, 1979). The frequency–time distribution of aftershocks was typical with a rate of occurrence decaying hyperbolically in time. In contrast, the 1987 event of magnitude 4.3, which occurred at the same mine, was followed by a few small aftershocks with the largest two aftershocks of magnitude 2.1, differing considerably from the level of seismicity observed during 3 months preceding the main shock (Gibowicz *et al.*, 1989). The daily release of seismic moment from aftershocks during the first month after the main event was found to be twice smaller than that from the tremors preceding the event. It seems, in general, that aftershocks to seismic events in mines are not as ubiquitous as those in natural seismicity.

10.5 Synthetic Seismograms

A seismogram can be considered as the response of the source–receiver path system to disturbances at the source. If the transfer functions are known, then a seismogram can be synthesized from a signal assumed at the source. It has been demonstrated in a multitude of papers that synthetic seismograms can be computed rather realistically, and a number of alternative procedures have been proposed.

The first approach are deterministic methods for predicting ground motion. The basic form of the displacement for a given seismic wave is described by equation (10.18). For a more detailed representation the seismic source is considered as elastic rebound along a fault produced by moving dislocations. Seismograms are then calculated from a Green function representation of the medium displacement. Green's functions for various types of faulting have been constructed, and numerous papers using this approach have been published [for references, see, for example, Bolt (1987)]. The procedure is a kind of complicated curve fitting assuming different values for the source parameters to find the best fit with the recorded seismic waves.

An effective method of constructing Green's functions to calculate synthetic seismograms of complex earthquakes is the use of small earthquake seismograms as empirical Green functions, introduced by Hartzell (1978). Assuming that a small earthquake has a simple source, its seismogram can be considered as an empirical Green function and deconvolved from the seismogram of a larger event by spectral division (e.g., Mueller, 1985; Frankel et al., 1986; Dan et al., 1990; Mori and Frankel, 1990). The observed seismogram $u_m(t)$ of a larger event can be expressed as the convolution of its source time function $S_m(t)$, the impulse response of the path $P(t)$, the recording site $R(t)$, and the instrument $I(t)$:

$$u_m(t) = S_m(t)*P(t)*R(t)*I(t). \qquad (10.30)$$

Assuming that the source time function $S_a(t)$ of an adjacent small event can be considered as a delta function $\delta(t)$, the waveform $u_a(t)$ of this event represents the impulse response of the path, site, and instrument or the empirical Green function for a source of that particular focal mechanism. Thus the focal mechanisms of the main shock and the Green function event are assumed to be similar. The deconvolution is performed by spectral division in the frequency domain.

$$\frac{u_m(\omega)}{u_a(\omega)} = \frac{S_m(\omega)}{S_a(\omega)} = S_m(\omega), \qquad (10.31)$$

where ω is the angular frequency. The quotient spectrum can be transformed back to the time domain, providing the source time function of the larger earthquake.

The second approach is the numerical solution of the equations of motion. This procedure takes into account more realistic structure around the source and near the surface. Numerical calculations are made by using many point sources on the fault and summing their contributions at an observation point to obtain synthetic seismograms that are close to the observed records.

As an example, the approach of Kikuchi and Kanamori (1982) should be mentioned. They developed a numerical method to deconvolve complex body waves into a multiple-shock sequence. Under an assumption that all subevents of a multiple shock have identical fault geometry and depth, the far-field source time function is obtained as a superposition of ramp functions. The height and onset time of the ramp functions are determined by matching the synthetic waveforms with the observed seismograms in the least-squares sense. The individual subevents are then identified by pair of ramp functions or discrete trapezoidal pulses in the source time sequence. The method can be used for the analyses of single and multistation data. As a test of the method Kikuchi and Kanamori (1982) analyzed teleseismic long-period P

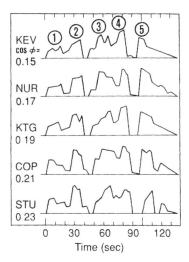

Figure 10.16 Far-field source time functions of the 1976 Guatemala earthquake. The functions are obtained from the records of five stations. Five major subevents are identified and marked by the numbers 1–5 [From Kikuchi and Kanamori (1982, Fig. 5).]

waves from the 1976 Guatemala earthquake. Far-field source time functions of this earthquake are shown in Fig. 10.16, reproduced from their paper. Five major subevents were identified. The method provides an effective tool for systematic analyses of multiple events.

Time-domain modeling of long-period body waves, such as that proposed by Kikuchi and Kanamori (1982), is concerned mainly with the gross complexities of the fault time history rather than with the minute details of the source function. Such details can be learned only from short-period and strong motion waveforms. A good example of the differences in source time functions, determined from the three sets of data, comes from the studies of the 1968 Borrego Mountain, California, earthquake. A comparison of the time functions for this earthquake determined from the long-period body waves (Burdick and Mellman, 1976), short-period body waves (Ebel and Helmberger, 1982), and strong motion data (Heaton and Helmberger, 1977) is shown in Fig. 10.17, reproduced from Ebel and Helmberger (1982).

During the past decade a number of numerical techniques have been developed for waveform modeling at regional and local distances, and the Earth models used are more and more complex to model real data (e.g., Hartzell and Heaton, 1983) and to understand the complexity of wave propagation in realistic structures (e.g., Campillo *et al.*, 1983). The various techniques and models were compared by Herrmann and Wang (1985), where relevant references can also be found. They considered time histories

TIME FUNCTIONS

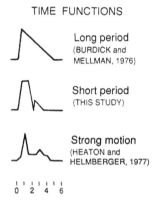

Long period
(BURDICK and
MELLMAN, 1976)

Short period
(THIS STUDY)

Strong motion
(HEATON and
HELMBERGER, 1977)

0 2 4 6

Figure 10.17 Comparison of the source time functions of the 1968 Borrego Mountain, California earthquake determined from the long-period body waves, short-period body waves, and strong motion data. The time scale at the bottom is in seconds. [From Ebel and Helmberger (1982, Fig. 12).]

of the eight dislocation source and two explosion source Green functions for simple Earth models, computed using several procedures. They found that each numerical technique for synthetic seismograms has imperfections, but peak amplitudes agree to within 10 percent and waveform shapes are almost identical.

So far only a few papers related to waveform modeling of seismic events in mines have been published. An interesting example of the application of waveform modeling to seismic events induced by mining has recently been published by MacBeth and Redmayne (1989). The work is an extension of previous waveform modeling of local and regional records (MacBeth and Panza, 1989) and is based on the modal summation of Panza (1985), used to construct synthetic seismograms, summing the contributions from a large number of higher-mode surface waves. The source of seismic events generated in the Midlothian coalfield in Scotland has been studied. The computation of synthetic seismograms permits distinction between the effects of different Earth structures, source depths, source orientation, and types of events. The hypocenters of the tremors appear to be closely grouped around the mine workings. Three types of seismic events have been found: normal faulting tremors, high-angle strike-slip tremors, and events with a focal mechanism resembling a single force, possibly corresponding to a collapse in mine workings.

A method of direct deconvolution for the retrieval of a source time function from *SH*-wave records has been proposed by Niewiadomski and Meyer (1986). The deconvolution operation in the time domain is a typical example of an ill-conditioned problem, and special methods are needed to solve such a problem. A technique providing reasonable results is the regularization method of Tikhonov (e.g., Tikhonov and Arsenin, 1979), which was used by Niewiadomski and Meyer (1986). The algorithm of direct deconvolution was applied to retrieve the source time function of the Lubin, Poland, mine tremor of June 20, 1987 (local magnitude $M_L = 4.3$) from broadband seismograms recorded at a distance of about 70 km from the source (Gibowicz et al., 1989). The source time function appears to be complex, composed of two subevents separated in time by a fraction of a second. This is an example demonstrating the complex nature of large seismic events induced by mining.

An efficient technique of Johnson (1974) for calculating seismograms from point seismic sources in a homogeneous half-space was used by McGarr and Bicknell (1990) to model the records of two tremors located in the Vaal Reefs gold mine near Klerksdorp in South Africa. Data from a single three-component surface station, supplemented by a seismic location network, were used. They found that a comprehensive description of the source process can be obtained by a trial-and-error calculation of synthetic seismograms until the observed seismograms are matched. Their experience shows that locally recorded broadband seismograms contain enough information to constrain the seismic moment, the source duration, and the fault-plane solution of seismic events.

10.6   Faults and Fractals

There are two general approaches to heterogeneity in the rupture process. The first is the development of physical models based on the heterogeneity of fault surface strength in a general sense, such as the barrier model and the asperity model. The second approach is purely geometric and is based on the concept of fractals introduced by Mandelbrot (1967) in a geologic context, which is a general model of the geometry of irregular systems. Mandelbrot (1967) noted that the length of a rocky coastline increased as the length of the measuring rod decreased according to a power law, and he associated the power with a fractal (fractional) dimension. This observation is based on the fact that the rocky coastline is scale-invariant. Photographs of a rocky coastline taken at different altitudes cannot be distinguished unless a scale is introduced. The scale invariance of geologic phenomena is well known—an object with a scale is always included whenever a photograph of a geologic feature is taken.

The fractal approach is used to characterize sets that exhibit irregularity regardless of the scale at which the set is examined. The self-similar figures are termed fractals. Natural fractals include trees, rivers, coastlines, and the shapes of mountains. A fractal approach to earthquake faulting has been applied in a number of studies (e.g., King, 1983; Smalley et al., 1985; Scholz and Aviles, 1986; Aviles et al., 1987; Okubo and Aki, 1987). The basic notions of the fractal concept have been described in some detail by King (1983), Okubo and Aki, (1987), and Turcotte (1989) and only a brief recapitulation of the idea is given here.

The notions of dimension are central in the development of fractals. Two of the different definitions of dimension—topological and fractal dimension —must be considered to identify a set as being a fractal set. Fractal dimension is a measure of a set's complexity or irregularity, and once it is determined, it provides a quantitative means of characterizing the fractal nature of that set. The notion of fractal dimension is best illustrated by an example given by Mandelbrot (1977) setting forth the length of the coastline of Britain. The question has no simple answer, since the length increases indefinitely as the interval at which it is measured decreases. A plot of the logarithm of the length of a coastline against the logarithms of the length of the yardstick used to measure the coast is a straight line. The slope of this line is $1 - D$, where D is the fractal dimension of the coastline. Since the length of a coastline increases more rapidly for larger D, then larger values of D can be directly associated with more complicated curves or coastlines.

Thus in general, if the number of objects N_i with a characteristic linear dimension r_i satisfies the relation

$$N_i = \frac{C}{r_i^D},$$ (10.32)

a fractal distribution is defined with the fractal dimension D and a constant C (e.g., Turcotte, 1989). Similarly, if the number of objects N with a characteristic linear dimension greater than r satisfies the relation

$$N = \frac{C}{r^D},$$ (10.33)

a fractal distribution is defined. The frequency–size distributions for fragments, ore deposits, oil fields, islands, fault systems, and earthquakes often satisfy this relation. The fractal dimension D provides a measure of the relative importance of large objects against small objects. Although relations (10.32) or (10.33) are mathematically valid over an infinite range, in physical situations there are always upper and lower limits on the applicability of the fractal distribution.

To obtain the fractal dimension of a fault system, the box-counting method (Mandelbrot, 1983) is usually used. A fault system enclosed in a square area is considered, which is divided into square boxes with a side length r. Then the number $N(r)$ of boxes that the fault line enters is plotted against r on a double logarithmic scale. If the graph is almost linear, which means that relation (10.33) holds well, its slope is equal to $-D$ and the fractal dimension can be obtained directly.

In any seismic region it is found that the number of earthquakes N with a magnitude greater than M satisfies the relation (Gutenberg and Richter, 1954)

$$\log N = a - bM, \tag{10.34}$$

where a and b are constants. The b value is widely used as a measure of regional seismicity. Its variations have been qualitatively linked to the heterogeneity of the material (Mogi, 1962), to the state of stress (Scholz, 1968), to changes in fracture mechanism in laboratory-scale rock fracture experiments (Meredith and Atkinson, 1983), and to the underlying physical processes of time-varying applied stress and crack growth under conditions of constant strain rate (Main et al., 1989). Aki (1981a) has shown that relation (10.34) is equivalent to the definition of a fractal distribution.

The seismic moment M_0 of an earthquake is often related to its magnitude by

$$\log M_0 = cM + d, \tag{10.35}$$

where c and d are constants, and there is a theoretical basis for taking $c = \frac{3}{2}$ (Kanamori and Anderson, 1975). The seismic moment can be approximated by (Kanamori and Anderson, 1975)

$$M_0 = C_1 r^3, \tag{10.36}$$

where $r = A^{1/2}$ is the linear dimension and A is the area of the fault rupture. By combination of relations (10.34), (10.35), and (10.36), the following relation is obtained:

$$\log N = -2b \log r + C_2, \tag{10.37}$$

where

$$C_2 = \frac{bd}{1.5} + a - \frac{b}{1.5} \log C_1. \tag{10.38}$$

Relation (10.37) can be rewritten as

$$N = C_2 r^{-2b}. \tag{10.39}$$

A comparison with definition (10.33) shows that

$$D = 2b. \tag{10.40}$$

Thus the fractal dimension of regional or local seismicity is simply twice the value of the parameter b.

Seismicity has fractal structures in space, time, and magnitude distributions expressed by the fractal dimension D, Omori's exponent p, and the b value, respectively. Correlation among these scaling parameters was considered by Hirata (1989). He calculated the fractal dimension and the b value for earthquakes in the Tohoku, Japan, region and found that Aki's formula (10.40) is not supported by his results, which show a negative correlation ($D = 2.3 - 0.73b$) between the b value and the fractal dimension D.

In general, the value of b does not depart greatly from the standard value close to 1. The size of large earthquakes, however, is significantly underestimated by the extrapolation of the size distribution of small earthquakes occurring on the same fault (e.g., Schwartz and Coppersmith, 1984; Davison and Scholz, 1985). This results from the different scaling relations for large and small earthquakes, which belong to different fractal sets (Scholz and Aviles, 1986). There is also evidence that significant departures from relation (10.32) are present for earthquakes with magnitude smaller than about 3, implying a breakdown of self-similarity at the small end as well (e.g., Aki, 1987). Scaling relations of earthquake sources are further discussed in Chapter 11.

Scholz and Aviles (1986) examined the fractal geometry of natural fractures and faults over the entire bandwidth from 10^{-5} to 10^5 m, relevant to earthquake processes. They found that the surfaces are fractal over this entire band but the fractal dimension D is a function of spatial frequency, or wavelength. Both abrupt and gradual transitions in D are observed and over some bandwidths the surfaces are smooth enough to become Euclidean. From these results Scholz and Aviles (1986) deduced that the asperity distribution on faults obeys an inverse power law and that fault roughness scales with spatial wavelength. The constants in these relations, however, depend on D and are valid over only limited spatial bands. They also showed that these results, combined with some simple physical models of the rupture process, provide a unifying framework for explaining a variety of seismic phenomena, arising from the fractal, strictly geometric, nature of faults. Specifically, these are the frictional instability, the generation of strong ground motions, and earthquake scaling relations.

No better example can be given than the distinction between large and small earthquakes. From the abrupt change in fractal dimension observed on the San Andreas fault system at a wavelength of about 10–12 km, it follows that earthquakes with dimensions larger than this could not be expected to be self-similar with smaller earthquakes. They should belong to different fractal sets. This dimension divides earthquakes into two families: large earthquakes that rupture the entire seismogenic layer and small earthquakes that do not (Scholz and Aviles, 1986). Thus this distinction, based purely on geometry, agrees with the observation that large earthquakes obey scaling laws different from those followed by small events (e.g., Archuleta *et al.*, 1982; Scholz, 1982; McGarr, 1986; Shimazaki, 1986).

Large and small earthquakes are self-similar within themselves, but the two sets are not self-similar with each other. From this it follows that the size distribution of large earthquakes should not cojoin the size distribution of small earthquakes. This is known to be the case and was termed a "fractal tear" by Scholz and Aviles (1986). The significance of their result, that the fractal function is an accurate representation of the topography of fractures and faults over the entire bandwith relevant to the earthquake rupture process, is that it must be incorporated into any physical model of seismic phenomena. In particular, any physical model must contain the scaling and self-similarity properties present in the geometry.

Chapter 11 | Seismic Spectra and Source Parameters

Waveform modeling in the time domain involves rather complex techniques and cannot be applied routinely. The Fourier transform of seismic records or time series into the frequency domain, on the other hand, does not change the content of the records, provided that both the amplitude and phase spectra are considered. In practice, only amplitude spectra are usually used for the determination of source parameters, and this procedure limits to some extent the amount of information that can be retrieved from the spectrum.

Spectral analysis has become a standard technique used in studies of small earthquakes. Early attempts to apply the spectral theory of a seismic source to seismic events induced by mining (Smith *et al.*, 1974; Spottiswoode and McGarr, 1975; Gibowicz *et al.*, 1977; Hinzen, 1982) have shown that simple source models in the form of a circular dislocation (Brune, 1970, 1971; Madariaga, 1976) or a rectangular fault (Haskell, 1964; Savage, 1972) can be successfully used for the interpretation of seismic spectra and the determination of source parameters of mine tremors.

Most seismic source theories, based on kinematic or quasidynamic dislocation models, predict that the far-field displacement spectrum should remain constant at low frequencies and become inversely proportional to some power of frequency at higher frequencies (Aki, 1967; Brune, 1970; Randall, 1973; Madariaga, 1976). Thus three independent parameters specify the far-field displacement spectrum: the low-frequency level, the corner frequency defined as the intersection of the low- and high-frequency asymptotes, and the slope coefficient controlling the rate of high-frequency decay of the spectrum.

In this chapter, the modern methods and techniques used for the determination of seismic source parameters are described in some detail. A number of primary or input parameters are considered, mostly in the frequency domain and some additional parameters in the time domain as well. Over 20 source parameters can be calculated from these input parameters. The most important parameter describing the strength of the source is the seismic moment or moment magnitude. Other magnitude scales, based either on coda-wave duration or on maximum amplitudes, are introduced to provide a basis for routine assessment of the source strength when no spectral analysis is undertaken. The determination of seismic energy and radiation efficiency is also considered. Several parameters describe the source dimensions and four independent estimates of stress release during the occurrence of seismic events are used.

11.1 Spectral and Time-Domain Parameters

The events selected for the determination of their source parameters are expected to be recorded on three-component sensors. The rock average densities and *P*- and *S*-wave velocities in the source area and at each seismic sensor must be evaluated in the first instance. For underground networks in mines, these quantities are often approximately the same in the source area and in the sensor vicinity, especially at short distances. *P*-wave velocities are usually available from event location procedures and then *S*-wave velocities, if they are not available, can be calculated by accepting the standard value of $\frac{1}{4}$ for the Poisson ratio.

The selected events must be located. Then the distance from the hypocentre, the azimuth to the epicenter measured from the north, and the angle of incidence measured from the vertical can be readily computed for each seismic station. Using the values of the azimuth and the angle of incidence, the traditional three-component seismograms N, E, Z can be rotated into the local ray coordinate system with one longitudinal component in the *P* direction and two transverse components in the *SV* and *SH* directions (e.g., Plesinger *et al.*, 1986). The rotation of components is recommended to cut down the number of seismic pulses and spectra to be processed for each station from a single event, from six (*P* and *S* waves on three components each) to three (one component for *P* wave and two components for *SV* and *SH* waves).

If triaxial sensors are installed in inclined boreholes, a situation occasionally found in deep mines, a simple geometric calculation of the angles between the source and particular components is not possible, unless the inclination and the azimuth of each component are specified. A much simpler approach is based on polarization analysis of direct *P*-wave pulses recorded on three components perpendicular to each other but ambiguously situated in space. Several techniques for polarization analysis are available, providing the required orientation of particle motion of *P* waves in terms of the azimuth and angle of incidence, including algorithms in the time and frequency domains (e.g., Montalbett and Kanasewich, 1970; Vidale, 1986; Park *et al.*, 1987).

The original seismograms provided by underground networks in mines represent either ground acceleration or ground velocity records. In general, the accelerometers are more often used in small-size networks covering selected areas in a given mine, where high-frequency small events are of special interest. The ground velocity sensors, on the other hand, are more widely used in extended networks covering large areas or even the whole mine. A numerical integration filter applied to the ground acceleration records provides the corresponding components of the velocity waveforms. A consecutive numerical integration of the ground velocity records provides, in turn, the displacement waveforms. Both these types of seismograms are needed for the extraction of source parameters.

The radiation coefficients of P and S waves are needed for the determination of seismic moment and seismic energy. The takeoff angle between the normal to the fault plane and the wave direction should be taken into account in the estimates of source dimension. These quantities can be computed for each seismic station from the fault-plane solution of a given event, whenever it is available. If the focal mechanism is not determined, the average radiation coefficients are usually accepted.

In almost all cases of practical significance, the spectral analysis is performed on limited portions of signals. The spectrum of the finite portion is a smoothed version of the spectrum of the original energy signal and the two spectra are not identical. If the portion $g_T(T)$ of an energy signal $g(t)$ is given in the interval $(-T/2, T/2)$ and T is too small, it may happen that the spectra do not even resemble each other. To diminish the deviation between the two spectra as much as possible, a taper $u(t)$ is introduced that is zero outside the interval $(-T/2, T/2)$ and then the selected portion of the signal can be written in the form (e.g., Meskó, 1984)

$$g_T(t) = g(t)u(t), \qquad (11.1)$$

where $u(t) = 0$ for $|t| > T/2$. The spectra of these functions are related by convolution

$$G_T(f) = G(f)^*U(f). \qquad (11.2)$$

The taper $u(t)$ should be chosen in such a way that the spectra $G_T(f)$ and $G(f)$ become close to each other in some sense. The smoothing effect caused by convolution of $G(f)$ with $U(f)$ become smaller if $U(f)$ is concentrated near $f = 0$. The central bulge of the spectral window is called the *main lobe*, and the subsidiary extrema are called *side lobes*. To minimize the smoothing effect, therefore, we have to concentrate the main lobe near $f = 0$ and to keep the amplitudes of the side lobes as small as possible. The most often used tapers of such a type are the Hanning and the Hamming tapers. The

Hanning taper is defined as follows (e.g., Meskó, 1984)

$$u(t) = \begin{cases} \dfrac{1}{2}\cos\dfrac{2\pi t}{T} & \text{for } |t| \le \dfrac{T}{2}, \\ 0 & \text{for } |t| > \dfrac{T}{2}, \end{cases} \qquad (11.3)$$

and its Fourier transform is

$$U(f) = \frac{T}{2}\operatorname{sinc}(fT) + \frac{T}{4}\left[\operatorname{sinc}(fT+1) + \operatorname{sinc}(fT-1)\right], \qquad (11.4)$$

where sinus cardinal sinc $x = (\sin \pi x)/(\pi x)$. The Hamming taper is described by

$$u(t) = \begin{cases} 0.54 + 0.46\cos\dfrac{2\pi t}{T} & \text{for } |t| \le \dfrac{T}{2}, \\ 0 & \text{for } |t| > \dfrac{T}{2}, \end{cases} \qquad (11.5)$$

and its Fourier transform by

$$U(f) = 0.54T\operatorname{sinc}(fT) + 0.23T\left[\operatorname{sinc}(fT+1) + \operatorname{sinc}(fT-1)\right]. \quad (11.6)$$

The coefficients of the Hamming taper are somewhat modified from those of the Hanning window, simply to decrease the amplitude of the highest side lobe, which is about 30 percent that for the Hanning taper. Other often used tapers are the triangle, Gaussian, and exponential tapers [for details see, e.g., Meskó (1984)].

The spectral analysis should be performed using interactive computer graphics allowing the choice and rapid change of signal and noise windows on various seismic channels. Windows containing the P pulse on the radial component and the first SV and SH pulses on the transverse components should be selected in each case. The time series should be tapered using a cosine taper (either Hanning or Hamming tapers) and then supplemented by zeros. The spectra can now be calculated by a fast Fourier transform (FFT) routine and corrected for instrumental effect by dividing them by the frequency response of the recording system.

The observed spectra must be corrected for attenuation and scattering effects along the travel path of seismic waves. Such corrections are of the utmost importance for the proper retrieval of source parameters of small earthquakes, even if they are recorded at close distances (e.g., Rovelli *et al.*,

1991). The uncorrected spectra are often characterized by the decay coefficient distinctly higher than the most often observed value of 2. In contrast, the corrected spectra of mine tremors are well described by a f^{-2} falloff (e.g., Gibowicz, 1990a). The attenuation rate of body waves is generally parameterized either by the average quality factor Q_c (Q_α for P and Q_β for S waves) along the ray path or by the attenuation operator t_c^* (t_α^* for P and t_β^* for S waves), described in Chapter 8. To correct for attenuation, the spectra are usually multiplied by the exponential term $\exp(\omega R/2cQ_c)$ or $\exp(\omega t_c^*/2)$, where R is the distance between the source and the receiver and c is either P- or S-wave velocity. When the values of Q_c are known even approximately, the corrections are straightforward. If not, they must be evaluated using the methods and techniques described in Chapter 8.

Various methods are used for the accurate and objective interpretation of seismic spectra to provide reliable estimates of source parameters. The simplest and most often used spectral model is that described by the low-frequency spectral level Ω_0 and the corner frequency f_0, above which the spectrum is assumed to fall off as a second power of frequency (Aki, 1967; Brune, 1970, 1971). For many spectra the corner frequency f_0 cannot be determined reliably from the intersection of two asymptotes (e.g., Brune *et al.*, 1979; Snoke *et al.*, 1983). For this reason, the approach proposed by Snoke (1987) can be used, replacing f_0 as a spectral observable by a parameter that can be determined reliably, providing a more robust procedure than that conventionally used. The integral J of the square of the ground velocity is such an observable, which is a direct measure of the energy flux of P or S waves as well. The energy flux of a plane wave is the product of the density, the wave velocity, and the integral of the square of the ground velocity (Bullen and Bolt, 1985).

If the limits of the spectral bandwidth of seismic instruments are f_1 and f_2, the integral J is given by (Snoke, 1987)

$$J = 2\int_0^\infty |V(\omega)|^2 \, df = 2\int_0^\infty |\omega U(\omega)|^2 \, df$$

$$= \frac{2}{3}[\Omega_0\omega_1]^2 f_1 + 2\int_{f_1}^{f_2} |\omega U(\omega)|^2 \, df + 2|\omega_2 U(\omega_2)|^2 f_2, \tag{11.7}$$

where $V(\omega)$ and $U(\omega)$ are the far-field ground velocity and displacement in the frequency domain. The first and third terms on the right-hand side are correction terms that have been estimated assuming a constant spectral amplitude Ω_0 for $f < f_1$ and a f^{-2} falloff for $f > f_2$. Numerical values of the integral are affected by resolution near zero frequency limited by the reciprocal of window length and by truncation at the Nyquist frequency (half of the sampling frequency). The effects of the bandwidth limitation on the estimates of source parameters can be significant when the corner frequencies are not in the middle of the selected frequency band (Di Bona and Rovelli, 1988).

The pulse shape of the far-field displacement in Brune's source model (Brune, 1970) is given by (Snoke, 1987)

$$u(t) = \Omega_0 \frac{t}{\tau^2} H(t) \exp(-t/\tau), \tag{11.8}$$

where τ is the rise time of the pulse and $H(t)$ is the Heaviside function. The displacement amplitude in the frequency domain for this pulse is given by

$$|U(\omega)| = \frac{\Omega_0}{1 + (\omega/\omega_0)^2}, \tag{11.9}$$

where $\omega_0 = 1/\tau$ is the circular corner frequency. Integral (11.7) of this spectrum is

$$J = \tfrac{1}{4}\Omega_0^2 (2\pi f_0)^3. \tag{11.10}$$

From the low-frequency level Ω_0 and the integral J, therefore, the corner frequency f_0 can be calculated (Snoke, 1987)

$$f_0 = \left(\frac{J}{2\pi^3 \Omega_0^2} \right)^{1/3}. \tag{11.11}$$

The spectral plateau Ω_0 is estimated by inspection on a graphic screen within an iterative computer procedure, into which the calculation of J and f_0 is also implemented, and the spectra are automatically approximated by two straight lines.

Such an approach has been used successfully to estimate the source parameters of small mine tremors in the Ruhr basin in Germany (Gibowicz et al., 1990) and of very small seismic events, with moment magnitude smaller than about -2, induced by the excavation of a shaft in granite in Manitoba, Canada (Gibowicz et al., 1991). It should be noted that corner frequencies calculated from underground records are much more reliable than those obtained from surface records (Malin et al., 1988). Examples of P-wave (A) and SH-wave (B) displacement spectra and their approximation by two straight lines for the event of March 23, 1987 from Heinrich Robert coal mine, Ruhr basin, recorded at a distance of about 5 km, are shown in Fig. 11.1. The event is characterized by moment magnitude $M = 1.9$ and the corner frequencies 5.5 and 4.8 Hz of P and SH waves, respectively. The instrumental response is flat for frequencies above 0.8 Hz, and the Nyquist frequency is 50 Hz. Another examples of acceleration waveforms (A), recorded at a distance of 79 m at the Underground Research Laboratory in Manitoba, and the corresponding P-wave (B) and S-wave (C) displacement spectra from the event of May 23, 1988 with moment magnitude $M = -1.9$

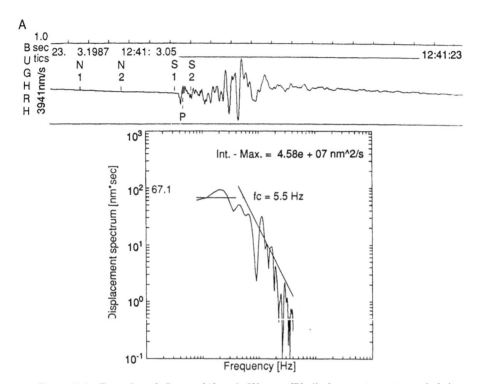

Figure 11.1 Examples of *P*-wave (A) and *SH*-wave (B) displacement spectra and their approximation by two straight lines for the event of March 23, 1987 with moment magnitude $M = 1.9$ from Heinrich Robert coal mine in Germany, recorded at a distance of about 5 km. The selection of signal and noise windows is shown by vertical dashed bars on the longitudinal and transverse component records at the top of each spectrum, where 1-s bars are also indicated. The low-frequency level, the energy flux, and the corner frequency are given on each spectrum.

are shown in Fig. 11.2. The corner frequencies of *P* and *S* waves are 1780 and 1240 Hz, respectively. The instrumental response is flat for frequencies above 0.5 kHz and below 5 kHz, and the Nyquist frequency is 20 kHz.

An alternative set of spectral parameters has been proposed by Andrews (1986). As did Snoke (1987), he used the integral *J*, but instead of Ω_0, his second parameter is the cumulative displacement squared *K* given in the form by Snoke (1987) as

$$K = 2\int_0^\infty |U(\omega)|^2 \, df$$
$$= 2|U(\omega_1)|^2 f_1 + 2\int_{f_1}^{f_2} |U(\omega)|^2 \, df + \frac{2}{3}|U(\omega_2)|^2 f_2, \qquad (11.12)$$

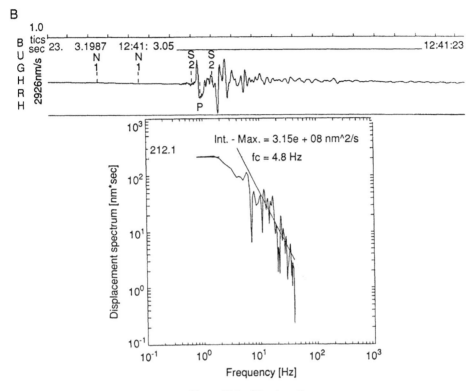

Figure 11.1   (Continued)

where $U(\omega_1)$ is used in the lower-limit correction term because Ω_0 is not an observable in this parametrization. The integral K for the Brune spectrum is

$$K = \tfrac{1}{4}\Omega_0^2(2\pi f_0),$$ (11.13)

and the low-frequency spectral level is related to the integrals J and K, trough relations (11.10) and (11.13), as (Andrews, 1986)

$$\Omega_0 = 2\left(\frac{K^3}{J}\right)^{1/4}.$$ (11.14)

Similarly, in this parametrization the corner frequency of the Brune spectrum is given by

$$f_0 = \frac{1}{2\pi}\left(\frac{J}{K}\right)^{1/2}.$$ (11.15)

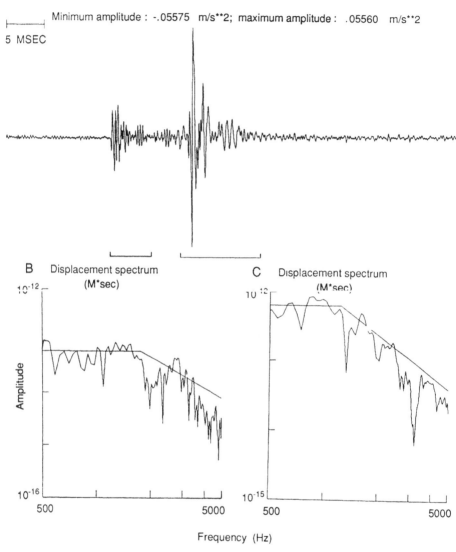

Figure 11.2   An example of acceleration waveforms (A), recorded at a distance of 79 m at the Underground Research Laboratory in Manitoba, Canada, and the corresponding P-wave (B) and S-wave (C) displacement spectra for the event of May 23, 1988 with moment magnitude $M = -1.9$. The time scale in milliseconds and the amplitude range in meters per square second are shown at the top of the figure. The time windows selected for spectral analysis are indicated by horizontal lines below the waveforms of P and S waves.

Snoke (1987) finds from a number of tests that the corner frequency f_0 estimated from formula (11.11) is consistent with what an observer might pick for the corner frequency for a simple event, whereas relations (11.14) and (11.15) often produce values of Ω_0 and f_0, which no observer would pick. Furthermore, experiments with synthetic spectra show that estimates of f_0 based on the (Ω_0, J) parametrization are more stable against changes in the falloff rate of the spectrum than estimates of the spectral level Ω_0 and corner frequency f_0 based on the (K, J) parametrization.

Thus, two independent parameters either Ω_0 and J or J and K are calculated directly from the spectra, leading to an automatic computation either of f_0 or Ω_0 and f_0. In either case our input parameters in the frequency domain, which are further used to estimate source parameters, are Ω_0, J, and f_0 from P, SV, and SH waves. Conventionally, the low-frequency level Ω_0 of S waves, recorded at a given station, is calculated vectorially from the SV and SH components, the integral J of S waves is the sum of SV and SH waves, and the corner frequency f_0 of S waves is the mean value from SV and SH waves. The corner frequency of P waves is expected to be greater than that of S waves; their ratio greater than one is an intrinsic property of earthquake source (Hanks, 1981). It could happen, however, that the apparent value of the ratio of the corner frequency of P wave over that of S wave is smaller than one when the observations are made at low values of the angle between the normal to the fault plane and the takeoff direction of seismic waves (Gibowicz et al., 1990). The ratio therefore can be taken as an indicator of the value of this angle when the fault-plane solution is not available.

If the original records are in the form of accelerograms, one more spectral parameter can be estimated. This is the maximum frequency f_{max} (Hanks, 1982), also called the *cutoff frequency* or *limiting frequency*, observed on acceleration spectra of S waves as the frequency above which the spectra show a sharp decrease with increasing frequency. The frequency f_{max} is also occasionally defined more formally as the frequency corresponding to the half-power level, that is, $P(f) = |F(f)|^2/\Delta t$, where $F(f)$ is the Fourier spectrum and Δt is the length of time window. To determine the frequency f_{max} in practice, either this definition is used or a more liberal approach is applied. In such a case, the acceleration spectra of S waves should be approximated by two straight lines corresponding to the values of Ω_0 and f_0 found from the displacement spectra and then f_{max} can be estimated by inspection on a graphic screen. The accepted value of f_{max} is the mean value from the estimates found from SV and SH waves. Its physical meaning will be described later.

Source parameters are conventionally found from the spectra of each record of an event and then averaged. A better method proposed by Andrews (1986) is to invert the record spectra to find separate station and event spectra, although this approach does not account for directional effects of the source. Assuming that each record spectrum is the product of a station-response spectrum and an event spectrum, we have

$$S_{i(k)}(f)E_{j(k)}(f) = R_k(f), \qquad (11.16)$$

where $R_k(f)$ is the seismic spectrum corrected for geometric spreading and attenuation of the kth record, $S_i(f)$ is the impulse–response spectrum of station i, $E_j(f)$ is the source spectrum of event j, $i(k)$ is the station number of record k, and $j(k)$ is the event number of record k. Equation (11.16) is essentially an extension to all records in a set of data of the method of Mueller (1985), who used the record of a small event as an impulse response to deconvolve the record of a larger event by finding a quotient in the frequency domain.

For K records, I stations, and J events, formula (11.16) describes a system of K equations to determine $I + J$ unknowns for each frequency band. There is one undetermined degree of freedom: if all station spectra are multiplied by an arbitrary function of frequency and all event spectra divided by the same function, the fit is not changed. This can be fixed by prescribing any one of the station spectra equal to one (Andrews, 1986). The system of equations is linearized by taking logarithms

$$\log S_{i(k)} + \log E_{j(k)} = \log R_k, \qquad (11.17)$$

where the frequency argument is suppressed, since the system (11.17) should be applied separately in each frequency band. The matrix has K rows and $I + J$ columns, and each row has only two nonzero elements. This matrix equation can be solved by the singular value decomposition method (e.g., Lawson and Hanson, 1974).

Simultaneous inversion for the quality factor Q and spectral parameters can be performed using seismic spectra. Preliminary estimates of Q, however, are needed for a general approximation of the displacement spectrum with a single corner frequency by four parameters (Boatwright, 1978)

$$\log U(f) = \log \Omega_0 - 0.43\pi(t/Q)f - \tfrac{1}{2}\log\left[1 + (f/f_0)^{2\gamma}\right], \quad (11.18)$$

where γ is the high-frequency decay of the spectrum and t is the travel time. The parameters Ω_0, f_0, γ, and Q can be estimated by an iterative linear inversion of spectral data, expanding relation (11.18) into a Taylor series around some trial estimate $x^* = (\Omega_0^*, f_0^*, \gamma^*, Q^*)$. For details, see Del Pezzo et al. (1987). In general, however, it is not possible to invert simultaneously for f_0, γ, and Q because of a severe trade-off between Q and γ, as might be expected from band-limited data (e.g., Boatwright, 1978; Del Pezzo et al., 1987). To limit this trade-off and at the same time to employ bounds for the attenuation, Q is assigned a number of values different from those of other estimates, and a set of inversion runs is obtained in the analysis of each earthquake. The method proved to be stable and converged well.

A modification of this method was used by Fehler and Phillips (1991) for simultaneous inversion for Q and spectral parameters of microearthquakes induced by hydraulic fracturing in granitic rock. Assuming $\gamma = 2$, they simultaneously fit many spectra to determine Ω_0 and f_0 for each event and one value of frequency-independent Q for all source–receiver paths. They also found that introducing some constraint on the variations in stress drop of a set of earthquakes improves the estimates of Q, f_0, and Ω_0. A more complex approach to the combined inversion for the three-dimensional Q structure and source parameters, using microearthquake spectra, was proposed by Scherbaum (1990) and applied by Scherbaum and Wyss (1990) to P-wave spectra to study the spatial distribution of attenuation in the Kaoiki, Hawaii, source volume. The observed spectra are interactively fit by spectral models consisting of a source spectrum with an assumed high-frequency decay, a single-layer resonance filter to account for local site effects, and a whole-path attenuation along the ray path. The individual Q values along each path are then used as Q starting values for a nonlinear iterative inversion of source parameters and a new Q value. Subsequently, the new Q values are used to reconstruct the next Q model, which again provides starting values for the next nonlinear inversion, and the process is repeated until no further improvement of the results is indicated.

A number of primary parameters measured in the time domain, supplementing spectral parameters, are used to estimate some source parameters. The slope $\dot{u}(t)/t$ of the initial S-wave pulse on the ground velocity waveforms is proportional to the dynamic stress drop reported by Boatwright (1980). The value of the slope is the vectorial sum of the values from SV and SH waves; $\dot{u}/t = [(\dot{u}/t)_{SV}^2 + (\dot{u}/t)_{SH}^2]^{1/2}$. The initial slope of velocity waveforms should be corrected for attenuation using the correction introduced by Choy and Boatwright (1988). The correction has been calculated theoretically by convolving a set of Futterman (1962) attenuation operators (introduced in Chapter 8) with a simple velocity pulse shape. Figure 11.3, reproduced from Choy and Boatwright (1988), shows the decrease in the measured initial slope plotted against the ratio $t^*/\tau_{1/2}$ of the attenuation operator to the measured pulsewidth. To correct the measured slope of ground velocity, it is sufficient to measure the initial pulsewidth, estimate the ratio $t^*/\tau_{1/2}$, and divide the measured slope by the decrease caused by the attenuation, as read from Fig. 11.3.

The pulsewidth or the pulse duration on the ground velocity waveforms, corresponding to the coherent rupture phase (Boatwright, 1984a), is the rise time of displacement. The pulsewidth should be corrected for attenuation by subtracting t^* (e.g., Boatwright, 1984b), and its final value is the mean value

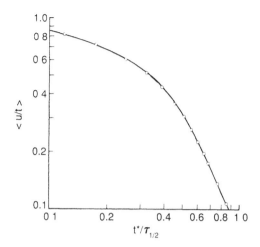

Figure 11.3 Decrease of the measured initial slope of the velocity waveform as a function of the ratio $t^*/\tau_{1/2}$ of the attenuation operator to the measured pulse width. [From Choy and Boatwright (1988, Fig. 14).]

form SV and SH waves. The rise time of displacement pulse can be used for an independent estimate of the source dimension (Boatwright, 1980).

The root-mean-square acceleration a_{rms}^{β} averaged over the duration of the shear wave arrival, or in fact over the faulting duration, is an important indicator of dynamic stress drop (Hanks and McGuire, 1981). The rms acceleration is measured from the onset of S wave within a window equal to the reciprocal of corner frequency. The final value of a_{rms}^{β} is the vectorial sum of the values from the SV and SH waves. The root-mean-square acceleration a_{rms}^{α} of P waves can also be measured in a similar manner, although this is a less significant parameter than a_{rms}^{β}. The ratio of the S-wave to P-wave rms acceleration defines the ratio of the average S-wave to P-wave radiation pattern, which in turn is a function of the change in rupture velocity (Boatwright, 1982).

Two more parameters in the time domain are introduced to serve as approximate routine measures of the event strength in terms of magnitude (described in Section 11.3). The magnitude can be based on either the recorded maximum aplitude of ground motion or the duration of records. The maximum amplitude A_{max} of S waves should be searched within a time window containing the S-wave group on two horizontal components and the maximum value of $A_{max} = (A_{SV}^2 + A_{SH}^2)^{1/2}$ from both the components is calculated. The duration of records should be taken as the mean value from three components. This parameter is measured from the onset of the first P-wave arrivals until a certain amplitude level chosen arbitrarily against the noise level is achieved; a proper choice is a matter of experience. This approach can be automated by calculating the amplitude ratio from two moving windows, one behind the other. The final value of the record duration is obtained by adding the travel time of P waves (from the origin to the sensor) to the value found from the records; the duration of radiated seismic waves, used for the calculation of magnitude, is defined as the time interval from the origin time to the end of recorded vibrations.

11.2 Seismic Moment

The seismic moment M_0, also called the *scalar moment*, in contrast to the moment tensor (introduced in Chapter 9), is a measure of earthquake strength defined in terms of parameters of the double-couple shear dislocation source model. It is the most reliable and useful measure of the strength of a seismic event. M_0 is expressed as follows (e.g., Aki and Richards, 1980)

$$M_0 = \mu \bar{u} A, \tag{11.19}$$

where μ is the shear modulus at the source, $\bar{u}$ is the average displacement across the fault, and A is the fault area. Occasionally, seismic moment can be calculated from field data using relation (11.19), whenever the average displacement across the fault is accessible to measurement and the source area can be determined either from the distribution of aftershocks or from the distribution of underground damage in mines. In practice, seismic moment is most often estimated from seismic records in either the time or frequency domain. For small seismic events, such as mine tremors, the spectral parameter Ω_0 is commonly used for a reliable determination of seismic moment.

The low-frequency level Ω_0 of the far-field displacement spectrum either of P or S waves is directly related to the seismic moment

$$M_0 = \frac{4\pi\rho_0 c_0^3 R \Omega_0}{F_c R_c S_c}, \tag{11.20}$$

where ρ_0 is the density of source material, c_0 is either P-wave velocity or the S-wave velocity at the source, R is the distance between the source and the receiver, F_c accounts for the radiation of either P or S waves, R_c accounts for the free-surface amplification of either P- or S-wave amplitudes, and S_c is the site correction for either P or S waves. If the focal mechanism is not determined, the rms averages of radiation coefficients over the whole focal sphere $R_\alpha = 0.52$ and $R_\beta = 0.63$ (Boore and Boatwright, 1984) can be used for the calculation of seismic moment. The free-surface effect is expressed by a factor of 2 for *SH* waves, whereas for *SV* and *P* waves it is a function of the angle of incidence and frequency and has to be specified individually. A free-surface correction can be neglected (i.e., assumed to be one) for sensors located in underground mines, especially for those located in boreholes. Similarly, site corrections are also often neglected, although near-surface site response is an important factor in source studies. The method of Andrews (1986), described in Section 11.1, can be used to find site-response spectra and site corrections for estimates of M_0.

From the theory of Fourier transform it follows that the low-frequency level Ω_0 is equal to the time integral of the displacement pulse in the time domain and this fact provides a reasonable check on the reliability of Ω_0 estimates in specified units. Durations and areas of displacement pulses are two independent parameters in the time domain, and they can be compared with the values of corner frequency and low-frequency spectral level in the frequency domain (e.g., Rovelli *et al.*, 1991). Here a more general problem of considerable practical interest is touched on, that is, the problem of the relation between the time-domain amplitude of a signal and its spectral amplitude. The conventional procedure for relating time-domain amplitude to spectral amplitude is to postulate that (Houston and Kanamori, 1986)

$$U(\omega_0) \propto A(t)C_D^m, \qquad (11.21)$$

where $A(t)$ is the maximum amplitude in the time domain, $U(\omega_0)$ is the spectral amplitude at the angular frequency ω_0 where $A(t)$ is measured or defined, and C_D is some measure of the duration of the signal. The exponent m can be estimated empirically when both time-domain and spectral amplitudes are available. Its value is usually between 0 and 1.

11.3 Magnitude

The seismic moment is fundamentally superior to any magnitude scale, since it quantifies a parameter of the commonly accepted earthquake source model. The only limitation for M_0 is the difficulty in properly processing seismic records on a large routine scale to determine the size of a seismic event, preferably in real time or at the time of its location. For this purpose, a variety of magnitude scales, in most cases defined in terms of amplitudes recorded over a particular limited spectral band, has been proposed. Magnitude by definition quantifies, therefore, the energy radiated over a fixed frequency band. Since the frequency distribution of radiated seismic energy changes with earthquake size (e.g., Aki, 1967), magnitude scales suffer intrinsic limitations such as saturation and discrepancies between various scales (e.g., Hanks and Kanamori, 1979). Thus, while M_0 is fundamentally and intrinsically superior, magnitude is a pervasive measure of earthquake strength.

The most popular measure of earthquake strength for small events is local magnitude M_L introduced by Richter (1935). He set up a "magnitude scale" of earthquakes, in which the magnitude of an earthquake is defined as the logarithm of the maximum amplitude A_0 measured in micrometers, traced on a seismogram recorded by a standard Wood–Anderson seismograph at a distance of 100 km from the epicenter. Reduction of amplitudes observed at various distances to the expected amplitudes at the standard distance of 100 km is made up by empirical functions obtained under the assumption that the ratio of the maximum amplitudes at two given distances is independent of the azimuth and is the same for all earthquakes considered

$$M_L = \log A(\Delta) - \log A_0(\Delta), \qquad (11.22)$$

where A is the maximum trace amplitude at distance Δ. This magnitude scale has also been used to quantify mine tremors in Poland (Gibowicz, 1963) and in South Africa (e.g., Spottiswoode and McGarr, 1975).

Later, the empirical functions were extended by Gutenberg and Richter (1956) to enable independent magnitude estimates from body- and surface-wave observations. These magnitude scales, denoted by M_s and originally defined for surface waves having 20 s period and by m_b defined for body waves with 1-s period recorded at teleseismic distance, are found to be a strong function of wave frequency. M_s in particular tends to an upper limit for great earthquakes. The moment magnitude M_w, defined by Kanamori (1977) for great earthquakes and based on seismic moment M_0, provides a more uniform scale.

In recent years, moment magnitude M (equal to M_w used for large earthquakes), instead of seismic moment itself, became a frequently used measure of the earthquake strength, simply as a matter of convenience. This has been formally defined by Hanks and Kanamori (1979) as

$$M = \tfrac{2}{3} \log M_0 - 6.0, \qquad (11.23)$$

where M_0 is in N·m (newtons by meters). Thus both M_0 and M are the strength parameters of equal rank. Relation (11.23) is simply a definition, as any magnitude scale is, with constants that correspond more or less to observation; its virtue is that it is uniformly valid in M_0 (Hanks and Boore, 1984).

The empirically found moment–magnitude relations have always been written as a linear relation between $\log M_0$ and magnitude

$$\log M_0 = a M_m + b, \qquad (11.24)$$

where M_m, in general, can be any magnitude and a and b are constants.

Relation (11.24) must be specified for a limited range of M_0, because it will always fail for a sufficiently large variation in M_0 as a result of the size-dependent frequency characteristics of the source excitation and the finite record bandwidth of any time-domain amplitude-based magnitude (Hanks and Boore, 1984).

To introduce a magnitude scale into everyday practice at a given mine, the time-domain input parameters, the maximum amplitude A_{max} and the record duration must be calibrated against the tremor strength (seismic moment or moment magnitude), hypocentral distance and site effects. For this, the values of seismic moment from a set of selected tremors representative for a given mine are needed.

The local magnitude M_A of a given seismic event, based on the maximum amplitude A^j_{max} recorded at station j, can be defined as

$$M_A = \log A^j_{max} + B \log R^j + C + D^j, \qquad (11.25)$$

where B, C, and D^j are constants. The constant B accounts for geometric spreading and attenuation of S waves along the source–receiver distance R, and its value should be between 1 and 1.5. The constant C is a "calibration" adjustment to the amplitude level corresponding to moment magnitude M, and the constants D^j account for the site effects at each receiver. To calculate constants in relation (11.25) from a set of i seismic events with magnitude M^i, the following set of equations is obtained by equating M_A to M

$$M^i = \log A^{ij}_{max} + B \log R^{ij} + C + D^j, \qquad (11.26)$$

which can be solved by least-squares regression analysis (e.g., Draper and Smith, 1981).

The coda-duration magnitude M_D based on the record duration is more often used magnitude than that based on the amplitudes (e.g., Bakun, 1984; Michaelson, 1990). Magnitude M^i_D can be defined as

$$M^i_D = B \log \tau^{ij} + CR^{ij} + D + E^j, \qquad (11.27)$$

where τ^{ij} is the time when the amplitude of the coda envelope decays to the cutoff criteria specified for a given system and corresponds to the lapse-time duration, which is the record duration plus the travel time of P waves; B is a constant close to 2, C is a constant close to zero, D is a "calibration" constant to the level of moment magnitude M, and E^j are constants accounting for the site effects. The constants are calculated from the set of linear equations which are obtained by replacing M_D by M in equations (11.27).

11.4   Seismic Energy

The radiated seismic energy represents the total elastic energy radiated by an earthquake. The seismic energy describes the potential for earthquake damage to artificial structures (buildings, etc.) better than the seismic moment, although the seismic moment provides better description of the overall size of an earthquake (e.g., Boatwright and Choy, 1986). The seismic energy is often used as a measure of the strength of seismic events in mines; at least this is a common practice in Polish mines. In the early days, when only analog records were available, simple techniques for estimating the seismic energy radiated in body waves was used. The energy flux was usually estimated from the peak motion, the dominant period, and the duration of the body-wave arrivals. Then the connection of the magnitude with the seismic energy was sought, usually as a linear relation between the magnitude and the logarithm of energy. The same methods were also used in studies of seismic events induced by mining (e.g., Gibowicz, 1963).

In the seventies, the radiated energy was often calculated from measurements of the seismic moment, corner frequency, and spectral falloff (e.g., Hanks and Wyss, 1972; Randall, 1973; Gibowicz et al., 1977), although this approach contains considerable uncertainties. Estimates of radiated seismic energy from direct measurements of the body-wave energy flux have demonstrated, on the other hand, that such an approach returns robust estimates of the energy (e.g., Snoke et al., 1983; Boatwright and Choy, 1986), provided that the radiated seismic waves are sufficiently sampled over the focal sphere (Boatwright and Fletcher, 1984), since the energy flux radiated in a specific direction in a specific wave type depends on the geometry of the rupture growth through focusing or directivity.

Neglecting directivity, Boatwright and Fletcher (1984) derived the following relations between the energy radiated in the P or S waves and the energy flux contained in the P- or S-wave arrivals:

$$E_c = 4\pi\rho_0 c_0 \langle F_c \rangle^2 \left(\frac{R}{F_c R_c} \right)^2 J_c, \qquad (11.28)$$

where $\langle F_P \rangle^2 = 4/15$ and $\langle F_S \rangle^2 = 2/5$ are the mean-square radiation pattern coefficients (Aki and Richards, 1980). The loss of energy from attenuation and scattering is usually accounted for in the calculation of the energy flux. There are two critical sources of error in relation (11.28): focal mechanism and directivity effects. Small errors in the radiation pattern coefficients, especially if the coefficients are small, can lead to large errors in estimates of the seismic energy. A common method of reducing these effects is to use a lower bound, or "water level," for the radiation pattern. Actual seismic sources can also display a significant amount of directivity.

The directivity function $[1 - \Delta V_r(\cos \psi)/c]^{-1}$, where ΔV_r is the average change in rupture velocity and ψ is the angle between the direction of a seismic ray leaving the source and the direction of rupture propagation, was introduced by Ben-Menahem (1961). If the angle ψ is small, then the recorded ground motion may be considerably increased in amplitude. This effect is called directivity. Although Boatwright and Boore (1982) have shown that large directivity effects occur at high frequencies, there is not much more empirical evidence for these effects. The difficulty in separating the effects of directivity from other factors responsible for the large variations in ground-motion amplitudes partly explains this situation. Furthermore, directivity effects are very large only for small values of ψ, and small values of ψ are commonly observed only over small portions of a fault rupture (Joyner, 1991).

To minimize the errors associated with these effects, the estimates of the energy flux, corrected for attenuation and scattering, should be arithmetically averaged before correcting for the radiation patterns (Boatwright and Fletcher, 1984). Then the resulting estimate of the seismic energy, neglecting free-surface effects, is

$$E_c = 4 \pi \rho_0 c_0 \langle F_c \rangle^2 \frac{\sum_n J_{c_n} R_n^2}{\sum_n F_{c_n}^2}, \qquad (11.29)$$

where the summation index runs over the measurements of the energy flux at n stations. The factor $\sum F_{c_n}^2 / 4 \pi \langle F_{c_n} \rangle^2$ is the radiation pattern coefficient appropriate to the sampling of the data corresponding to the distribution of takeoff angles. If the focal mechanism is not known, this factor becomes the number of measurements divided by 4π. This technique is appropriate for the uniform sampling of the focal sphere. When the focal sphere is not sampled densely, at a few takeoff angles only, relation (11.29) must be modified by including appropriate weights in both summations (Boatwright and Fletcher, 1984).

The values of the total seismic energy $E = E_P + E_S$ of small mine tremors displayed against the seismic moment often show that the energy for a given seismic moment can vary as much as by a factor of 20 (e.g., Gibowicz et al., 1990). The ratio of S- to P-wave energy is an important indicator of the type of focal mechanism responsible for the generation of seismic events in mines. There is definite evidence from natural earthquakes that the energy radiated in P waves is a small fraction of that in S waves, with the ratio E_S/E_P ranging between 10 and 30 (e.g., Boatwright and Fletcher, 1984; Boatwright and Quin, 1986). In contrast, it was found that the ratio E_S/E_P for small mine tremors in the Ruhr Basin, Germany, ranges from 1.5 to 30 (Fig. 11.4); for two thirds of the events this ratio was smaller than 10 (Gibowicz et al.,

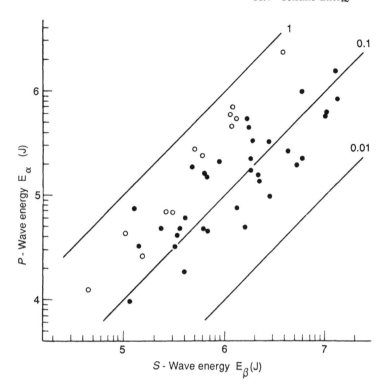

Figure 11.4 Logarithm of *P*-wave energy E_α versus logarithm of *S*-wave energy E_β for small mine tremors in the Ruhr Basin, Germany. The values of the ratio E_α/E_β are indicated by straight lines [From Gibowicz *et al.* (1990, Fig. 12).]

1990). A similar result was also reported from the Underground Research Laboratory in Manitoba, Canada, where extremely small seismic events induced by the excavation of a shaft in granite were observed (Gibowicz *et al.*, 1991). The observed energy depletion in *S* waves could possibly be explained by a non-double-couple focal mechanism of some mine tremors, enriching the energy radiated in *P* waves, and implying that tensile failures, or at least shear failures with tensile components, are often generated in mines.

The radiation efficiency of *P* and *S* waves is another source parameter, introduced by Boatwright and Quin (1986) and defined as

$$\varepsilon_c = \frac{\mu E_c}{\sigma_{rms} M_0},$$
(11.30)

where σ_{rms} is the dynamic stress drop calculated from the root-mean-square acceleration, described in Section 11.6. The total radiation efficiency is the sum of the radiation efficiency of each wave type $\varepsilon = \varepsilon_P + \varepsilon_S$ and it seems to depend on the average rupture velocity in a sense that it increases with increasing velocity. The radiation efficiency therefore becomes an interesting parameter, readily calculated, which could provide an insight into the problem of whether slow or fast rupture processes are dominant in seismic events induced by mining.

The seismic efficiency η of mine tremors, that is, the ratio of the seismic energy E to the total energy E_T released during rupture processes ($E = \eta E_T$), seems to be very low. McGarr (1976) has estimated the seismic efficiency in a gold mine in South Africa by comparing the total energy released by the closure of mine excavation with the observed seismic energy radiated during the same time interval and he has found the value of about 0.24 percent. A comparison of high shear stresses driving the ruptures in the mine (40–70 MPa in the source area) with seismic stress drops has led to the values of seismic efficiency from 0.26 to 3.6 percent (McGarr et al., 1979).

The strain energy released during faulting is consumed in heat, seismic waves, and microstructural defects formed during crushing. New surfaces are the principal defects produced at shallow crustal depths. Fault gauge from mining induced shear fractures at a depth of 2 km in a mine in South Africa has been studied by Olgaard and Brace (1983) to determine the energy allocated to creating new surfaces. They have found that the surface energy could vary from 1 to 10 percent of the total energy released during a mine tremor. Thus most of the energy released during faulting is converted into heat.

11.5   Source Dimensions

While calculation of the seismic moment from relation (11.20) is independent of the source model, estimates of the source dimension are heavily model-dependent. The radius r_0 of the circular fault is inversely proportional to the corner frequency f_c of either P or S wave

$$r_0 = \frac{K_c \beta_0}{2\pi f_c},   (11.31)$$

where K_c is a constant depending on the source model and β_0 is the S-wave velocity in the source area. For the simplest source model of Brune (1970, 1971), represented by a circular dislocation with instantaneous stress release, the constant $K_S = 2.34$ and is not dependent on the angle of observation, and only S waves are considered. This model has been used in innumerable studies to estimate the source size of natural earthquakes. Similarly, in mine seismicity studies, following the results from natural earthquake studies, the source size has been estimated by some authors using Brune's model. In Polish mines, however, it was found that whenever the size and geometry of underground damage caused by rockbursts could be estimated, its radius was considerably smaller than that predicted by Brune's model (Gibowicz *et al.*, 1977; Gibowicz, 1984; Gibowicz *et al.*, 1989). It was also recently found that the Brune model provides unrealistic estimates of the source size of seismic events in French coal mines (Revalor *et al.*, 1990).

The use of Brune's model in mine seismicity studies is vigorously advocated by McGarr (e.g., McGarr, 1984, 1991), who argues that the discrepancies found between the smaller size of underground damage caused by rockbursts and the larger source size predicted by Brune's model, could readily be explained in terms of his inhomogeneous faulting model (McGarr, 1981). Only localized regions within the seismic source would experience a genuine reduction in stresses although displacement occurs over a broader zone. The underground damage would then be related to the area of asperity failure and not to the whole source area.

Whatever the reality behind large mine tremors might be, McGarr's arguments could not be upheld in the case of small seismic events induced by the excavation of a shaft in granite at the Underground Research Laboratory in Manitoba, Canada (Gibowicz *et al.*, 1991). In this case, the corner frequencies of both P and S waves from a single tremor were highly variable, implying that they depend on the angle of observation. The reciprocals of corner frequencies often corresponded to a single pulse with maximum amplitudes in the time domain, implying that the corresponding area is connected with a genuine stress drop. The source radii calculated from the average values of S-wave corner frequencies using Brune's model were unrealistically large in terms of shaft geometry and visual observations of rock failures around the shaft.

The quasidynamic model of the circular fault of Madariaga (1976), on the other hand, provides reasonable results in good agreement with independent observations in mines (e.g., Gibowicz, 1984; Gibowicz *et al.*, 1989; Gibowicz *et al.*, 1990). Madariaga (1976) studied a plane circular model of faulting with fixed rupture velocity. The model is quasidynamic since the effective stress at the fault is specified. The coefficient K_c is a function of the angle θ of observation, that is, the angle between the normal to the fault and the takeoff direction of P or S waves, and this fact must be taken into account when the number of observations is limited (e.g., Gibowicz *et al.*, 1990).

The variation of the coefficient K_c as a function of azimuth θ for three values of the rupture velocity are shown in Fig. 11.5, reproduced from Madariaga (1976). Two angles θ are provided by fault-plane solutions for each receiver, corresponding to two nodal planes. The values of the source radius r_0 can be calculated from relation (11.31) for two sets of angles θ for

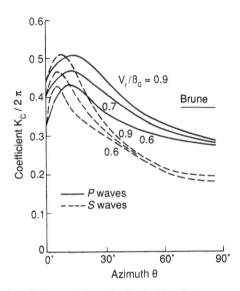

Figure 11.5   Variation of the coefficient $K_c/2\pi$ for P and S waves as a function of azimuth θ for three values of the rupture velocity $V_r/\beta_0 = 0.6$, 0.7 and 0.9. The constant value of $K_S = 2.34$ corresponding to the source model of Brune is also marked. [Modified from Madariaga (1976, Fig. 10).]

both P and S waves, using graphic relations given in Fig. 11.5. The values showing smaller scatter against the average value can then be accepted as the most probable values of the source radius, estimated separately from P and S waves but corresponding to the same nodal plane. Thus under favorable conditions, the corner frequencies of P and S waves can specify a fault plane out of two nodal planes provided by standard focal mechanism determinations (Gibowicz *et al.*, 1991). If the focal mechanism is not determined, the source radius could be calculated by using the average values of the coefficients $K_P = 2.01$ and $K_S = 1.32$ provided by Madariaga (1976) and the mean values of corner frequencies f_c averaged from various receivers.

   An independent check on the values of source radius is provided by the set of observations of the pulse duration τ_r of S waves. The pulse duration of the ground velocity or the rise time of the displacement, expected from a circular rupture, is a simple function of the source radius, rupture velocity V_r and angle θ (Boatwright, 1980; 1984a)

$$\tau_r = \frac{(12 - 13\xi)}{16 V_r} r_t,$$ (11.32)

where $\xi = V_r \sin\theta/\beta_0$ and the source radius is marked by r_t to distinguish it from the radius r_0 calculated from the corner frequency. The difference between the estimates of source size determined from the signal-duration measurements and the spectral measurements increases as the rupture complexity increases.

The rupture velocity between $0.6\beta_0$ and $0.9\beta_0$ is usually assumed in the estimates of source dimensions (e.g., Madariaga, 1976). There are some indications, on the other hand, that mine tremors are probably slower events than natural earthquakes of the same magnitude (Gibowicz et al., 1989). The average change in rupture velocity ΔV_r, caused by the abrupt acceleration or deceleration of rupture front, can be monitored by the ratio of the root-mean-square acceleration in the S waves to that in the P waves. Boatwright (1982) has shown that the ratio of the rms acceleration in the S waves to the rms acceleration in the P waves provides a direct estimate of the ratio of the average S-wave radiation pattern to the average P-wave radiation pattern, which in turn is a function of the change in rupture velocity.

An approximation of far-field displacement spectra by two asymptotes is usually adequate for simple seismic events, whereas complex events, composed of several subevents, can introduce considerable uncertainty into the estimates of source parameters. Examples of simple and complex mine tremors in a gold mine in South Africa have been described by McGarr et al. (1981) and a complex major seismic event in a copper mine in Poland has been described by Gibowicz et al. (1989). Two examples of ground motion and displacement spectra from such mine tremors are shown in Figs. 11.6 and 11.7, reproduced from McGarr et al. (1981). Simple ground motion and its displacement spectrum for an $M_L = 0.72$ event in a South African mine are shown in Fig. 11.6, and complex ground motion and its spectrum for an $M_L = 1.45$ event, composed of at least three major subevents, are presented in Fig. 11.7. The spectrum of a complex event is rather difficult to analyze in terms of Brune's (1970) or Madariaga's (1976) models; its corner frequency is not as well defined as that of a simple event.

The appearance of an intermediate-frequency part, inversely proportional to the frequency, on some spectra implies partial stress drop during a seismic event according to the Brune (1970) model or a fault geometry different from the circular one. In the latter case two corner frequencies f_1 and f_2 found from the spectrum can be used to evaluate a rectangular fault model (Haskell, 1964; Savage, 1972), which assumes simultaneous slip across the fault width W and bilateral rupture along the fault length L at a constant rupture velocity. For the rupture velocity $V_r = 0.9\beta_0$, Haskell's model gives the following relations

$$L = 3.6\beta_0/2\pi f_1 \tag{11.33}$$

and

$$W = 4.1\beta_0/2\pi f_2, \tag{11.34}$$

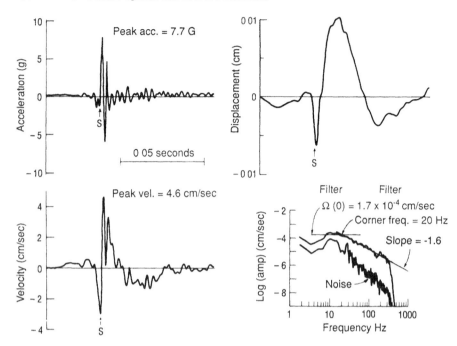

Figure 11.6 Simple ground motion and its displacement spectrum for the horizontal component of event 6 of April 21, 1978 with $M_L = 0.72$ at East Rand Propriety Mines. South Africa. [From McGarr *et al.* (1981, Fig. 5).]

where the fault surface $A = LW$ corresponds approximately to the surface $A = \pi r_0^2$ for r_0 calculated from Brune's model.

Although elongated sources are possible for large shallow earthquakes that rupture the upper crust to long distances along the fault (Scholz, 1982, 1990), there is no clear mechanism for producing elongated faults for most small earthquakes (Brune *et al.*, 196). There seems to be, on the other hand, positive evidence that elongated rectangular faults are reasonable source models for some seismic events generated in Polish deep mines (e.g., Gibowicz *et al.*, 1977; Gibowicz and Cichowicz, 1986; Gibowicz *et al.*, 1989). It might be expected that mining excavations could create favorable conditions for producing such faults.

Partial stress-drop events might occur when the stress release is not uniform and coherent over the whole fault plane. These complexities would occur as a result of asperities or barriers on the fault or complex fault geometry (Brune *et al.*, 1986). The corresponding displacement spectra are shown in Fig. 11.8, reproduced from Brune (1976). For a very small partial stress drop $\epsilon = 0.01$ there is a broad frequency band over which the spectrum falls off as ω^{-1}, and has relatively much more high-frequency energy than a complete stress drop event ($\epsilon = 1$) of the same seismic moment.

The partial stress drop model provides one of the explanations for low-stress-drop events. Since the earliest days of spectral analysis of small earthquakes it was noted that along with events with stress drops of the order of 10 MPa, many events occur with a stress drop as low as 0.01 MPa. The Anza, California, digital array has provided new high-resolution data on the high- versus low-stress-drop events recorded in the same area. Figure 11.9, reproduced from Brune *et al.* (1986), shows shear-wave spectra for the high-stress-drop event of 11.9 MPa and the low-stress-drop event of 0.02 MPa from the same cluster, recorded at the same station. Often the high-stress-drop event is represented by a simple pulse with a well-defined corner frequency and relatively steep high-frequency falloff, whereas many of the low-stress-drop events are more complex in both the time and frequency domains. Brune *et al.* (1986) have speculated that seismic events can be divided into partial stress-drop events ($\Delta\sigma < 10$ MPa), complete stress drop

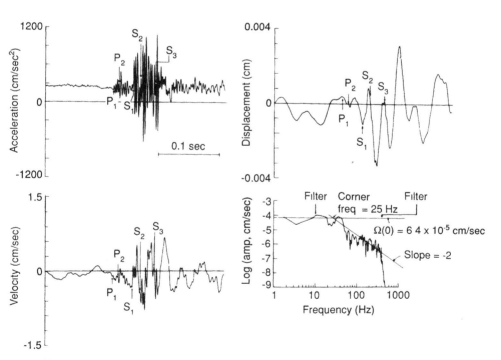

Figure 11.7 Complex ground motion and its displacement spectrum for the horizontal component of event 3 of April 21, 1978 with $M_L = 1.45$ at East Rand Proprietary Mines, South Africa. [From McGarr *et al.* (1981, Fig 6).]

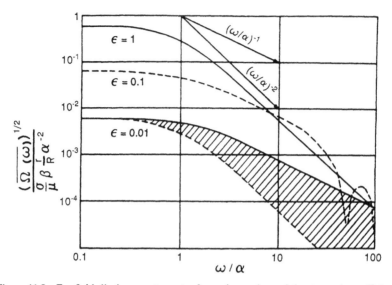

Figure 11.8   Far-field displacement spectra for various values of the stress-drop efficiency ϵ. The crosshatched area represents the part of the spectra that is important in distinguishing the partial stress-drop model ($\epsilon < 1$) and complete stress-release model ($\epsilon = 1$). (From Brune (1976, Fig. 4).]

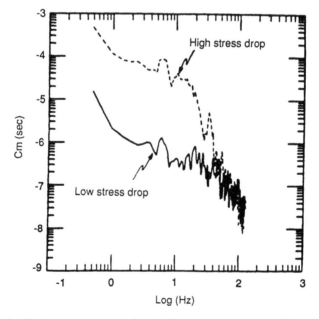

Figure 11.9   Displacement spectrum for the high-stress-drop event of May 27, 1983 ($\Delta\sigma =$ 11.9 MPa) and for the low-stress-drop event of July 20, 1984 ($\Delta\sigma = 0.02$ MPa) recorded at KNW station of the Anza, California, telemetered digital array. [From Brune *et al.* (1986, Fig. 5).]

events ($\Delta\sigma \cong 10$ MPa), and "overstress" stress-drop events with stress drops greater than the absolute effective stress, which can be maintained for a very long time ($\Delta\sigma > 30$ MPa).

11.6 Stress-Release Estimates

At least four different estimates of stress release during earthquakes are used in observational practice at present. The one that represents stress change most accurately is the static stress drop defined as the average difference between the initial and final stress levels over the fault plane. Only the dynamic stress drop (also called the *effective stress*), the difference between the initial stress and the kinetic friction level on the fault, can be determined from seismic data, as well as the *apparent stress*, a quantity based on the radiated energy and seismic moment, which has been proposed as a measure of stress release (e.g., Snoke *et al.*, 1983).

When a complete stress release is assumed, the stress drop can be calculated from the relation (Brune, 1970, 1971)

$$\Delta\sigma = \frac{7}{16} \frac{M_0}{r_0^3}, \tag{11.35}$$

and it represents the uniform reduction in shear stress acting to produce seismic slip over the circular fault. This stress drop is taken as an approximation of the static stress drop and is often termed the *Brune stress drop* $\Delta\sigma = \Delta\sigma_B$ for $r_0 = 2.34\beta_0/2\pi f_s$; it is the most widely used estimate of stress release. Although stress drops show considerable variability from event to event, for most mine tremors they are found to fall within the range from 0.01 to 10 MPa, similar to the ranges for natural earthquakes. Independent estimates of the state of stress within the source region in South African gold mines indicate that stress drops are only small fractions of the total shear stress producing faulting (McGarr *et al.*, 1975, 1979; Spottiswoode and McGarr, 1975).

The apparent stress is (Wyss and Brune, 1968)

$$\sigma_a = \frac{\mu E}{M_0} = \eta\bar{\sigma}, \tag{11.36}$$

where μ is the shear modulus of the source material, E is the radiated energy, η is the seismic efficiency, and $\bar{\sigma} = (\sigma_1 + \sigma_2)/2$ is the average shear stress acting on the fault, in which σ_1 is the stress level before and σ_2 is the stress level after the occurrence of an earthquake. Madariaga (1976) has demonstrated that the apparent stress is proportional to the dynamic stress drop, but does not represent an actual stress difference. If the P-wave contribution to the seismic energy and the azimuthal dependence of the energy flux are neglected, the Brune stress drop is a constant multiple of the apparent stress (Snoke, 1987). For some mine tremors the energy of P waves cannot be neglected and the apparent stress becomes an independent parameter (Gibowicz *et al.*, 1990, 1991).

Two of the estimates of dynamic stress drop are calculated from records of the ground velocity and ground acceleration. The dynamic stress drop σ_d of Boatwright (1980, 1981) is calculated from the initial S-wave slope $\dot{u}/t$ of the far-field velocity waveforms using the relation

$$\sigma_d = \frac{(\rho_0 \rho \beta)^{1/2} \beta_0^{5/2}}{V_r^3} \frac{R}{F_S} (1 - \xi^2)^2 \left| \frac{\dot{u}}{t} \right|, \qquad (11.37)$$

where ρ_0 and β_0 are the density and shear-wave velocity, respectively, at the source, and ρ and β are the corresponding quantities at the receiver; V_r is the rupture velocity; R is the source–receiver distance or the geometric spreading factor, including the amplification at the free surface in a more general case; F_S is the S-wave radiation pattern; and $\xi = V_r \sin\theta / \beta_0$, in which θ is the angle between the normal to the fault and the takeoff direction of the shear wave, and the vector sum of $\dot{u}/t$ from two horizontal components is used. The dynamic stress drop σ_d strongly correlates with the apparent stress (Boatwright, 1982).

The second estimate of dynamic stress drop is calculated from measurements of the rms acceleration a_{rms}^S averaged over the duration of the shear wave arrival. To invert these measurements Hanks and McGuire (1981) used the relation

$$\sigma_{rms} = \frac{2.7 \rho_0 R}{0.85} \left(\frac{f_s}{f_{max}} \right)^{1/2} a_{rms}^S, \qquad (11.38)$$

where f_s is the shear-wave corner frequency and f_{max} is the high-frequency limit of the recorded acceleration; the factor of 0.85 accounts for free-surface amplification, an rms radiation pattern of 0.6, and equal partitioning of energy onto both horizontal components. A rather pleasant surprise resulting from strong ground-motion studies is the remarkable observational stability of a_{rms}^S from one station to the next, and this results in highly stable estimates of dynamic stress drops (Hanks, 1984).

If the rupture process is simple, with a constant rupture velocity throughout the rupture area, the three estimates of stress drop ($\Delta\sigma, \sigma_d, \sigma_{rms}$) will be approximately the same. If the rupture geometry includes asperities or barriers, the three estimates will differ because σ_d is most sensitive to the stress release of the first subevent and both $\Delta\sigma$ and σ_{rms} stress drops are influenced by the stress release from all the subevents in the faulting process (Boatwright, 1984b). Complexity in the rupture process is most probably significant for a majority of earthquakes with magnitude greater than 4, and misleading estimates of source parameters may result from the application of relations based on the assumption of a simple event (Madariaga, 1979; Rudnicki and Kanamori, 1981; Boatwright, 1984a, 1984b). There is no simple procedure to correct for complexity. As a first step, the degree of complexity should be quantified, possible along the lines suggested by Boatwright (1984a), who determined the waveform complexity from the ratios of signal duration, energy flux, and seismic moment in the entire waveform to those corresponding to the most prominent pulse.

The average displacement $\bar{u}$ across the fault plane is an important source parameter, often used for the source characterization. The displacement $\bar{u}$ across the circular fault can be calculated directly from relation (11.19), which for the source area $A = \pi r_0^2$ becomes

$$\bar{u} = \frac{M_0}{\mu \pi r_0^2}. \tag{11.39}$$

In converting seismic comment or average displacement and fault dimensions to stress drop $\Delta\sigma$, a relation of the form

$$\Delta\sigma = C\mu(\bar{u}/l) \tag{11.40}$$

is usually used (Kanamori and Anderson, 1975; Boore and Dunbar, 1977), where l is a measure of fault dimension and C is a numerical factor related to the shape of the fault. For simple geometries, C can be found analytically; otherwise, numerical solutions are required (Boore and Dunbar, 1977; Parsons et al., 1988). For a circular fault of radius r_0 embedded in an infinite medium

$$C = \frac{7\pi}{16}, \tag{11.41}$$

$$l = r_0; \tag{11.42}$$

the value used in relations (11.35) and (11.39). For a two-dimensional fault of width W in an infinite medium and displacement along the long axis

$$C = 4/\pi \tag{11.43}$$

$$l = W \tag{11.44}$$

(Knopoff, 1958). The shape factor C for the same two-dimensional fault near the free surface is available only when the fault breaks to the surface. In this case the factor $l = W$ applies if

$$C = 2/\pi \qquad (11.45)$$

(Kanamori and Anderson, 1975).

Finite-element numerical results for the shape factor C have been presented by Parsons *et al.* (1988) for rectangular strike-slip and two-dimensional strike-slip and dip-slip faults at various depths of burial in an isotropic medium and under a uniform stress drop. All faults were oriented vertically, except a two-dimensional fault that was dipped at 10°. For rectangular strike-slip faults, the factor C ranges from 2.04 for the surface breaking case to 2.55 for the infinitely buried case for $L/W = 1$, from 1.26 (surface breaking) to 1.51 (infinitely buried) for $L/W = 2$, and from 0.65 (surface breaking) to 1.28 (infinitely buried) for $L/W \to \infty$ which is the two-dimensional case, where L and W are the fault length and width, respectively. Vertical deep-slip faults are stiffer than corresponding strike-slip faults and the factor C varies from 0.67 (surface breaking) to 1.70 (infinitely buried) for $L/W \to \infty$. The factor C is considerably reduced by shallow surface breaking dips; C equals 0.48 for a two-dimensional dip-slip fault with a 10° surface-breaking dip. These different values of the shape factor C might be of interest for the estimation of stress drops in mine seismicity studies whenever elongated source-related faults are expected.

The effect of directivity on the stress parameter determined from ground-motion observations has been considered by Boore and Joyner (1989). They calculated this effect for a range of rupture velocities and three directions of rupture propagation along a fault dipping 45°, and the ranges of takeoff angles were taken to simulate ground-motion observations at regional and close distances. Their results suggest that stress parameters determined for earthquakes depend on the amount of directivity incorporated into the observations and not accounted for in the model used to derive these parameters. Observations from different types of waves or from different distributions of stations can lead to apparent variations in the stress parameters, even if the actual stress parameter is constant. The differences can be quite large; well over a factor of 2.

11.7 Scaling Relations

It has been shown that the far-field displacement spectrum is characterized by frequency-invariant amplitudes below the corner frequency and a decay of amplitudes at frequencies above the corner frequency. The corner period, the reciprocal of the corner frequency, is proportional to the source duration and is related to the source dimensions. A source scaling relation describes the manner in which the source duration or the source dimension increases with increasing seismic moment. For a given seismic moment, smaller source dimensions would give rise to a shorter source duration, a higher corner frequency, and larger amplitudes above the corner frequency. This in turn would mean an increase in stress drop.

In studies of large earthquakes occurring in seismic regions of the world, it has been found that stress drop is roughly independent of the seismic moment (Kanamori and Anderson, 1975), which means that M_0 is proportional to r_0^3 [relation (11.35)]. A constant-stress-drop scaling relation has been confirmed by innumerable studies and has become an accepted model for large and moderate earthquakes. Recently the constant stress drop pattern was reported for small volcanic events in Italy (De Natale *et al.*, 1987), for the aftershocks with magnitudes down to 2 near the Norwegian coast (Chael and Kromer, 1988), and for microearthquakes with magnitudes less than about 2 observed at Anza, California (Frankel and Wennerberg, 1989). The constant-stress-drop model implies a self-similar rupture process regardless of the scale of the seismic events.

The constancy of stress drop has been found for some seismic events in gold mines in South Africa (Spottiswoode, 1984) and in coal mines in Poland (Gibowicz *et al.*, 1977).

In contrast to these results, there is growing evidence of a breakdown in similarity between large and small earthquakes (e.g., Chouet *et al.*, 1978; Archuleta *et al.*, 1982; Haar *et al.*, 1984; Archuleta, 1986; Fletcher *et al.*, 1986; Dysart *et al.*, 1988; Glassmoyer and Borcherdt, 1990). Generally, a marked decrease in stress drop with decreasing seismic moment, for seismic moment below about 10^{13} N · m ($M < 3$), is reported. In Fig. 11.10, reproduced from Fletcher *et al.* (1986), the seismic moment (in dyn · cm; 1 N · m = 10^7 dyn · cm) as a function of the source radius, bounded by the contours of constant stress drop (in bars; 1 MPa = 10 bar), is shown for Anza, California, small earthquakes recorded by a wideband array. The source radii for these events are roughly constant over four orders of magnitude in seismic moment. This apparent constancy leads to a strong dependence of stress drop on the seismic moment. Another manifestation of this apparent breakdown in similarity is the divergence of the scaling of peak acceleration and ground velocity from that expected from theoretical considerations following the similarity relation (McGarr, 1986).

It seems that the most convincing evidence of the breakdown in scaling relations for small seismic events comes from studies of mine-induced seismicity, based on underground seismic networks where recorded waveforms

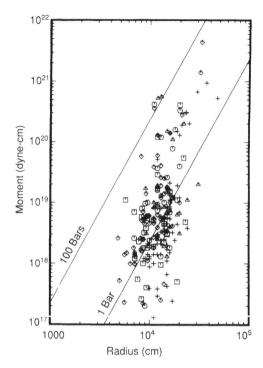

Figure 11.10   Seismic moment as a function of source radius for Anza, California, earthquakes Data are plotted by source regions marked by various symbols. [From Fletcher *et al.* (1986, Fig. 1), copyright by the American Geophysical Union.]

are not practically affected by free-surface effects and where site effects are not expected to be drastic.

Stress drop decreasing with decreasing seismic moment has been definitely observed for small mine tremors with the moment from about 10^{11} to 10^{13} N · m in the Polish copper mines (Gibowicz, 1985). In South Africa more than 100 mine tremors with magnitude from 0 to 3 were recorded on the surface and at depths of 1768 m and 3048 m within the Western Deep Levels gold mine (Bicknell and McGarr, 1990). It was found that the breakdown in scaling principles persists with the underground data recorded within a few hundred meters of the source and in the source spectra. A seismic monitoring system was installed 200 m ahead of an operational stope in the same mine, and an on-line estimate of the source parameters of microtremors was performed (Cichowicz *et al.*, 1990). The seismic moment ranged from 10^4 to 10^7 N · m, the corner frequency of S waves ranged from 700 to 2500 Hz, and

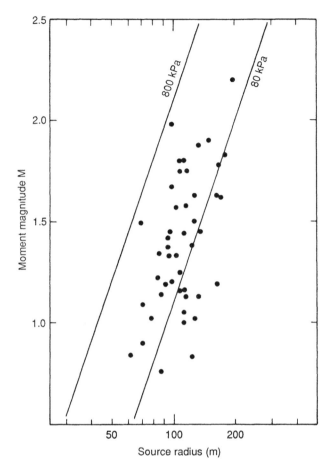

Figure 11.11 Moment magnitude as a function of source radius for small seismic events at the Heinrich Robert coal mine in Germany. [Modified from Gibowicz *et al.* (1990, Fig. 10).]

the stress drop ranged from 0.01 to 1.4 MPa. It was found that the corner frequency is independent of the seismic moment for events with seismic moment smaller than about $5 \cdot 10^5$ N $\cdot$ m.

The evidence for scaling relations for small seismic events with the seismic moment 10^{11}–10^{12} N $\cdot$ m at the Heinrich Robert mine in Germany suggests that the stress drop is moment dependent (Gibowicz *et al.*, 1990). This evidence is shown in Fig. 11.11, where moment magnitude versus source radius is presented. The observed corner frequencies ranged from 4 to 14 Hz. The idea of a characteristic earthquake advocated by Aki (1984, 1988), in which asperities and barriers along the fault are considered as its physical basis, was adopted to the notion of a characteristic mine tremor of the barrier type (Takeo, 1983; Aki, 1984), in which the slip varies by more than one order of magnitude whereas the length of the fault is stable. The characteristic fault length at the Heinrich Robert mine is about 200 m, corresponding exactly to the width of mine longwalls in the area. Thus it is possible to speculate that the source dimension of induced seismic events could be related to the geometry of mining operations (Gibowicz *et al.*, 1990).

The nonsimilar behavior of small earthquakes has been widely interpreted as a source effect, involving either an upper limit to the radiated frequency, that is, the presence of a characteristic fault length, or the dependence of stress on the seismic moment. The change in spectra scaling can also be explained by attenuation effects or by any process that limits high frequencies, whether it is due to the source, the propagation path, the local site, or the recording instrument. This effect of finite bandwidth has been recognized for some time (e.g., Hanks and McGuire, 1981; Hanks, 1982; Hanks and Boore, 1984). Boore (1986) has shown that a moment-independent filter that attenuates high frequencies, regardless of its origin, produces marked changes in the scaling expected from the usual analysis of self-similar models. A conciusive siudy of the causes of the removai of high frequencies from the radiated field will require recordings at varying depths in the Earth and at a variety of surface sites for a set of seismic events. Such observations are possible in studies of seismicity in mines, as proposed by Hanks (1984), in which stations located at considerable depth are removed from the free surface and from near-surface effects strongly influencing the recorded motions and leading to biased source parameters.

The problem of the nonsimilar behavior of small seismic events is identical with and directly related to that of the origin of the high-frequency band-limiting of the radiated shear acceleration field or, in other words, the upper limit frequency of shear acceleration spectra that Hanks (1982) called "f_{max}." An example of the acceleration spectrum with such a limit, for the San Fernando, California, earthquake of 1971 with magnitude $M_L = 6.4$, is shown in Fig. 11.12, reproduced from Hanks (1982), in which f_{max} is about 10 Hz.

The origin of f_{max} is controversial and sometimes confusing because it could be attributed to source, path, or recording site effects or to a combination of all three. There are strong arguments that f_{max} is the source property (e.g., Papageorgiou and Aki, 1983a, 1983b, 1985; Aki, 1984, 1987, 1988; Archuleta, 1986; Papageorgious, 1988) and may represent a minimum source dimension. There is some theoretical justification for this view as well. If this interpretation is correct, then stress drop would decrease without limit for earthquakes at the minimum source radius, since seismic moment is not found to have a lower observational limit (Scholz, 1990). There are, however, equally strong arguments that f_{max} is controlled by local recording site

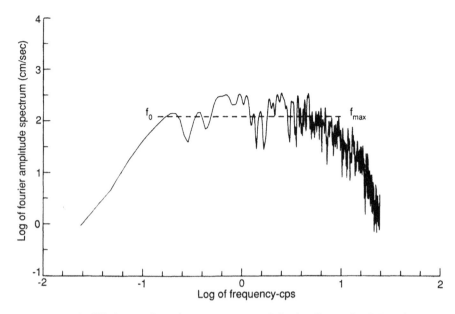

Figure 11.12 Whole-record acceleration spectrum of the San Fernando, California, earthquake of February 9, 1971, $M_L = 6.4$, at Pacoima Dam. The frequencies f_0 and f_{max} are estimated as indicated. [From Hanks (1982, Fig. 2).]

conditions (e.g., Hanks, 1982; Anderson and Hough, 1984; Cranswick *et al.*, 1985; Hauksson *et al.*, 1987; Frankel and Wennerberg, 1989). Hanks (1982) has argued that since high frequencies are strongly attenuated in the immediate vicinity of the Earth's surface, there is a maximum frequency f_{max} of seismic waves that may be detected by surface instruments.

Papageorgiou and Aki (1983a) interpreted the observed f_{max} in terms of the cohesive zone of extent L at the crack tip acting as a low-pass filter with a cutoff frequency equal to V_r/L $(= f_{max})$, where V_r is the spreading velocity of the rupture front. The frequency f_{max} of large earthquakes may be determined by the width of fault zone. The slip near a crack tip associated with a major earthquake may be smeared out over a distance comparable to the fault-zone width, and high-frequency waves may be smoothed out (Aki, 1987). Another very interesting observation is that the frequency f_{max} from large earthquakes is roughly the same as the corner frequency of small earthquakes when the latter become nearly constant and independent of seismic moment below a magnitude of about 3 (e.g., Archuleta, 1986; Aki, 1987, 1988). The corner frequency of small earthquakes with the fault length smaller than the fault width may be primarily controlled by the fault-zone width.

The other fundamental scaling relation for earthquakes is expressed in their frequency–seismic moment distribution. In any region it is found that during a given period of time the number $N(M)$ of earthquakes with moment magnitude M obeys a relation

$$N(M) = aM^{-b}, \tag{11.46}$$

where a and b for a given set of observations are constants, although in general they are variable in space and time. This relation is historically known as the Gutenberg–Richter or Ishimoto–Aida relation. This type of distribution is observed for widely diverse phenomena and is often referred to as the *law of Pareto*. A power-law size distribution of this type is typical of fractal sets and arises simply from the self-similarity of earthquakes (e.g., Scholz, 1990).

There is evidence that significant departures from relation (11.46) are present for earthquakes with magnitude smaller than about 3, implying a breakdown of self-similarity at the small end as well. Aki (1987) studied the frequency–magnitude relation for earthquakes with magnitude range from $-1/3$ to 4 recorded at a borehole station in the Newport-Inglewood fault zone in California. He found a clear departure of the observed frequency–magnitude relation for earthquakes with $M < 3$ from the linear relation (on a logarithmic scale) extrapolated from the data for $M > 3$. The observed number of earthquakes with $M \cong 0.5$ was almost 10 times smaller than that expected from the extrapolation of empirical relation for earthquakes with magnitudes larger than 3. Aki (1987) concluded that the observed frequency–magnitude relation departs from self-similarity for earthquakes with magnitude smaller than about 3. This departure coincides with the widely observed departure of the seismic moment-corner frequency relation from self-similarity.

Chapter 12 | Statistical Assessment of Seismic Hazard in Mines: Statistical Prediction

Although the occurrence of mine tremors is not strictly a random process, a statistical approach to the analysis of seismic events in mines provides a reasonable basis for seismic hazard assessment. In this chapter two fundamental types of statistical approach are considered.

The first approach is the assessment of seismic hazard, understood as an estimation of the mean probability (over space and time) of the occurrence of a seismic event with a certain magnitude within a given time interval. Different models can be applied under real mining conditions. It should be noted that this approach is valid only when seismic activity exhibits stationarity and independence (Lomnitz, 1966). Despite these limitations, the defined concept of seismic hazard has some important advantages and is worth serious attention. The probability that rockbursts will occur in particular mining areas in given time intervals provides information that is needed for the estimation of expected losses. Such evaluations are usually simple and fast and do not require advanced computational techniques.

The second approach is the continuous evaluation of the hazard as a function of time or the amount of extracted rocks. For this purpose certain models of stationary hazard evaluations are used and applied within the moving time windows. Thus, the process is time dependent through the time dependence of its parameters. We refer to such an approach as a *time-dependent hazard* or, equivalently, a *statistical prediction*. We introduce two models: the first based on a theoretical dependence between the radiated seismic energy and the amount of extracted rocks (Kijko, 1985), and the second based on empirical dependence of the number of events and their energy on the distance from a longwall face (Kijko and Syrek, 1988). The selection of material in this chapter reflects the author's personal interests and is far from a complete coverage of the statistical methods that can be applied in seismicity studies in mines.

12.1 The Gutenberg–Richter Frequency–Magnitude Distribution

Many observations indicate that seismic events induced by mining follow the same rules as those obeyed by natural earthquakes. A prominent place in the analysis of both types of seismicity belongs to the frequency–magnitude relation introduced by Gutenberg and Richter (1944, 1954), and already described in Chapter 11 [relation (11.46)], which on a logarithmic scale takes the following form

$$\log n = a - bm, \tag{12.1}$$

where n is the number of earthquakes with magnitude m, and a and b are parameters. Relation (12.1) may be interpreted either as being a cumulative relationship, if n is the number of events with magnitude equal or larger than m in a given time interval, or as being a density law, if n is the number of events in a certain small magnitude interval around the value of m. The parameter a is a measure of the level of seismicity, whereas the parameter b, which is typically close to 1, describes the relative number of small and large events in a given interval of time. Once the parameters a and b in relation (12.1) are determined, practical implications for the mining industry would follow. For example, it was found at one of East Rand gold mines in South Africa that over a 100-day period in an area involving approximately 300 m of face advancing at an average rate of 20–30 cm per day, the parameters a and b were 2.19 and 0.63, respectively (McGarr, 1984). If the events with $M_L = 2.0$ are taken as a minimum size of the tremors that might be expected to cause damage, then the Gutenberg–Richter relation (12.1) shows that 8 or 9 cases of disruption to the mining operations may occur every 100 days.

There are numerous papers devoted to the frequency–magnitude relation. They all show that the Gutenberg–Richter relation holds for virtually all magnitude ranges, in all locations and at all times (Rundle, 1989). Only in some special cases does the log–linear relation (12.1) break down. Criticism of relation (12.1) is consistently centered around its unsatisfactory behavior in the high-magnitude range (e.g., Lomnitz, 1974). The tendency to overestimate the likelihood of strong event occurrence thus appears to be built into the formula (Lomnitz-Adler and Lomnitz, 1979). As a result, some seismologists have proposed an alternative approach by introducing either maximum magnitude m_{max} (Cornell, 1968) or a certain modification of relation (12.1) (e.g., Merz and Cornell, 1974). According to Aki (1987), the nonlinearity can also appear for very small events. Taylor et al. (1990) attribute the nonlinearity of the frequency–magnitude relation to either observational or source effects.

Great efforts have been devoted to the investigation of the physical significance of relation (12.1) and its parameters. From the rock deformation experiments of Scholz (1968) it follows that b-value variations are directly related to the stress conditions. He found that the parameter b varied in a characteristic way: it decreased as stress was increased. Scholz (1968) believes that the state of stress rather than the heterogeneity of rocks plays a more important role in determining the value of b. Somewhat similar results were also found from the study of earthquakes (Gibowicz, 1973a, 1973b). The parameter b depends also on the rheology and structure of the material, and especially on the presence of defects. The parameter b therefore can be considered, more generally, as a parameter controlling the medium capability of releasing the accumulated energy.

Time variations of the coefficient b were found during earthquake sequences in New Zealand (Gibowicz, 1973a; Imoto, 1987) and for earthquakes in California (Wyss and Lee, 1973). The variations preceded the 1976 Tangshan earthquake in China (Li *et al.*, 1978) and the 1976 Friuli earthquake in Italy (Cagnetti and Pasquale, 1979). These changes were also observed in laboratory experiments modeling the fault movements (e.g., Weeks *et al.*, 1978; Main *et al.*, 1989; Meredith *et al.*, 1990). In most cases a decrease in b values before slips on cut surfaces, followed by an increase after the event, were found. Characteristic changes of b values are considered, among the other precursory phenomena, in models of physical processes occurring in the areas of pending earthquakes (e.g., Mjackin *et al.*, 1975), and time variations of b are routinely observed for earthquake prediction in Kamchatka (Fedotov *et al.*, 1976) and in Japan (Shibutani and Oike, 1989).

It should be noted that precursory phenomena before seismic events induced by mining were also observed. Anomalous seismicity changes (an increase followed by a decrease) were recorded prior to moderate rockbursts in a deep silver mine in northern Idaho (U.S.A.) (Brady, 1977), in coal mines in Poland (Gibowicz, 1979), and in gold mines in South Africa (Brink, 1990).

Several efficient statistical procedures for the evaluation of parameter b are available. To demonstrate the maximum-likelihood method, we will describe the classic procedure of the b parameter evaluation introduced by Aki (1965) and Utsu (1965), performed under the assumption that magnitudes are continuous and unlimited from the top.

If the magnitudes of seismic events are assumed to be independent, identically distributed random variables, the frequency–magnitude Gutenberg–Richter relation (12.1) can be written in the form

$$f(m) = \begin{cases} 0 & \text{for } m < m_{min}, \\ \beta \exp[-\beta(m - m_{min})] & \text{for } m \geq m_{min}, \end{cases} \qquad (12.2)$$

or

$$F(m) = \begin{cases} 0 & \text{for } m < m_{min}, \\ 1 - \exp[-\beta(m - m_{min})] & \text{for } m \geq m_{min}, \end{cases} \qquad (12.3)$$

where $f(m)$ and $F(m)$ are the probability density and cumulative distribution functions of magnitude m, respectively; m is considered as a continuous variable that may assume any value above the threshold value m_{min}; and $\beta = b \ln(10)$. If the magnitudes are considered to be independent, then the joint probability density for the set of N magnitudes m_i $(i = 1, \ldots, N)$ is equal to the product of the individual probability densities $f(m_i)$. The maximum-likelihood estimate of the parameter β is the value of $\hat{\beta}$ for which the likelihood function $L(\cdot)$, proportional to the joint probability density, is maximum. The maximum likelihood condition can be written as

$$L(\beta|m_1, \ldots, m_N) = \text{const} \prod_{i=1}^{N} f(m_i|\beta) = \max, \qquad (12.4)$$

or equivalently

$$\sum_{i=1}^{N} \frac{\partial}{\partial \beta} \ln f(m_i|\beta) = 0. \qquad (12.5)$$

Taking the logarithm of $f(m|\beta)$ and calculating its derivative, we can write the maximum-likelihood condition (12.5) as

$$\frac{N}{\beta} - \left(\sum_{i=1}^{N} m_i - N m_{min} \right) = 0, \qquad (12.6)$$

from which the classic Aki-Utsu maximum likelihood estimate of β follows (Aki, 1965; Utsu, 1965)

$$\hat{\beta} = \frac{1}{\langle m \rangle - m_{min}}, \qquad (12.7)$$

where the sample mean magnitude $\langle m \rangle = \sum_{i=1}^{N} m_i / N$. From the central limit theorem (Eddie et al., 1982) it follows that for sufficiently large N, $\hat{\beta}$ is approximately normally distributed about its mean value equal to (12.7) with the standard deviation equal to

$$\hat{\sigma}_\beta = -\left(\frac{\partial^2 \ln L}{\partial \beta^2} \right)^{-1/2} = \hat{\beta}/\sqrt{N}. \qquad (12.8)$$

The standard deviation of $\hat{b}$ is obtained by dividing $\hat{\sigma}_\beta$ by $\ln(10)$.

Another more precise approximation of (12.7), based on the β distribution of Utsu (1966), was given by Zhang and Song (1981) as

$$\hat{\sigma}_\beta = \hat{\beta} N / \left[(N-1)(N-2)^{1/2} \right], \qquad (12.9)$$

which for the large number N of events is equivalent to the estimation (12.8) of Aki (1965). For an estimate of time varying seismic hazard, $\hat{\sigma}_\beta$ is more useful when obtained from the formalism involving time-dependent β. Shi and Bolt (1982) showed that for large samples N and slow temporal changes of β, the standard deviation of $\hat{\beta}$ is

$$\hat{\sigma}_\beta = \hat{\beta}^2 \left\{ \sum (\langle m \rangle - m_i)^2 / [N(N-1)] \right\}^{1/2}, \tag{12.10}$$

where the sum is taken from $i = 1$ to $i = N$. The problem of β varying slowly with time was also studied by Guttorp and Hopkins (1986). They show that under an additional assumption that the number of seismic events in time is a Poisson process and that the event magnitudes are independent of the number of events, the approximate standard deviation of $\hat{\beta}$ is

$$\hat{\sigma}_\beta = \frac{\sinh(\Delta m \, \hat{\beta}/2)}{\Delta m/2}, \tag{12.11}$$

where Δm is the magnitude interval corresponding to the accuracy of magnitude determination.

The maximum-likelihood procedure has several desirable statistical properties. It provides more robust estimates and is asymptotically unbiased. Moreover, in the limit of large samples, the variance of the maximum-likelihood estimate is at least as small as any other unbiased estimator (Eadie et al., 1982).

Since its first derivation in 1965, the Aki–Utsu formula (12.7) has been successfully used in numerous studies in various areas with entirely different patterns of seismicity. Nevertheless, this approach has significant shortcomings, such as the assumption that magnitudes are unbounded from the top and that they are continuous variables.

The belief that there must be an upper limit to the earthquake magnitude has been expressed by many seismologists (e.g., Yegulalp and Kuo, 1974; Weichert, 1980). Knopoff and Kagan (1977) have demonstrated that if the Gutenberg–Richter relation is valid for a given set of earthquakes, then the upper bound to the magnitude must be present. Assuming that relation (12.1) describes the cumulative Gutenberg–Richter relation and that the seismic energy–magnitude relation is of the form

$$\log E = c + dm, \tag{12.12}$$

where c and d are constants, Knopoff and Kagan (1977) showed that the total amount of energy released by earthquakes in a time unit is

$$\hat{E}_{\text{TOTAL}} = \int_{E_{\min}}^{E_{\max}} E \, dN = \text{const} \; E^{1-b/d} \Big|_{E_{\min}}^{E_{\max}}. \tag{12.13}$$

For typical values of $b \cong 1$ and $d \cong 1.5$ and for $E_{max} \to \infty$, the total amount of released seismic energy E_{TOTAL}, therefore, tends to infinity. This fact Knopoff and Kagan (1977) called "the E_{max} catastrophe."

Thus it is clear that the upper bound of E_{max} (or equivalently m_{max}) must be introduced. It is easy to show that if the upper limit of magnitude is taken into account, the density and cumulative probability distribution functions (12.2) and (12.3) take the form (Page, 1968)

$$f(m) = \begin{cases} 0 & \text{for } m < m_{min} \text{ and } m > m_{max}, \\ \dfrac{\beta \exp[-\beta(m - m_{min})]}{1 - \exp[-\beta(m_{max} - m_{min})]} & \text{for } m_{min} \leq m \leq m_{max}, \end{cases}$$

$$(12.14)$$

and

$$F(m) = \begin{cases} 0 & \text{for } m < m_{min}, \\ \dfrac{1 - \exp[-\beta(m - m_{min})]}{1 - \exp[-\beta(m_{max} - m_{min})]} & \text{for } m_{min} \leq m \leq m_{max}, \quad (12.15) \\ 1 & \text{for } m > m_{max}. \end{cases}$$

Following the procedure described by relations (12.4) and (12.5), the maximum-likelihood equation for β is (Page, 1968)

$$\frac{1}{\beta} = \langle m \rangle - m_{min} + \frac{(m_{max} - m_{min})\exp[-\beta(m_{max} - m_{min})]}{1 - \exp[-\beta(m_{max} - m_{min})]}. \quad (12.16)$$

The exact evaluation of β from equation (12.16) requires the knowledge of m_{min} and m_{max} and can be obtained only by recursive solutions. Nevertheless, a simple approximation of β is possible. It is clear that β obtained from the solution of equation (12.16) is smaller than β estimated from the Aki–Utsu formula (12.7), in which the presence of m_{max} is ignored. Thus, a proper estimation of β can be performed by using simple formula (12.7) and then by correcting the result for the expected bias. Consequently, with an accuracy to the second term of the Taylor expansion of (12.16), the approximate β value becomes

$$\hat{\beta} = \hat{\beta}_0(1 - \kappa_{max}), \quad (12.17)$$

where

$$\kappa_{max} = \hat{\beta}_0 \frac{(m_{max} - m_{min})\exp[-\beta_0(m_{max} - m_{min})]}{1 - \exp[-\beta_0(m_{max} - m_{min})]}, \quad (12.18)$$

and $\hat{\beta}_0$ is the Aki–Utsu estimation of β [relation (12.7)].

The second restriction in the Aki–Utsu formula (12.7) comes from the assumption that the magnitudes are continuous. It is well known that magnitudes are determined to a fraction of the magnitude unit: to 0.1 of the unit at the best. The magnitudes of historical earthquakes, for example, are known with an accuracy of only 0.5 of the unit. The fact that the discretization of magnitude values may affect the estimates of the parameters in frequency–magnitude relations was first noted by Utsu (1966), who proposed a correction for formula (12.7). An extensive study of the effects of discretization of the magnitude axis and the effect of grouping of magnitudes into certain magnitude intervals was performed by Bender (1983). She showed that the maximum-likelihood estimation of β requires the solution of the equation

$$\frac{q}{1-q} - \frac{nq^n}{1-q^n} = \sum_{i=1}^{n} \frac{(i-1)k_i}{N}, \tag{12.19}$$

where $q = \exp(-\beta\Delta m)$, $n = (m_{max} - m_{min})/\Delta m$ and is equal to the number of magnitude intervals of with Δm, k_i $(i = 1, \ldots, n)$ are the numbers of events with magnitudes in the ith interval, and $\Sigma k_i = N$ is the total number of events with magnitudes complete above the threshold m_{min}. Bender (1983) also showed how to correct the continuous magnitudes in the Aki–Utsu and Page formulas for the bias following magnitude discretization.

For the case of $m_{max} \rightarrow \infty$, the Bender's correction to the Aki–Utsu formula (12.7) can be readily approximated as

$$\hat{\beta} = \hat{\beta}_0 \kappa_\Delta, \tag{12.20}$$

where

$$\kappa_\Delta = \hat{\beta}_0 \frac{\Delta m}{2} \frac{1 + \exp\left(-\hat{\beta}_0\Delta m\right)}{1 - \exp\left(-\hat{\beta}_0\Delta m\right)}, \tag{12.21}$$

and $\hat{\beta}_0$ is calculated according to formula (12.7), in which the sample mean magnitude $\langle m \rangle = \Sigma k_1 m_1/N$, where $m_i = m_{min} + i\,\Delta m/2$ is the center of the ith magnitude interval and summation is taken from $i = 1$ to $i = n$. Correction (12.21) was introduced for the first time by Utsu (1971), who showed that formula (12.7) provides β estimates systematically smaller when the magnitude interval Δm is large. For $\Delta m \rightarrow 0$, the correction factor $\kappa_\Delta \rightarrow 1$ and relation (12.20) becomes equal to formula (12.7).

Other alternative solutions were given by Guttorp and Hopkins (1986) and Tinti and Mulargia (1987), who showed that the maximum-likelihood estimator of β for grouped data coincides with the solution obtained by both the method of moments and the maximum entropy:

$$\hat{\beta} = \frac{1}{\Delta m} \ln\left(1 + \frac{\Delta m}{\langle m \rangle - m_{min}} \right). \tag{12.22}$$

Formula (12.22) is superior to the Bender relation (12.19) because its solution does not require recursive procedures.

One more simple correction to formula (12.7) is possible. Zhang and Song (1981) and Ogata and Yamashina (1986) showed that relation (12.7) provides a biased estimation of β and that the bias is significant when the number of events is small or β is large. The bias can be removed using the following relation:

$$\hat{\beta} = \hat{\beta}_0 \frac{N}{N-1}. \tag{12.23}$$

This correction can play a significant role when the $\hat{\beta}$ value is scanned in the time domain with the time window of constant length. Each time the window covers a small number of events, the bias of β evaluation becomes significant.

Thus far we have focused mainly on the estimation of the b parameter in unlimited and truncated from the top, continuous and discrete Gutenberg–Richter relations. In all cases it was assumed that the observed magnitudes are correct and that their uncertainties result only from magnitude discretization. Magnitude uncertainties, however, whether resulting from discretization or from observational errors, can be considered in an entirely different way (Tinti and Mulargia, 1985; Bender, 1987; Kijko and Sellevoll, 1992). The results of theoretical and numerical studies show that significant magnitude uncertainty strongly affects the parameter a, whereas the estimators usually employed for the calculations of b remain valid. The effects of magnitude errors, following normal distributions with the standard deviation up to 0.2 of the magnitude unit, can be neglected (Kijko and Sellevoll, 1992). The problem of magnitude uncertainty is discussed further in Section 12.6.

An example of the application of relation (12.7) to an assessment of time-dependent seismic hazard in mines is shown in Fig. 12.1, reproduced from Lasocki (1990). The data are related to the area of longwall 11 in seam 507 at Bobrek coal mine, Upper Silesia, Poland. The analysis covers the period from the beginning of longwall in August 1986 to April 21, 1987. The number of seismic events (altogether 339 events with energy equal or larger than 10^4 J) varied significantly in time with the average rate $\lambda = 1.15$ events per day. The probability of occurrence of a seismic event with the energy $E_0 \geq 3 \cdot 10^5$ J was calculated as follows (Lasocki, 1990). Assuming that the energy-magnitude relation is described by formula (12.12), the Gutenberg–

Richter relation (12.1) takes the form

$$\log n = A - B \log E, \tag{12.24}$$

where $A = a - c/d$, $B = b/d$, and the maximum-likelihood estimation of B is

$$\hat{B} = \frac{\ln 10}{\langle \log E \rangle - \log E_{min}}. \tag{12.25}$$

In (12.25) $\langle \log E \rangle$ is the sample mean of the logarithm of energy of considered events and E_{min} is the threshold value above which the catalog of events is complete; in this case, it is assumed to be equal to 10^4 J. Assuming further that the occurrence of seismic events is Poissonian in time and stationary within time interval $(t - \Delta T, t)$, where ΔT is long enough to observe data for a reliable estimation of parameters, the probability of occurrence of at least one event with the energy $E \geq E_{min}$ between t and $t + \Delta t$ is $1 - \exp(-\lambda \Delta t)$, where λ is the mean activity rate defined as a mean number of events with energies $E \geq E_{min}$ per a time unit. Thus, if the process is time-dependent through a time dependence of its parameters $B(t) \equiv B$ and $\lambda(t) \equiv \lambda$, but can be considered stationary for any time interval $\langle t - \Delta T, t \rangle$, the probability of occurrence of an event with the energy greater than or equal to a given energy E_0 ($E_0 \geq E_{min}$), within a time interval $(t, t + \Delta t)$, is

$$P\left[E \geq E_0 | (t, t + \Delta t)\right] = 1 - \exp\left[\lambda_0(t) \Delta t\right]. \tag{12.26}$$

In relation (12.26) $\lambda_0(t)$ is the rate of occurrence of seismic events with the

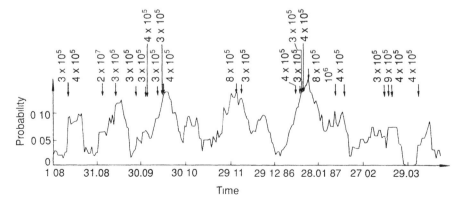

Figure 12.1   Time-dependent probability of the occurrence of seismic events with the energy $E \geq 3 \cdot 10^5$ J for the longwall 11 area at the Bobrek coal mine, Poland. The time of the occurrence of observed events is marked by arrows. [From Lasocki (1990, Fig. 4.13)]

energy $E \geq E_0$ and is equal to $\lambda(t)[1 - F(E_0|t)]$, where $F(E_0|t)$ is the probability cumulative distribution function of the seismic energy E equal to

$$F(E|t) = 1 - \left(\frac{E_{\min}}{E}\right)^{B(t)}. \tag{12.27}$$

In this way, the probability (12.26) can be estimated for any time t and predicted for any time interval $(t, t + \Delta t)$ from the events recorded within the interval $(t - \Delta T, t)$. The results of such an estimation of the probability (12.26) are shown in Fig. 12.1. For this case $\Delta t = \Delta T = 10$ days was accepted. The arrows indicate the time of occurrence and the energy of strong seismic events with energy $E \geq E_0 = 3 \cdot 10^5$ J. The majority of these events occurred when the estimated probability was either increasing or at high level. The only exception are two events (September 26, 1986 and April 5, 1987) that were not preceded by the increased probability.

12.2   Extreme Distributions of Magnitudes

Catalogs of seismic events are often incomplete at low magnitudes, and this can bias predicting capabilities of some results based on the frequency–magnitude relation (12.1). Largest events, on the other hand, are better known, are more homogeneous in time, and are more accurately determined than the small events. Usually they dominate the principal effects such as energy or seismic moment release, or any damage that might occur. It can be therefore expected that the methods that preferentially analyze the strongest events are at least as useful as the methods relying on complete files.

The theory of extreme values can be summarized as follows. Let X be a random variable with a certain cumulative probability function $F(x)$. The probability that x will be the largest among n independent samples $X_1, \ldots, X_n$ from the same distribution $F(x)$ is (Benjamin and Cornell, 1970)

$$F_{\max}(x) = P(X_1 \leq x, X_2 \leq x, \ldots, X_n \leq x) = [F(x)]^n, \tag{12.28}$$

which is the exact distribution function of the largest values. According to Gumbel (1962), there are only three distributions of extremes, asymptotically equivalent to relation (12.28), and each distribution assumes specific behavior of the absolutely largest value of the variable x.

The first asymptotic distribution of the largest values, known also as *Gumbel I distribution*, is defined as

$$F_{\max}^1(x) = \exp\{-\exp[-\alpha(x - u)]\}, \tag{12.29}$$

where α and u are the distribution parameters, $\alpha > 0$, and x is unlimited from both sides. The second asymptotic distribution of the largest values, known as *Gumbel II distribution*, is defined as

$$F_{max}^{II}(x) = \exp\left[-\left(\frac{u - x_{min}}{x - x_{min}}\right)^k\right],$$
(12.30)

where $k > 0$, $u > x_{min} \geq 0$, and x_{min} is the lower limit of x. The third asymptotic distribution of the largest values, *Gumbel III distribution*, is defined by

$$F_{max}^{III}(x) = \exp\left[-\left(\frac{x_{max} - x}{x_{max} - u}\right)^k\right],$$
(12.31)

where $k > 0$, $u < x_{max}$, and x_{max} is the upper limit of x.

The extreme-value statistics was first applied in seismology by Nordquist (1945), who demonstrated that the largest earthquakes in California are in good agreement with Gumbel I extreme distribution. The major break-through in application of extreme-value statistics was made by Epstein and Lomnitz (1966), who have shown that Gumbel I asymptote can be derived directly from assumptions that seismic events are generated by a simple Poisson process and that they follow the frequency–magnitude relation (12.1).

Epstein and Lomnitz (1966) followed the fact that for independent identically distributed random variables the maximum of the sequence of a fixed size n is distributed by relation (12.28), where n is random with the probability distribution $P(n)$, and the probability that the maximum of the sequence will not exceed m is equal to $P(n)[F(m)]^n$. Thus, if the number n of seismic events per time interval Δt follows the Poisson distribution

$$P(n|\Delta t) = \frac{(\lambda \Delta t)^n \exp(-\lambda \Delta t)}{n!}$$
(12.32)

and magnitude m is a random variable with the Gutenberg–Richter cumulative probability distribution function (12.3), the cumulative distribution of the largest magnitude during time interval Δt is

$$F_{max}(m|\Delta t) = \sum_{n=0}^{\infty} P(n|\Delta t)[F(m)]^n$$

$$= \sum_{n=0}^{\infty} \frac{(\lambda \Delta t)^n \exp(-\lambda \Delta t)}{n!}[F(m)]^n.$$
(12.33)

Denoting $\Lambda = \lambda \Delta t F(m)$, and since $\sum_{n=0}^{\infty}\Lambda^n/n! = \exp(\Lambda)$, the distribution of the largest magnitudes m becomes

$$F_{\text{max}}(m|\Delta t) = \exp\{-\lambda \Delta t[1 - F(m)]\}, \qquad (12.34)$$

or equivalently

$$F_{\text{max}}(m|\Delta t) = \exp[-\lambda \Delta t \exp(-\beta m)]. \qquad (12.35)$$

Formula (12.35) is just Gumbel I distribution (12.29) for $\beta = \alpha$ and $\lambda \Delta t = \exp(\alpha u)$. After estimating λ and β, we can obtain several useful characteristics of seismic hazard. For example, the mean return period, which is the mean number of time intervals required for an event with magnitude not less than m to be observed, by definition equal to

$$Rp(m) \equiv 1/\lambda(m), \qquad (12.36)$$

takes the form $\exp(-\beta m)/\ln[F_{\text{max}}(m)]$.

The work of Epstein and Lomnitz (1966) has initiated an "avalanche" of applications of extreme values methods for the earthquake hazard estimation. An extensive list of references on this subject can be found, for example, in a work of Campbell (1982). Moreover, by deriving the first Gumbel distribution from the commonly accepted rules related to the earthquake occurrence, Epstein and Lomnitz (1966) have established a formal basis for the applications of this distribution in earthquake statistics. Soon after publication of their work, however, a number of authors noticed that the application of the Gutenberg–Richter frequency–magnitude relation with unbounded argument results in unbounded distribution of extremes. Since, from a physical point of view, there must exist an upper bound to the earthquake magnitude [see equation (12.13)], Gumbel I distribution of extreme magnitudes has only approximate character. To avoid this contradiction, some authors use Gumbel III distribution for which an upper bound for the magnitude is introduced. Although Gumbel III distribution has been applied in seismology rather successfully, it cannot be derived from the common rules of the earthquake occurrence.

The extreme distributions were applied for the first time to mine tremors by Oczkiewicz and Szukalski (1974). They studied the events that occurred between September 1972 and January 1974 in the Lubin Copper Basin, Poland, and they found that the strongest event to be expected during the next five years, with a probability of 0.99, would have an energy of $6 \cdot 10^8$ J. On March 24, 1977, however, a large tremor with the energy $2.5 \cdot 10^{10}$ J ($M_L = 4.5$) occurred. This means that the prediction of the strongest event based on a too short period of observation, and not corrected for the trend of increasing release of seismic energy, could not be successful. More reliable estimation of seismic hazard for the same area, based on the monthly largest

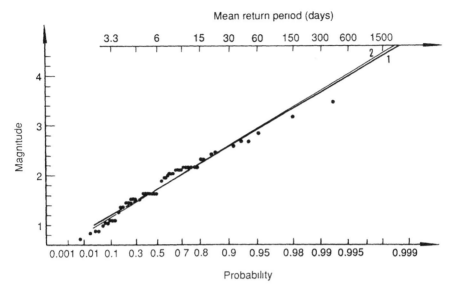

Figure 12.2 Gumbel III distribution of maximum magnitudes of seismic events observed every 3 days at the Wujek coal mine, Poland, from September 3, 1977 to April 19, 1978. Curve 1 is the distribution calculated by the least-squares method, and curve 2 is the distribution determined by the maximum-likelihood method [From Dessokey (1984, Fig 2)]

events during 8 years of observation and using Gumbel III distribution, was performed by Kijko *et al.* (1982).

A comprehensive comparative study of the application of different models of extreme distributions and different fitting techniques for seismic hazard evaluation in mines was carried out by Dessokey (1984) and Idziak *et al.* (1991). An example of the application of Gumbel III distribution at Wujek coal mine in Poland is shown in Fig. 12.2, reproduced from Dessokey (1984). Over a thousand seismic events that occurred during the mining of a closing face in seam 501 from September 3, 1977 to April 4, 1978 were used in the analysis. The length of the face was approximately 1000 m. The recorded events were analyzed to determine seismic hazard and to estimate mean return periods for the strongest events.

In addition to Gumbel I and Gumbel III distributions, mining applications of Gumbel II distribution are also known. Zuberek (1989) has shown that the maximum amplitudes of acoustic emission signals recorded during uniaxial compression of sandstone samples are consistent with the Gumbel asymptotic distribution of the second type. The same model was applied by Idziak *et al.* (1991) in their analysis of the distribution of strong seismic events with the energy about 10^6 J, which occurred between 1977 and 1987 in the Upper Silesia Coal Basin, Poland.

Gumbel I distribution was used by Lasocki (1990) to the assessment of time-dependent seismic hazard at Szombierki coal mine in Poland. He assumed that the occurrence probability of strongest seismic events with the energy $E \geq E_0$ during the next time interval Δt is $P[E \geq E_0|(t, t + \Delta t)] = 1 - F_{max}(E_0)$, where $F_{max}(E_0)$ is described by formula (12.34), in which the magnitude distribution $F(m)$ was replaced by the energy distribution (12.27). Following the assumption that the process within the time window ΔT is stationary, the maximum-likelihood evaluation of the parameters $B(t) = B$ and $\lambda(t) = \lambda$ is obtained from solutions of the system of two equations (Lasocki, 1990)

$$\begin{cases} \dfrac{n}{\lambda(t)} - \dfrac{n}{\exp[\lambda(t) - 1]} = \sum_{i=1}^{n} \left(\dfrac{E_{min}}{E_i} \right)^{B(t)}, \\[3ex] \lambda(t) = \dfrac{n + \sum\limits_{i=1}^{n} \ln\left(\dfrac{E_{min}}{E_i} \right)^{B(t)}}{\sum\limits_{i=1}^{n} \left(\dfrac{E_{min}}{E_i} \right)^{B(t)} \ln\left(\dfrac{E_{min}}{E_i} \right)^{B(t)}}, \end{cases} \tag{12.37}$$

where E_i is the energy of the strongest event observed during the time window ΔT preceding time t. Thus, the sequence of probabilities $P[E \geq E_0|(t, t + \Delta t)]$ for every Δt can be calculated.

12.3 Bimodal Distributions of Mine Tremors

An introduction to this topic is given in Section 2.3. In this section the bimodal distribution, its mathematical models, and possible explanations, is described in some detail.

During the study of the recurrence of strong seismic events in Polish mines with the use of extreme distributions (e.g., Kijko et al., 1982, 1987; Dessokey, 1984; Drzezla et al., 1986; Idziak et al., 1991), an interesting feature was noticed: the empirical distributions of extreme seismic energies is more complex than might be expected from the most general theoretical considerations, such as the distributions of Gumbel (1962). This feature was observed in underground coal mines of Upper Silesia, Poland, at Doubrava coal mine of the Ostrava-Karvina Coal Field, Czechoslovakia, and in copper ore mines of the Lubin Copper Basin, Poland. The application of standard modification methods (truncation or an assumption that the distribution parameters are

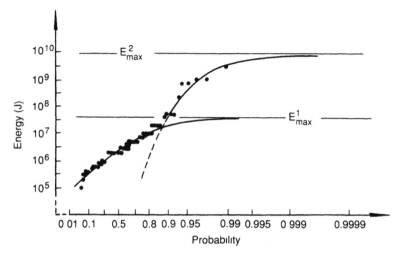

Figure 12.3 Probability distribution of the monthly maximum seismic energy release at the Szombierki coal mine, Poland, from January 1975 to June 1982. [From Kijko *et al.* (1987, Fig. 2).]

random variables) did not change the character of empirical distributions, and no satisfactory fit to the data was found. A closer look at the empirical distributions suggests that they are of bimodal character (Fig. 12.3).

The bimodal distribution results from the mixing of random variables generated by two different phenomena. This statement invokes a fundamental question of what those two (or possibly more) mechanisms are generating seismic events in mines. The problem is of practical importance in everyday mining practice as well. This concerns, in particular, the high-energy component of the distribution corresponding to the largest events. Possible physical and mathematical models generating bimodal distributions of seismic energy, therefore, are considered in the following, and some hypotheses to explain the observed distributions are formulated.

12.3.1   Mathematical Models

Three basic mathematical models for bimodal distribution are discussed below. Unfortunately, the statistics of observed events cannot be used to discriminate between these three models. In all three cases, the theoretical distributions fit empirical data quite well. Of course, the derived distribution functions in models I, II, and III are not the only possible distributions describing the observations. The number of possible modifications of the models presented below is practically unlimited.

Model 1 This model was proposed by Stankiewicz (1989), who assumed that at any moment the state of the rockmass can be described by a stress that increases linearly in time in the absence of a seismic event

$$x = x_0 + \alpha t,$$ (12.38)

and drops to a certain value from the interval (x_0, x) if the event occurs. We take x_{max} as the maximum value of stress: once this value is reached, the event occurs immediately. We assume also that $p_1(x)dt$ is the probability that if at a given time t the system is in the state x, an event will occur at the time dt, and $p_2(x|y)$ is the conditional probability that after the event the system will pass from the state x to y. Then $P(x,t)$, the probability that at the instant t the system is in the state x, will obey Kolmagorov equation (Molchan, 1984; Stankiewicz, 1989)

$$p_1(x)P(x,t) + \alpha \frac{\partial P(x,t)}{\partial x} + \frac{\partial P(x,t)}{\partial t}$$

$$= \int_x^{x_{max}} p_2(x|y)P(y,t)p_1(y)\,dy.$$ (12.39)

Let us assume that $p_2(x|y) = 1/y$, which means that the event is followed by the stress drop to a certain value in the interval (x_0, x) with the probability described by a uniform distribution, and $p_1(x) = p_1 = $ const, which means in turn that the probability of event occurrence does not depend on the amount of stress in the rock if the stress does not exceed its maximum x_{max}.

Let us also accept that the released seismic energy is connected with the stress drop through a power function

$$E = A(\Delta x)^\delta.$$ (12.40)

Furthermore, we assume that seismic events can be generated by two different mechanisms, between which a simple interaction is observed. To induce the seismic event generated by the second mechanism, a certain threshold value of stress $x_{max}^{(1)}$ must be exceeded. If the value of x prior to the occurrence of event exceeds $x_{max}^{(1)}$ ($x_{max}^{(1)} < x \le x_{max}$), an interaction takes place, and the energy released by the event is

$$E = \eta A(\Delta x)^\delta.$$ (12.41)

Our mathematical formalism is general and does not require any physical assumptions regarding the nature of the interaction, which could be, for example, the coupling between the events induced by mining and of tectonic origin or the events generated by two major geologic features such as layers with different strength and faults or dikes.

Substituting the probabilities $p_1(x)$ and $p_2(x|y)$ into equation (12.39) and using relations (12.40) and (12.41), the stationary solution of equation (12.39), expressed as the energy distribution, is obtained (Stankiewicz, 1989; Glowacka et al., 1990)

$$F(\varepsilon) = 1 - D \begin{cases} V(\varepsilon) + V(\varepsilon_1)C[1 - \exp(-\gamma\kappa)]V(\varepsilon_{max} - \kappa), \\ V(\varepsilon_1)\{1 - C\exp[\gamma(\varepsilon_{max} - \kappa)] + \exp(\gamma\varepsilon_1)\} - V(\varepsilon_{max} - \kappa), \\ V(\varepsilon - \kappa) - V(\varepsilon_{max} - \kappa), \end{cases}$$

$$(12.42)$$

where the first relation is valid for $\varepsilon \leq \varepsilon_1$, the second for $\varepsilon_1 < \varepsilon \leq \varepsilon_2$, and the third for $\varepsilon > \varepsilon_2$, and $\kappa = \log(\eta)$, $\varepsilon = \log(E/E_{min})$, $\varepsilon_1 = \log(E_{max}^{(1)}/E_{min})$, $\varepsilon_2 = \varepsilon_1 + \kappa$, $\varepsilon_{max} = \log(E_{max}/E_{min})$, $E_{max}^{(1)} = A(x_{max}^{(1)\delta})$, $E_{max} = \eta A(x_{max}^{\delta})$, $\gamma = \ln(10)/\eta$, $V(\varepsilon) = \exp[(-C\exp(\gamma\varepsilon)]$, and the normalizing constant D is equal to $\{\exp(-C) + CV(\varepsilon_1)[1 - \exp(-\gamma\kappa)] - V(\varepsilon_{max} - \kappa)\}^{-1}$.

It is clear that, in general, the seismic energy distribution (12.42) has bimodal character and becomes unimodal only for $\kappa = 0$. An example of the application of relation (12.42) is shown in Fig. 12.4, reproduced from Stankiewicz (1989), where the frequency–energy distribution is presented for

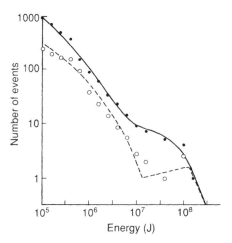

Figure 12.4 Frequency–energy distribution of 919 seismic events observed at the Nowy Wirek coal mine, Poland, from January 1978 to December 1985. The bimodal cumulative distribution is shown by a continuous line, and the density distribution is marked by a dashed line. [From Stankiewicz (1989, Fig. 31)]

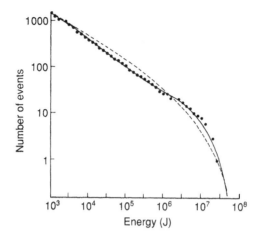

Figure 12.5 Frequency–energy distribution of 1658 seismic events recorded at the Doubrava coal mine, Czechoslovakia, from November 1982 to May 1986. The bimodal cumulative distribution is shown by a continuous line, and the truncated Gutenberg–Richter relation is marked by a dashed line. [From Glowacka *et al.* (1990, Fig. 1).]

919 seismic events recorded between 1978 and 1985 at Nowy Wirek coal mine in Poland. The bimodal density cumulative distribution is marked by a continuous line and the bimodal density distribution is shown by a dashed line. Another example is shown in Fig. 12.5, reproduced from Glowacka *et al.* (1990), where the frequency–energy distribution is presented for 1658 events recorded between November 1982 and May 1986 at Doubrava coal mine in Czechoslovakia. The bimodal cumulative distribution is marked by a continuous line, and the truncated Gutenberg–Richter distribution [equation (12.15)] is shown by a dashed line. In both the cases distribution (12.42) fits the data rather well.

Model II—High- and Low-Energy Seismicity Independent Processes Alternative formulas describing bimodal empirical distributions of seismic energy can be obtained in a much simpler way under an assumption that high- and low-energy seismic activities are mutually independent. Empirical distributions of the strongest events (Fig. 12.3) suggest that seismic events are generated by two mechanisms. The first is responsible for a low-energy component of the distribution and generates events with x [where x is either magnitude or log(energy)] from the interval $(x_{min}, x_{max}^{(1)})$. Accordingly, the second mechanism generates a high-energy component from the interval $(x_{min}, x_{max}^{(2)})$, where $x_{max}^{(1)} < x_{max}^{(2)}$.

Let λ_1 be the mean number of seismic events in a time unit, generated by the first mechanism in the interval $(x_{min}, x_{max}^{(1)})$. By analogy, let λ_2 be the mean number of seismic events of the second kind in the interval $(x_{min}, x_{max}^{(2)})$. Thus, if $F_1(x)$ and $F_2(x)$ are the distribution functions of the random variables x generated by the first and second mechanisms, respectively, the effective distribution function of x takes the form

$$F(x) = \begin{cases} \pi_1 F_1(x) + \pi_2 F_2(x), & x_{min} \leq x \leq x_{max}^{(1)}, \\ \pi_1 + \pi_2 F_2(x), & x_{max}^{(1)} \leq x \leq x_{max}^{(2)}, \end{cases} \quad (12.43)$$

where $\pi_1 = \lambda_1/(\lambda_1 + \lambda_2)$ and $\pi_2 = \lambda_2/(\lambda_1 + \lambda_2)$.

We assume that seismic events generated by both the first and second mechanisms are Poisson processes with the parameters λ_1 and λ_2, respectively (Kijko et al., 1987). From the assumption of independence it follows that the course of seismic activity of the first kind has no effect on the second kind, and vice versa. It is well known (e.g., Epstein and Lomnitz, 1966; Campbell, 1982) that if a random variable of the distribution $F(x)$ is a sequence of events with a Poisson distribution, then the distribution of extreme values of x is described by equation (12.34), in which $\Delta t = 1$ and λ is a parameter of the Poisson distribution. After simple transformations, we obtain from equations (12.34) and (12.43) the sought distribution of the strongest events in a general form

$$F_{max}(x) = \begin{cases} F_{max}^{(1)}(x) F_{max}^{(2)}(x), & x_{min} \leq x \leq x_{max}^{(1)}, \\ F_{max}^{(2)}(x), & x_{max}^{(1)} \leq x \leq x_{max}^{(2)}, \end{cases} \quad (12.44)$$

where $F_{max}^{(i)}(x) = \exp\{-\lambda_i[1 - F_i(x)]\}$ for $i = 1, 2$.

To be able to use the derived bimodal distribution (12.44), it is necessary to define the form of the distributions $F_1(x)$ and $F_2(x)$. We can assume that the magnitudes (or the energy logarithms) of seismic events generated by the first and second mechanisms are random variables with double-truncated exponential distributions (12.15)

$$F_i(x) = \frac{1 - \exp[-\beta_i(x - x_{min})]}{1 - \exp[-\beta_i(x_{max}^{(1)} - x_{min})]}, \quad (12.45)$$

where β_i are the parameters of distributions generated by the first ($i = 1$) and second ($i = 2$) mechanisms, respectively.

The derived distribution of the strongest seismic events is indeed of a bimodal nature, and its interpretation is rather clear. Assuming, for example, that seismicity of the first kind results from stresses generated by mining works, the second kind would be tectonic activity. This would mean that mining does not affect tectonic processes and that earthquakes would occur even if no mining is carried out in the area. The fit of the observed largest seismic events by the derived extreme distribution (12.44) is shown in Fig. 12.3.

Model III—High-Energy Seismicity Dependent on Low-Energy Component   It is possible to imagine that in a given area residual tectonic stresses are present, which would not generate earthquakes in the absence of mining. The mining works generate additional stresses that are released in the form of a low-energy component of seismic activity and that act as a "triggering mechanism" for tectonic stresses.

In a similar way, it could be supposed that a layered structure of the rockmass, with highly differentiated strength parameters, would generate seismic activity with two or more mutually correlated modes. These two situations can be described as follows. By analogy to model I, let λ_1 and λ_2 be the parameters of Poisson distribution of seismic events generated by the first and second mechanisms, respectively. If i events of the first kind are generated, then the number j of events of the second kind would be a Poisson distribution with the mean $i\lambda_2/\lambda_1$. Thus, the conditional probability of the occurrence of j events of the second kind is

$$P_2(j|i) = \exp(-i\lambda_2/\lambda_1)\frac{(i\lambda_2/\lambda_1)^j}{j!}, \qquad (12.46)$$

and the probability of occurrence of the event of this kind is

$$P(j) = \sum_{i=1}^{\infty} P_2(j|i)P_1(i)$$

$$= \exp(-\lambda_1)\frac{(i\lambda_2/\lambda_1)^j}{j!}\sum_{i=1}^{\infty}\frac{i^j}{i!}[\lambda_1\exp(-\lambda_1/\lambda_2)]^i, \qquad (12.47)$$

where $P_1(i)$ is the probability of occurrence of i first kind events, or according to our assumptions $P_1(i) = \exp(-\lambda_1)(-\lambda_1)^i/i!$.

Distribution (12.47) is known in the statistical literature as a Neyman distribution of type A. To determine the relevant distribution function $F_{max}(x)$ of extreme seismic events, we should define the probability of occurrence of extreme k events, without specifying their kind. On the basis of distributions $F_{max}^{(1)}(x)$, $F_{max}^{(2)}(x)$, $P_1(i)$ and $P_2(i)$, this probability is

$$F_{max(k)}(x) = \sum_{i=1}^{\infty} P_1(i)\left[F_{max}^{(1)}(x)\right]^i P_2(k-i|i)\left[F_{max}^{(2)}(x)\right]^{k-i}. \qquad (12.48)$$

Thus, after a series of simple transformations the sought cumulative distribution function can be written in the form (Kijko *et al.*, 1987)

$$F_{max}(x) = \sum_{k=0}^{\infty} F_{max(k)}(x)$$

$$= \exp\left\{-\lambda_1\left[1 - F_{max}^{(1)}(x)\right]\exp(-\lambda_1/\lambda_1)\left[1 - F_{max}^{(2)}(x)\right]\right\}. \quad (12.49)$$

Assuming that the number of second-kind events is small in comparison with the total number of observed seismic events, the cumulative distribution of extreme events takes the final form as follows:

$$\overset{\cdot\cdot}{F}_{max}(x) = F_{max}^{(1)}(x)\left[F_{max}^{(2)}(x)\right]^{F_1(x)}, \qquad x_{min} \leq x \leq x_{max}^{(2)}. \quad (12.50)$$

12.3.2 Possible Explanations of Bimodal Distributions

Although formal mathematical models of bimodal distributions are formulated, there is still no answer as to why the observed distributions of seismic events in mines are bimodal. In this respect, various hypotheses could be considered.

It can be assumed that the bimodality (or multimodality) is caused by inhomogeneous major geologic features, such as layers with different strength, faults or dikes, and seismicity observed in mines is induced entirely by mining excavations. In the course of mining, the process of stress release in a rockmass attains some steady state with certain energy characteristics. The perturbation of the stress field induced by adjacent mining or a substantial change in properties of the medium may activate a new process or alter the existing processes, generating seismicity. This may result in an increase or decrease of the number and energy of events originated from the same geologic features, and/or the expansion of seismic activity into new regions in the rockmass. The energy distributions of events generated, for example, in distant layers with different thickness and strength, would be characterized by different parameters that could not be described by simple unimodal distributions.

It could also be assumed that the low-energy component of the distribution is the result of the release of stresses generated by mining, whereas the high-energy component of the distribution is the result of the release of residual tectonic stresses accumulated in the rockmass.

It seems reasonable to suppose that the first hypothesis may hold in some mining districts, whereas for others the second hypothesis may be acceptable. Whatever explanation is offered, the so-called geologic factors definitely play the most important role in the generation of seismic events in mines. The most striking example of such a role displayed by tectonic instability and tectonic stresses is seismicity induced by surface mining, described in Section 1.2.

12.4   Seismicity and Rock Extraction

It is rather obvious that seismicity in mines is related to mining conditions and excavation methods. Several attempts to link seismic activity with various quantities characterizing mining works were mostly of empirical nature (e.g., Šklenar and Rudajev, 1975; Bober and Kazimierczyk, 1979; Roček and Skorepova, 1982). In the studies carried out in deep gold mines in South Africa (e.g., Cook, 1976; McGarr and Wiebols, 1977), the sum of released seismic energy per excavation area was assumed to be the criterion of rockburst threat. High values of released energy were interpreted as an increase in the number of rock cracks thus formed, which could lead to a rockburst.

The first serious approach to the formulation of a theoretical basis for the problem was made in the 1960s. Assuming that the change in potential energy of a system resulting from mining is equal to the product of weight of mined rocks and their depth, it was shown that no more than half of this energy can be stored as elastic energy (Cook, 1976). The rest of the energy must be released by a rock body in different forms, such as heat and friction. These considerations imply that the seismic energy is proportional to the volume of mined rocks. The same conclusions were obtained by Duvall and Stephenson (1965) from theoretical assessments. They showed that for a rockmass of circular cross section the amount of radiated seismic energy is proportional to the volume of mined rocks. A comprehensive study of how and what seismicity will result from a given amount of mining was carried out by McGarr (1976, 1984), and an analysis of energy changes accompanying underground mining was presented by Brady and Brown (1985).

In this section, a simple model describing the relationship between seismicity and excavated rocks, based on the theoretical solutions of Randall (1971) and McGarr (1976), is formulated and some examples of its applications to evaluate time dependent seismic hazard in mines are given.

Following Randall's (1971) theoretical results, McGarr (1976) has shown that mining-induced seismicity expressed by the sum of seismic moments $\sum M_0$ and the volume ΔV of convergence accompanying elastic deformation around a stope can be expressed as

$$\sum M_0 = k\mu|\Delta V|, \tag{12.51}$$

where μ is the rigidity modulus and k is a constant ranging from $\frac{1}{2}$ to $\frac{4}{3}$, but usually is close to 1.

Following McGarr's (1976) considerations leading to equation (12.51), and physical links between the volume of convergence accompanying deformation around the stope and the volume of fractured rocks as a result of seismic activity, Kijko (1985) proposed the relation

$$\sum M_0 = \text{const } \Delta V_f, \tag{12.52}$$

where ΔV_f is the volume of fractured rocks resulting from seismic activity. Assuming that the transverse cross section A_f of the fractured rocks is proportional to the transverse cross section A of the opening, $A_f = \theta A$, the volume ΔV_f of the fractured rocks can be expressed by the volume ΔV of a mined layer as $\Delta V_f = \theta \Delta V$, and from relation (12.52) we obtain

$$\Delta M_0 = \text{const } \theta \Delta V. \tag{12.53}$$

In general, the parameter θ would depend on the depth and mining technique, mechanical properties of mined rocks and rocks forming the opening, and on the type of support of the opening. The parameter θ should decrease when mining is carried out with backfilling. It should increase in the case of flexible hanging wall setting on a footwall, or in the case of the so-called bottom squeeze.

To be able to use equation (12.53) in mining practice, the values of seismic moments M_0 should be replaced by other parameters that are routinely determined, such as magnitude or seismic energy. It is generally accepted that local magnitude of small earthquakes is proportional to the logarithm of seismic moment (e.g., Randall, 1973; Gibowicz, 1975; Spottiswoode and McGarr, 1975)

$$\log M_0 = p + q M_L, \tag{12.54}$$

where p and q are constants. Although for simple models of seismic source and for events with small source size, the constant q is close to 1, its value here is not specified to preserve more general form for our solution.

From equations (12.53) and (12.54) we have

$$\sum 10^{q\,M_L} = \text{const } \theta \Delta V. \tag{12.55}$$

Relation (12.55) links seismic activity expressed in terms of tremor magnitude with the mined volume ΔV.

The strength of seismic events in mines is often assessed in terms of energy rather than magnitude. In such a case the magnitude in equation (12.55) should be replaced by the energy. This can be done by using empirical relation of type (12.12), valid for a given area. For seismic events obeying a constant-stress-drop scaling relation (see Section 11.7), the coefficient $d = q$. From relations (12.12) and (12.54) it follows that equation (12.55) now takes the form

$$\sum E^{d/q} = \text{const } \theta \, \Delta V. \tag{12.56}$$

The sum $\sum E$ of released seismic energy accompanying the excavation of a certain volume ΔV does not have to be considered deterministically as a constant, but could be understood as a random variable as well, estimated from the known cumulative distribution function of energy. Let us assume that the seismic events associated with the mined volume ΔV are described by the Gutenberg–Richter relation (12.1), where m is the local magnitude M_L bounded from the top by m_{max} and that the magnitude–energy relation is represented by equation (12.12). Since the number dN of tremors with the energy from the interval $(E, E + dE)$ can be expressed as $dN = \text{const } E^{-(b/q+1)} dE$, where $\text{const} = 10^{a+c\,b/d}/d \ln(10)$, the sum of released seismic energy is

$$\sum E = \text{const} \int_{E_{min}}^{L_{max}} E^{-b/d} \, dE$$

$$= \text{const}(1 - b/d)\left(E_{max}^{1-b/d} - E_{min}^{1-b/d} \right), \tag{12.57}$$

where E_{min} is the threshold energy above which all events induced by the mined volume ΔV are recorded. Taking into account relation (12.54) between the seismic moment M_0 and magnitude, and using the cumulative distribution function of magnitude truncated from both ends [formula (12.15)], for $d > b$ and $E_{max} \gg E_{min}$, the relation between the seismic moment release $\sum M_0$ and seismic energy $\sum E$ can be approximated as

$$\ln \sum M_0 = \frac{1}{B} \ln \sum E + \ln B \frac{c_2 c_3}{c_1}, \tag{12.58}$$

where B and c_3 are parameters characterizing the state of the rockmass, $B = (d - b)/(q - b)$, $c_1 = \text{const}(1 - b/d)$, $c_2 = (1 - b/d)10^{a+pb/q}/d \ln(10)$, and $\log c_3 = (p - cq/d)/(1 - b/q)$.

Assuming that during mining the parameter $\theta = \text{const}$, from relations (12.53) and (12.58) we finally obtain

$$\sum E = \text{const}(\Delta V)^B. \tag{12.59}$$

It should be noted that even if relation (12.12) between the magnitude and energy is of arbitrary character, equation (12.59) is still valid, but the quantity E is not necessarily the true released energy. The range of applicability of relation (12.59) is much wider than it could be expected from the original assumptions (Glowacka and Kijko, 1989). If the linear relation $\Delta V_f = \theta \Delta V$, for example, is replaced by a more general relation $\Delta V_f = \theta(\Delta V)^n$, where n is a constant, the original form of equation (12.59) does not change.

One limitation to the presented model is the assumption that stresses in the rockmass result exclusively from mining, whereas tectonic stresses are ignored. In mines where tectonic stresses are significant, the presented model can be successfully used only after certain modifications.

As an example, the evaluation of seismic hazard at Doubrava coal mine, Czechoslovakia, described by Glowacka et al. (1990), is briefly discussed. To simplify the notation the symbol E_Σ instead of ΣE and V instead of ΔV are used. The symbol ΔV here denotes the volume of rocks excavated during the specified time interval Δt.

We define the seismic hazard as a probability that the seismic energy ΔE_Σ, released as a result of the excavation of the volume ΔV, will exceed a given value E_0 during the next specified time interval Δt. According to relation (12.59), the energy expected to be released during the next time interval $\Delta t_i = t_{i+1} - t_i$ is

$$\Delta E_{\Sigma i} = CV_{i+1}^B - CV_i^B, \tag{12.60}$$

where the parameters B and C are calculated from the previous energy release. The released energy $\Delta E_{\Sigma i}$ obviously has a random character. We assume that $\Delta E_{\Sigma i}$ follows normal distribution, truncated from the left-hand side by 0, with the average value $\langle \Delta E_\Sigma \rangle$ and standard deviation σ_E. Thus, the probability $P[\Delta E_\Sigma > E_0|(V, V + \Delta V)]$ that the sum of seismic energy ΔE_Σ released within the excavated volume $(V, V + \Delta V)$ will exceed the specified value E_0 is

$$P\left[\Delta E_\Sigma > E_0|(V, V + \Delta V)\right]$$
$$= 1 - \frac{\sqrt{2}}{\pi \sigma_E} \int_0^{E_0} \exp\left(-\frac{1}{2}\sigma_E^2\right)\left(E - \langle \Delta E_\Sigma \rangle\right)^2 dE, \tag{12.61}$$

where $\langle \Delta E_\Sigma \rangle$ is the average expected sum of seismic energy released during the time interval Δt and calculated according to formula (12.60) in which the parameters B and C are estimated from the whole previous run of the energy release E_Σ and from the volume V of mined rocks.

The described algorithm was applied in the Doubrava coal mine, Czechoslovakia (Glowacka et al., 1990), where 1658 seismic events were recorded from November 1982 to May 1986. The energy of the strongest event was estimated to be $4 \cdot 10^7$ J. The observed frequency–energy distribution is of bimodal nature (Fig. 12.5). Relation (12.61) therefore was rewritten in the form

$$E_\Sigma = E_{\Sigma 1} + E_{\Sigma 2}, \tag{12.62}$$

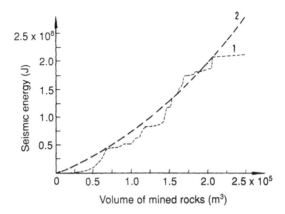

Figure 12.6   Cumulative seismic energy versus volume of mined-out rocks in weekly intervals at the Doubrava coal mine, Czechoslovakia. Curve 1 shows the observed dependence, whereas its approximation by relation (12.85) is shown by curve 2. [From Glowacka *et al.* (1990, Fig. 3).]

where

$$\begin{cases} E_{\Sigma 1} = C_1 \cdot V^{B_1} & \text{for } E < E_{\max}^{(1)}, \\ E_{\Sigma 2} = C_2 \cdot V^{B_2} & \text{for } E > E_{\max}^{(1)}, \end{cases}$$

and $E_{\max}^{(1)} = 8 \cdot 10^5$ J. The parameters C_1, C_2, B_1, and B_2, like those in relation (12.59), were continuously updated in the course of mining and seismic energy release. The results are displayed in Fig. 12.6, where curve 1 shows the observed sum of released seismic energy against the mined out volume in weekly intervals, and curve 2 is calculated from relation (12.62).

In the second example we demonstrate how the probability of the strong event occurrence can be assessed when several independent methods of seismic hazard evaluation are simultaneously employed. This approach was described by Glowacka and Lasocki (1992). In our case the resulting probability is combined from two sources. The first is the probability of the seismic event occurrence based on changes of the parameters a and b from the Gutenberg–Richter relation during the mining (Lasocki, 1990); this approach is discussed in Section 12.1. The second is the probability of the release of a certain amount of seismic energy resulting from the excavation of a certain amount of rocks; the approach discussed in the previous example.

Let p_0 denote the probability of the occurrence of an event with the energy not less than a certain threshold E_0 within the specified time interval Δt. Such a probability is continuously updated in the course of mining activity according to the formula

$$p_0 = n(E \geq E_0)/N, \tag{12.63}$$

where $n(E \geq E_0)$ is the number of time intervals Δt during which an event with the energy $E \geq E_0$ has occurred and N is the total number of time intervals Δt in the time period under consideration. Let p_i $(i = 1, 2)$ denote the conditional probabilities of event occurrence with $E \geq E_0$, within a time interval Δt in which a certain premonitory effect a_1 is observed. In our case, according to formula (12.26), $a_1 = (B, \lambda)$, where B and λ are the parameters determining the probabilities of event occurrence during mining works. Similarly, according to formula (12.60), $a_2 = (B, C)$. Assuming that the probabilities p_1 and p_2 are determined independently (which is only partly true because both approaches require partly the same input data), the resulting probability p for small Δt, when probabilities p_0, p_1, and p_2 are small, can be approximated as (Aki, 1981a)

$$p = p_0 \frac{p_1}{p_0} \frac{p_2}{p_0}, \tag{12.64}$$

where p is the resulting synthesis of the probabilities p_0, p_1 and p_2.

An evaluation of the resulting probability p was performed for longwall 11/507 in the Bobrek coal mine, Poland (Glowacka and Lasocki, 1992). Four out of seven large seismic events were preceded by the increased probability. Three out of seven events were not predicted because the probabilities p_2 were close to zero during the final stage of the longwall existence.

12.5 The Number and Energy of Seismic Events in the Stope Area

Stresses under real mining conditions cannot be determined in a credible way, nor can an assessment of the stability of the rockmass's complex system be made even if the value of stresses are known (Drzezla *et al.*, 1988). Statistical approaches to the analysis of the space–time–energy distribution of seismic events is a certain solution, when the time-dependent statistical distributions can be created and time-dependent seismic hazard assessed.

In this section a simple approach to this problem is considered, in which the dependence of the number and energy of seismic events on the distance from the stope face is analyzed. By monitoring this distance during mining it is possible to assess, to some extent, seismic hazard and its trends. To clarify this concept, a few examples of such an approach are briefly described.

Example 1. The dependence of the number and energy of seismic events on the distance from an active longwall face was described by Syrek and Kijko (1988). Altogether, 843 events with energy $\geq 10^4$ J were analyzed. The tremors occurred in three selected areas of Wujek coal mine in Poland between February 9, 1979 and May 4, 1981. Events used in the analysis occurred within the longwall panel or were situated ≤ 50 m on either side from the panel, 150 m ahead and 220 m behind the longwall face. The events were counted within the moving window 10 m wide, and then two curves were calculated: the number of events N and the sum of their energies E_Σ as a function of distance from the longwall face. The mining was carried out at depths of about 500–700 m, in 2.5–3.0-m-high longwalls, in seams of similar thickness (5.5–8.0 m). The number N and energy E_Σ of seismic events as a function of distance from the longwall face were shown in Fig. 2.2. Both curves are of similar character, reaching the highest values in the vicinity of the longwall face. The maximum of curve N is exactly at the face, whereas the maximum of curve E_Σ is at a distance of ~ 12 m from the longwall face. Both curves show also a steady, almost symmetric, decrease along the panel, ahead and behind the face. Theoretical values of vertical stresses σ as a function of distance from the face were evaluated to compare their pattern with the distribution of the number and energy of seismic events (Fig. 2.2). The stresses were calculated as a superposition of analytic solutions of Gil (1991) for a viscoelastic and plastic model of the rockmass. There is a good correlation between the maxima of the two curves E_Σ and σ about 12 m ahead of the longwall face. The two curves are also of a similar shape ahead of the face, along the longwall panel.

Example 2. In example 1 the seismic energy was calculated over large longwall panels for several months. Thus, the energy curves display the averaged distributions of seismic activity and are not able to monitor the current changes in seismicity. If the distribution of seismic energy and the state of stress are related, then a change in stresses, and notably a change in the distance between the maximum of stress and the longwall face, should affect the energy distribution as well. Consequently, a change in the character of energy distribution, and especially in the position of the maximum energy against the distance from an active face, should indicate a change in the state of seismic hazard. The number N of events, the seismic energy E_{DB} released by distressing blasting, and the distance X_0 between the maximum energy and the face for longwall VIII in the Wujek coal mine are compared in Fig. 12.7, reproduced from Syrek and Kijko (1988). The seismic events were analyzed during a period in which the longwall advanced 60 m. The energy was summarized over a consecutive 60-m segment of the panel, the origin of which was displaced by 20 m relative to the previous segment. In this manner, a series of distributions for consecutive 60-m windows, shifted each time by 20 m, were obtained. Longwall VIII was mined in the years 1983–1984 with

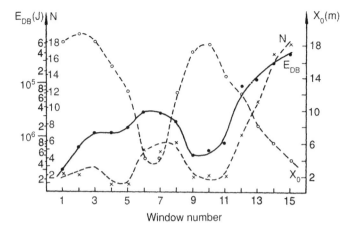

Figure 12.7   Relations between the number N of seismic events, energy E_{DB} released by destressing blasting, and distance X_0 between the maximum energy and the longwall face for the longwall VIII area at the Wujek coal mine, Poland. [From Syrek and Kijko (1988, Fig. 5)]

hydraulic filling, and 194 events with energy $\geq 10^4$ J were recorded. In the course of the longwall advance two periods of enhanced rockburst hazard occurred, which caused damage to mining workings. Both periods were well marked on the curve of strong events (N) and the curve of energy released by destressing blasting (E_{DB}). Curves N, E_{DB} and X_0 could provide a measure of the state of seismic hazard. Curve X_0 in Fig. 12.7 behaves "symmetrically" in relation to curves E_{DB} and N. The minimum values on the curve X_0 correspond to the maxima of the curves E_{DB} and N. The correlation between the three curves is rather obvious. The described correlations also indicate that the parameter X_0 might be of great value for estimating the state of seismic hazard.

12.6   Statistics with Incomplete and Uncertain Data

It often happens that as a result of changes in detection capability of a seismic network, caused by changes in instrumentation and station coverage, the catalogs of seismic events can differ in different intervals of time. This time-dependent incompleteness of catalogs makes them unsuitable for the use in seismic hazard evaluation, based on the existing procedures. In this section a procedure is described that facilitates the use of catalogs with different threshold of completeness and incorporates the uncertainty of earthquake magnitude determination. The method was originally developed for earthquake hazard evaluation, based on the largest historical earthquakes and time-dependent complete files of instrumentally recorded events (Kijko and Sellevoll, 1989, 1992). The proposed procedure is a maximum-likelihood method for estimating the hazard parameters: the maximum regional magnitude m_{max}, the seismic activity rate λ, and the b parameter of the Gutenberg–Richter relation.

We assume the Poisson occurrence of seismic events with the activity rate λ [equation (12.32)] and accept the doubly truncated Gutenberg–Richter distribution of magnitude m [equations (12.14) and (12.15)], in which m belongs to the domain (m_{min}, m_{max}) and m_{min} is the known threshold value of magnitude. The desired seismicity parameters are $\theta = (\beta, \lambda)$ and m_{max}. Following Tinti and Mulargia (1985), let us assume that the observed (apparent) magnitude is distorted by an observational stochastically independent error ε. If the error is assumed to be normally distributed with the standard deviation σ, the properly normalized density and cumulative probability functions of the apparent magnitude become, respectively (Kijko and Sellevoll, 1992)

$$\tilde{f}(m|m_{min}, \sigma) = f(m|m_{min}, \sigma)/[1 - F(m_{min}|m_{min}, \sigma)] \tag{12.65}$$

$$\tilde{F}(m|m_{min}, \sigma) = [F(m|m_{min}, \sigma) - F(m_{min}|m_{min}, \sigma)]/[1 - F(m_{min}|m_{min}, \sigma)]. \tag{12.66}$$

where

$$f(m|m_{min}, \sigma) = \beta A(m)/(A_1 - A_2)C(m|m_{min}, \sigma), \tag{12.67}$$

$$F(m|m_{min}, \sigma) = [A_1 - A(m)]/(A_1 - A_2)D(m|m_{min}, \sigma), \tag{12.68}$$

$$C(m|m_{min}, \sigma) = \frac{e^{\gamma^2}}{2}\left[\text{erf}\left(\frac{m_{max} - m}{\sqrt{2}\,\sigma} + \gamma\right) + \text{erf}\left(\frac{m - m_{min}}{\sqrt{2}\,\sigma} - \gamma\right)\right], \tag{12.69}$$

$$D(m|m_{min}, \sigma) = \left\{A_1\left[\text{erf}\left(\frac{m - m_{min}}{\sqrt{2}\,\sigma}\right) + 1\right] + A_2\left[\text{erf}\left(\frac{m_{max} - m}{\sqrt{2}\,\sigma}\right) - 1\right]\right.$$

$$\left. - 2C(m|m_{min}, \sigma) \cdot A(m)\right\}/2[A_1 - A(m)], \tag{12.70}$$

$A_1 = \exp(-\beta m_{min})$, $A_2 = \exp(-\beta m_{max})$, $A(m) = \exp(-\beta m)$, $\text{erf}(\cdot)$ is the error function (Abramowitz and Stegun, 1970), $\gamma = \beta\sigma/\sqrt{2}$, and $m \geq m_{min}$.

It may be verified that for m inside the interval (m_{min}, m_{max}), the correction function $C(m|m_{min}, \sigma)$ may be well approximated by a constant equal to $\exp(\gamma^2)$, and

$$\lim_{\sigma \to 0} C(m|m_{min}, \sigma) = 1, \quad \text{and} \quad \lim_{\sigma \to 0} D(m|m_{min}, \sigma) = 1. \tag{12.71}$$

Relations (12.71) are in full agreement with our intuitive expectations: the less the random errors perturb the real magnitude, the more the apparent magnitude distributions $f(m|m_{\min}, \sigma)$ and $F(m|m_{\min}, \sigma)$ appear to correspond to $f(m)$ and $F(m)$.

Finally, assuming that the model in which the density function vanishes below the cutoff magnitude $m_{\min}$ is unrealistic (in practice the transition occurs gradually), the relation between the apparent activity rate $\tilde{\lambda}(m)$ and the "true" one takes the form

$$\tilde{\lambda}(m) = \lambda(m)\frac{e^{\gamma^2}}{2}\left[1 + \mathrm{erf}\left(\frac{m_{\max} - m}{\sqrt{2}\,\sigma} + \gamma\right)\right].\qquad(12.72)$$

Let us assume that our catalog of seismic events can be divided into s subcatalogs (Fig. 12.8). Each of them has its span T_i and is complete starting from the known magnitude $m^{(i)}_{\min}$. For each subcatalog i, let $\mathbf{x}_i = |x_{ij}, \sigma_{ij}|$ be the apparent magnitude and its standard deviation with $j = 1, \ldots, n_i$, where n_i denotes the number of earthquakes in each complete subcatalog and $i = 1, \ldots, s$. If the size and number of seismic events are not mutually dependent, the likelihood function of the sought parameters θ and $L_i(\theta|\mathbf{x}_i)$ is the product of two functions $L_\beta(\beta|\mathbf{x}_i)$ and $L_\lambda(\lambda|\mathbf{x}_i)$. According to relation (12.4)

$$L_\beta(\beta|\mathbf{x}_i) = \mathrm{const}\prod_{j=1}^{n_i}\tilde{f}\left(x_{ij}|m_i, \sigma_{ij}\right),\qquad(12.73)$$

and following an assumption that the number of earthquakes per unit time is

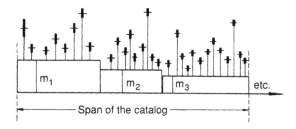

Figure 12.8 Illustration of the data that can be used to obtain seismic hazard parameters by the procedure described in the text. The approach permits the combination of the parts of the catalog with variable threshold of completeness $(m_1, m_2, m_3, \ldots)$ and uncertain magnitudes. It is assumed that the observed magnitude is the true magnitude distorted by a random error normally distributed with the mean equal to zero and the standard deviation σ (marked by thick vertical bars).

a Poisson random variable, $L_\lambda(\lambda|\mathbf{x}_i)$ can be written as

$$L_\lambda(\lambda|\mathbf{x}_i) = \text{const} \cdot \exp\left[-\bar{\lambda}\left(m_{\min}^{(i)}\right) \cdot T_i\right]\left[\lambda\left(m_{\min}^{(i)}\right) \cdot T_i\right]^{n_i}, \quad (12.74)$$

where const is a normalizing factor, the apparent activity rate is as defined by relation (12.72), $\lambda(m_{\min}^{(i)}) = \lambda[1 - F(m_{\min}^{(i)}|m_{\min})]$, and $m_{\min}$ is the "total" threshold value of magnitude. The only condition in the choice of this value is that $m_{\min}$ cannot exceed the threshold magnitude value of any part of the catalog. Relations (12.73) and (12.74) define the likelihood function of the sought parameters for each complete subcatalog i, where $i = 1,\ldots,s$. The joint likelihood function based on the whole catalog is given by

$$L(\boldsymbol{\theta}|\mathbf{x}) = \prod_{i=1}^{s} L_i(\boldsymbol{\theta}|\mathbf{x}_i). \quad (12.75)$$

The maximum-likelihood estimate of $\boldsymbol{\theta}$ is the value of $\hat{\boldsymbol{\theta}}$ that maximizes the likelihood function (12.75). Our likelihood function does not provide satisfactory evaluation of $m_{\max}$. From the formal point of view, the maximum likelihood estimate of $m_{\max}$ is simply the largest observed seismic event magnitude $X_{\max}$. This follows from the fact that the likelihood function (12.75) decreases monotonically for $m_{\max} \to \infty$. A more realistic estimation of $m_{\max}$ can be carried out by introducing some additional condition or information.

At present, there is no generally accepted method to estimate the maximum magnitude $m_{\max}$. It can be estimated from empirical relationships between the magnitude and various tectonic and fault parameters (e.g., Wyss, 1979; Singh et al., 1980; Coppersmith and Youngs, 1989) or from extrapolation of frequency–magnitude curves and the use of strain rate or rate of seismic moment release (e.g., Papastamatiou, 1980; Anderson, 1986). Such an approach was applied by McGarr (1984) to evaluate the maximum possible magnitude of seismic events in mines. The value of $m_{\max}$ can also be evaluated from statistical procedures (e.g., Yegulalp and Kuo, 1974; Dargahi-Noubary, 1983; Kijko and Sellevoll, 1989, 1992). We describe two statistical procedures for the evaluation of $m_{\max}$, in which crucial role is played by the maximum observed magnitude $X_{\max}$.

Method 1. In this approach the condition for the evaluation of $m_{\max}$ is based on some properties of the end-point estimator of a uniform distribution. It is easy to show that if a random variable ξ follows uniform distribution in the range $(0, a)$, where a is an unknown, the unbiased estimation of a is equal to (Eadie, et al., 1982)

$$\hat{a} = \frac{n+1}{n}\xi_{\max}, \quad (12.76)$$

where ξ_{max} is the maximum observed value of ξ, $\xi_{max} = \max(\xi_1, \ldots, \xi_n)$, and n is the number of observations. Since the value of any cumulative distribution function $F(m)$ follows the uniform distribution within the interval $(0, 1)$, then by replacing ξ by $F(m)$, where $F(m)$ is the cumulative probability function of magnitude m, the following relation is obtained:

$$1 = \frac{n+1}{n} F(X_{max}), \tag{12.77}$$

where X_{max} is the maximum observed magnitude in the catalog. Assuming that $F(m)$ is the double-truncated Gutenberg–Richter magnitude distribution (12.15) and that $n = \lambda T$, where λ is the activity rate and T is the span of the catalog, the maximum possible magnitude is given by

$$m_{max} = -\frac{1}{\beta} \ln \left\{ A_1 - \left[A_1 - A(X_{max}) \right] \frac{\lambda T + 1}{\lambda T} \right\}, \tag{12.78}$$

where $A(X_{max}) = \exp(-\beta X_{max})$. Relation (12.78) is in full agreement with our intuitive expectations: the larger the period of observation, the less the estimated maximum possible magnitude m_{max} deviates from the maximum observed magnitude X_{max}.

Method 2.   In this approach the evaluation of m_{max} is obtained from the condition

$$X_{max} = \text{EXPECTED}(X_{max}|T), \tag{12.79}$$

that the largest observed magnitude X_{max} is equal to $\text{EXPECTED}(X_{max}|T)$, which is the largest expected magnitude in the span T of the catalog. After the application of the moment-generating function, the largest expected magnitude within the time interval T is given by (Kijko and Sellevoll, 1989)

$$\text{EXPECTED}(X_{max}|T)$$

$$= m_{max} - \frac{E_1(Tz_2) - E(Tz_1)}{\beta \exp(-Tz_2)} - m_{min} \exp(-\lambda T), \tag{12.80}$$

where $z_i = -\lambda A_i / (A_2 - A_1)$ and $E_1(\cdot)$ denotes an exponential integral function (Abramowitz and Stegun, 1970)

$$E_1(x) = \int_x^{\infty} \exp(-\zeta)/\zeta \, d\zeta. \tag{12.81}$$

Since for most real seismic data $Tz_1 > 1$ and $Tz_2 > 1$, $E_1(z)$ can be approximated as

$$E_1(x) = \frac{1}{x} \exp(-x) \frac{x^2 + a_1 x + a_2}{x^2 + b_1 x + b_2}, \tag{12.82}$$

where $a_1 = 2.334733$, $a_2 = 0.250621$, $b_1 = 3.330657$, and $b_1 = 1.681534$. Formula (12.82) is an approximation of the exponential integral function with the maximum error of $5 \cdot 10^{-5}$ for $1 \leq x \leq \infty$. By including condition (12.79) into equation (12.75), a set of equations determining the maximum-likelihood solution is obtained, which can be solved by an iterative procedure.

The estimation of seismic hazard parameters in the Klerksdorp gold mining district in South Africa is described as an example. The Klerksdorp gold field covers an area of about 300 km^2, and mining is carried out at an average depth of 2.3 km below the surface. This implies that about 500 km^3 of rock needs to be monitored for seismic activity (Gay et al., 1984; Heever, 1984; Scheepers, 1984). Mining in the district started in 1886, but on a large scale it started in 1952. For several years no major seismic activity was encountered. As mining went deeper, more mine-related seismic events started to occur. The extent of associated damage and injury was such that by 1969 it was decided to establish a permanent facility to monitor seismic activity.

For this reason the Chamber of Mines of South Africa, in collaboration with four mines in the area, established the regional seismic network in 1971. From 1971 the network had expanded considerably and by the beginning of 1989 it consisted of 30 working stations covering an area of about 250 km^2. Until 1989, the more advanced analyses were limited by analog-based technology of the recording system. This fact and the technical development of seismic monitoring systems prompted the four mines in 1989 to implement an upgrade of its seismic facilities. The new system is the Integrated Seismic System (ISS), a fully digital seismic and nonseismic data acquisition and processing network (Mendecki, 1990).

The catalog of seismic events used in our study has been compiled and homogenized by Glazer (1991). The catalog has been divided into three parts. The first part contains $n_1 = 514$ events for the period January 1, 1972 to December 31, 1984 and is complete at and above the threshold local magnitude $m_{min}^{(1)} = 3.0$. The second part contains $n_2 = 522$ events recorded from January 1, 1985 to August 30, 1990 with the threshold magnitude $m_{min}^{(2)} = 2.7$. The last part contains $n_3 = 523$ events observed from September 1, 1990 to December 31, 1991 and is complete from magnitude $m_{min}^{(3)} = 2.5$. It was assumed that for each data set the standard deviations of magnitude are equal to $\sigma_1 = 0.3$, $\sigma_2 = 0.2$, and $\sigma_3 = 0.1$, respectively. The maximum ob-

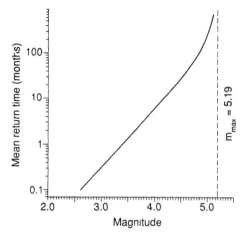

Figure 12.9 Mean return periods of seismic events for the Klerksdorp gold mining district in South Africa.

served local magnitude X_{max} was assumed to be equal to 5.0. An application of the estimation procedure for the parameters β and λ, in which m_{max} was determined according to relation (12.78), provides the following values: $\hat{\beta} = 2.90 \pm 0.07$, $\hat{\lambda} = 692.8 \pm 35.1$ (for $m_{min} = 2.0$), and $\hat{m}_{max} = 5.19 \pm 0.15$. The mean return periods of seismic events with a given magnitude for the discussed area are shown in Fig. 12.9. It is interesting to note that the replacement of condition (12.78) by (12.79) provides the following values: $\hat{\beta} = 2.90 \pm 0.07$, $\hat{\lambda} = 692.7 \pm 35.1$ (for $m_{min} = 2.0$), and $\hat{m}_{max} = 5.16 \pm 0.15$, which are only slightly different from the values estimated in the first approach.

Chapter 13 | Prediction and Prevention of Large Seismic Events

A number of statistical approaches to the prediction of mine tremors, based on statistical analysis of the space and time distribution of seismic events in a given area, are described in Chapter 12. About 20 years ago earthquake prediction became a leading expression for seismologists throughout the world. As a result, major efforts were undertaken in the research of earthquake source mechanism. Only then did it become apparent that each earthquake is different and no general rules could be found to describe its behavior deterministically by a few selected parameters. The same applies to rockbursts and mine tremors; their properties have to be specified for given mining and geologic environments.

Prediction of earthquakes is generally described as meaning the accurate forecasting of the place, size, and time of an impending earthquake, and is usually divided into three categories: long-term, intermediate-term, and short-term (e.g., Scholz, 1990). The first involves a prediction made years to decades in advance, the second involves a warning of a few weeks to a few years in advance, and the third with a warning of a few hours to a few weeks. Long-term prediction usually amounts to determining the recurrence time of earthquakes on a given fault segment and predicting the approximate time of the next event from the known time of the previous one. This problem, based on the repetition of the seismic cycle, a primary feature of the earthquake process, seems to be well established, and considerable progress has been made in this respect (Scholz, 1990). The hypothesis of seismic gap, an increased earthquake potential after a long quiet period, however, has been recently denied by Kagan and Jackson (1991). They argue that the data suggest that places of recent earthquake activity have larger than usual seismic hazard and that the seismic gap model is not supported by the history of strong earthquakes during the last 10–15 years.

Intermediate- and short-term predictions depend on the identification of precursory phenomena of various kinds that indicate that the stress loading process has reached an advanced stage. No consistently reliable phenomena precursory to earthquakes have been found. Although the evidence supports the existence of premonitory phenomena, it is difficult to indicate a single precursor that can be universally accepted. The first results of a project undertaken by the International Association of Seismology and Physics of the

Earth's Interior (IASPEI) to evaluate claims about precursors to earthquakes have recently been presented by Wyss (1991). Only three out of 28 nominations were accepted for the preliminary list of significant precursors that IASPEI is compiling. The aim of the project is to establish a consensus as to which phenomena may be real precursors, useful for earthquake prediction studies. This evaluation is of practical importance as some authors believe that hundreds of case histories of significant precursors exist, while others reject the entire evidence presented.

13.1 Precursory Phenomena

Similarly as for natural earthquakes, no consistently reliable phenomena precursory to mine tremors have been found. In this section various precursory phenomena to both earthquakes and seismic events in mines are described briefly, although the possibility that some of them are spurious cannot be ruled out.

The most frequently reported precursory phenomena involve patterns of seismicity. This probably results from the fact that seismicity is routinely monitored and constitutes the largest data set available to examination for evidence of precursors (Scholz, 1990). The seismic cycle, that is, the time interval between two major events on a specified fault segment, is accompanied by characteristic patterns in seismicity. The main event is followed by an aftershock sequence, which decays into a postseismic period of quiescence extending over the entire region surrounding the rupture zone and occupying about 50–70 percent of the recurrence period. This is followed by a general increase in background seismicity and then is followed sometimes by an intermediate-term quiescence. The succeeding major event is preceded by the immediate foreshocks, which are usually concentrated close to the hypocenter. A pronounced decrease in foreshock activity is often observed just before the final rupture.

This type of pattern has been reported many times. An outstanding example is the case of the 1978 Oaxaca, Mexico, earthquake ($M = 7.8$), whose successful prediction was based on a seismic quiescence (Ohtake et al., 1977, 1981). One out of three nominations accepted for the preliminary list of significant precursors of IASPEI (Wyss, 1991) is related to seismic quiescence before strong aftershocks reported by Matsu'ura (1986). She studied aftershocks of some large shallow earthquakes in Japan. Large aftershocks triggered their own aftershock sequences and were preceded by seismic quiescence expressed by an appreciable decrease in the rate of occurrence described by the modified Omori formula. The aftershock activity then recovered to the normal level shortly before the occurrence of the large aftershock. In general, the validation of a quiescence requires rigorous statistical testing and is a controversial subject.

Acoustic emission monitoring has been in the longest and most extensive use in mines (e.g., Hardy, 1981, 1984), and it is believed that this technique could provide a valuable indication of approaching rockbursts (e.g., Brady, 1977; Brink and Mountford, 1984; Leighton, 1984; Rudajev *et al.*, 1985; Zavyalov and Sobolev, 1988; Calder *et al.*, 1990). A relationship between the rate of microseismic events and the state of stress in the rock was verified in 1938 during a seismic study of pillars in a deep copper mine in Michigan (Bolstad, 1990). This led to the hypothesis that a relationship exists between microseismic activity and rock failure. The main problem, however, with microseismic techniques today is their reliability. Although a rapid increase in the number of microseismic events sometimes precedes a rockburst, at other times an identical pattern would result in a false alarm, and rockbursts frequently occur without any warning at all (e.g., Leighton, 1984; Bolstad, 1990). The acoustic emission technique in its classic form is based on counting the number of acoustic pulses and their cumulative energy. A waveform analysis of microseismic events could possibly provide more indicators of rockburst hazard.

Foreshocks are the most obvious precursory phenomena preceding earthquakes. Their occurrence has led to the famous prediction of the Haicheng, China, earthquake of 1975 (Wu *et al.*, 1978). This case was accepted for the preliminary list of significant precursors of IASPEI (Wyss, 1991). Foreshock activity varies greatly in individual cases—from single events to swarms. No relation between the size of the largest foreshock and the size of the main shock was found, other than that the foreshock is always smaller (Jones and Molnar, 1979). The main difficulty with applying foreshocks in earthquake prediction is their similarity to other earthquakes in a given area. The only distinction recognized so far is that foreshock sequences are characterized by smaller values of the parameter b of the frequency–magnitude relation than are aftershocks or other earthquakes (Suyehiro, 1966).

Anomalous crustal deformations preceding earthquakes have also been reported frequently. Many of these observations were obtained from individual tiltmeters or strainmeters and it is difficult, therefore, to determine if they resulted from local site effects or are of more regional character. Two notable examples, discussed by Mogi (1985) and Scholz (1990), however, were obtained from conventional geodetic surveys of regional extent. These are the Niigata, Japan, earthquake of 1964 ($M = 7.5$) and the Sea of Japan earthquake of 1983 ($M = 7.7$). Repeated leveling along the Honshu coast indicated that a broad region surrounding the rupture area of the Niigata earthquake was rapidly uplifted about 5 years prior to the earthquake. The anomalous crustal movement preceding the 1983 earthquake was very similar to the Niigata case. A broad uplift of several centimeters about 5 years before the earthquake was observed. It should be noted, however, that no decision was made against five nominations for the IASPEI preliminary list of significant precursors, which describe crustal movements and strain and tilt changes before three major Japanese earthquakes, including the 1983 earthquake in the Sea of Japan (Wyss, 1991).

Anomalous horizontal deformations of the Earth's surface, appearing from the beginning of January 1977, were observed at strain rosettes situated at a distance of 3 km from the epicenter of the Lubin, Poland, mine tremor of March 24, 1977 of magnitude 4.5 (Gibowicz *et al.*, 1979). The observed pattern of horizontal strain changes implies that the stress began to accumulate in the source area at least 3 months before the occurrence of the tremor, and that the area of stress accumulation was much larger than the source area itself. Measurements of the vertical and horizontal deformations of boreholes at the Rudna copper mine in Poland have shown that, under specific conditions of roof and pillar mining, the changes of the calculated vertical stresses could well be related not only to the local stress field but also to stress changes at a distance, associated with imminent seismic events (Siewierski *et al.*, 1989a, 1989b). Out of six strong rockbursts observed during preliminary deformation measurements, three were preceded by distinct stress increases.

Microgravimetric measurements in mines for the prediction of seismic events have been advocated by Fajklewicz (e.g., Fajklewicz and Jakiel, 1989). The measurements performed for 2 years at the Pstrowski coal mine in Upper Silesia, Poland, show that the time changes of gravity microanomalies are of a regular pattern. Local negative changes of the microanomaly are associated with approaching mine tremors within a radius of 60–100 m. Regional time changes of the gravity microanomaly are believed to signal the development of elastic strain in the whole investigated area and the approaching violent release of accumulated strain energy (Fajklewicz and Jakiel, 1989).

Seismic-wave propagation anomalies form another group of precursory phenomena. The first study of anomalous precursory changes in the propagation of seismic waves was done by Semenov (1969). He found a reduction in the ratio V_P/V_S of compressional and shear-wave velocities before a number of earthquakes near Garm in Tadjikistan, Commonwealth of Independent States (CIS). This result was confirmed for three small earthquakes at Blue Mountain Lake, New York (Aggarwal *et al.*, 1973). It as found that the ratio V_P/V_S decreased by 10–15 percent within an area surrounding the rupture and then recovered its normal value just prior to the earthquake. From these observations the dilatancy–diffusion theory of earthquake precursors, described in the next section, was developed. Subsequently, large number of papers were published, reporting premonitory velocity changes. This early optimism based on dilatancy–diffusion models was not upheld when accurate measurements of seismic velocity (e.g., McEvilly and Johnson, 1974; Kanamori and Hadley, 1975; Mogi, 1985) revealed no significant precursory changes before the occurrence of some earthquakes, and some earlier observations have been criticized.

A potentially better method is to study the shear coda waves, described in Chapter 8, that sample a given volume (Aki, 1985; Sato, 1988). Observations of coda Q^{-1} values indicate, in general, an increase in scattering and attenuation before large earthquakes within a substantial volume containing the rupture area of the earthquake (e.g., Novelo-Casanova et al., 1985; Jin and Aki, 1986; Peng et al., 1987; Tsukuda, 1988; Su and Aki, 1990), although there is a lack of consensus among authors as to the nature of effects, their possible physical mechanism, or even their existence as precursory phenomena, as opposed to a coseismic effect (Sato, 1988). Although the idea that crack density could change as a function of time and affect coda Q^{-1} changes seems reasonable, the method for detecting these changes is far from being established [see Wyss (1991) for comments by the IASPEI panel]. It appears that the reported changes in coda Q can likely be attributed to spatial variations in coda Q or to differences in the character of the S wave caused by focal mechanism changes or variations in the angle of incidence. For these reasons two nominations for coda Q^{-1} changes as precursory phenomena were not accepted for the IASPEI preliminary list of significant precursors, although the evaluating panel thought that it was definitely worthwhile to continue research on the problem of coda Q^{-1} precursors.

Interesting results on the estimation of the quality factor Q in a stope environment from coda-wave analysis have been reported from South Africa (Cichowicz and Green, 1989). The analysis of coda waves from microtremors occurring immediately in front of an advancing mine face, at the Western Deep Levels gold mine, provided an estimate of the size of the fracture zone induced by the stope. It was found that the rockmass contains a large proportion of fractured rocks at a distance of about 15–20 m from the stope. The quality factor was found to be about five times smaller in the stope fracture zone than outside of this zone. Cichowicz and Green (1989) hope to ascertain whether such fracturing observations could possibly be related to the likelihood of rockburst occurrence.

Shear-wave splitting, described in Chapter 7, is another anomaly in the propagation of seismic waves that could possibly be used as a precursor to large earthquakes. Shear-wave splitting is a result of propagation through the stress-aligned fluid-filled cracks known to exist in most rocks in the crust, forming extensive-dilatancy anisotropy or EDA (Crampin, 1987a). It is expected that the immediate effect of any change of stress before earthquakes would be to modify the geometry of the EDA cracks, and thus alter the behavior of the shear-wave splitting. Thus, the hypothesis is that stress changes before earthquakes can be monitored by analyzing shear-wave splitting (Crampin, 1987a). It is claimed (Crampin *et al.*, 1991) that this hypothesis has been confirmed by observations of changes in shear-wave splitting before and after the 1986 North Palm Springs earthquake, southern California, of magnitude $M = 6$, which can be interpreted as stress-induced changes in EDA-crack geometry (Peacock *et al.*, 1988; Crampin *et al.*, 1990). Similar effects have also been observed before and after small earthquakes at Enola, Arkansas, in 1984 (Booth *et al.*, 1990), and before and after hydraulic fracturing at the geothermal project in Rosemanowes Quarry in Cornwall in 1982 (Crampin and Booth, 1989). The discussion of how to measure and use *S*-wave splitting, however, is still in progress (Aster *et al.*, 1990). The nomination of shear-wave splitting as a precursor to earthquakes was not accepted for the IASPEI preliminary list of significant precursors (Wyss, 1991). The evaluating panel was not convinced that this precursor exists, since the observational evidence is rather weak and the presented material failed to meet the selection guidelines. The panel, however, thought that it is important to continue research on this problem since the idea that crack density and orientation could change as a function of time and effect a delay and polarization change of shear waves seems reasonable.

Changes in electric rock resistivity and, in some cases, of magnetic and electromagnetic fields, have been reported to occur before a number of earthquakes in the CIS, China, Japan, and the United States [see Mogi (1985) for a review]. There is typically a decrease in resistivity, often for several months before the earthquake. One of the most credible cases is the 1976 Tangshan, China, earthquake with $M = 7.8$, where the anomaly was observed at several sites. Several nominations for precursors of this type, however, failed to meet the selection guidelines for the IASPEI preliminary list of significant precursors (Wyss, 1991).

Continuous monitoring of electric resistivity changes seems to be a promising tool for the analysis of the variation of stresses and the occurrence of seismic events in mines. The observations conducted for several years at the Lubin copper mine in Poland by Stopinski (e.g., Stopinski and Dmowska, 1984) allowed preliminary assessment of resistivity levels corresponding to nonseismic, possibly seismic, and seismic states of stress. The measurements show that characteristic patterns of resistivity changes accompany approaching mine tremors, and also follow blasting operations and roof detachments.

Changes in pressure, flow rate, and chemical composition of groundwater, oil, or gas have been reported to occur before earthquakes. The hydrologic precursors, in which water-level changes were observed, were reviewed by Roeloffs (1988). A change in radon content of a well in the vicinity of the 1966 Tashkent, Uzbekistan, earthquake with $M = 5.5$ was reported as the first geochemical precursor by Ulomov and Mavashev (1971). A threefold increase in the radon content of the well water was observed at least one year before the earthquake. A number of radon observations were described by

Wakita *et al.* (1988). Continuous monitoring of radon concentration in groundwater in Japan in one case showed a short-term anomaly, radon concentration decrease, related to the 1978 Izu-Oshima-kinkai earthquake of magnitude $M = 7.0$, which occurred near the monitoring station, but no abnormal changes have been observed in other cases (Wakita *et al.*, 1988). The radon anomaly before the Izu-Oshima-kinkai earthquake coincided with changes in water levels, temperatures, and flow rates of several nearby water sources. This radon concentration decrease was accepted as a precursor for the IASPEI preliminary list of significant precursors (Wyss, 1991). Geochemical precursors and their possible mechanisms were reviewed by Thomas (1988). One of those mechanisms is an increase in reactive surface area caused by cracking.

Reports of radon gas emission as an earthquake precursor are numerous. To assess its possible use as a rockburst precursor, radon monitoring was performed at two gold mines in South Africa (McDonald, 1984). Although one of the major problems of underground monitoring of radon is the effect of fluctuating ventilation on the gas, fluctuations in the emanation of radon gas were detected and in a few instances were correlated with seismicity.

13.2 Dilatancy Models

The theory of dilatancy of rocks prior to rupture plays an important role in any discussion of the precursory phenomena of earthquakes. Many solids show an increase in volume during deformation. The term *dilatancy* meant originally the increase of volume of granular masses as a result of deformation. The term is now generally taken to describe the increase in volume as a result of application of a deviatoric stress (e.g., Scholz, 1990). Dilatancy was interpreted by Brace *et al.* (1966) as being caused by the development of pervasive microcracking within the rock, with a simultaneous increase in void space. The important aspect of dilatancy for earthquake prediction is its effect on the measurable physical properties of the Earth's crust.

There are two types of dilatancy model: volume dilatancy and fault-zone dilatancy models, described in some detail by Scholz (1990); his description is briefly followed here. In the volume dilatancy models it is assumed that dilatancy occurs in a volume of rock surrounding the fault zone, and in the fault-zone dilatancy models it is supposed that dilatancy occurs only within the fault zone itself.

The best known volume dilatancy model is the dilatancy-diffusion model proposed by Nur (1972) and expanded by Whitcomb *et al.* (1973) and Scholz *et al.* (1973). It assumes that dilatancy occurs within the stressed volume around an impending rupture zone and develops at an accelerating rate. The model is illustrated in Fig. 13.1, reproduced from Scholz *et al.* (1973), where

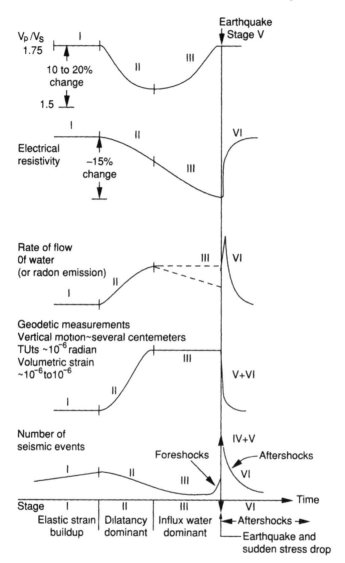

Figure 13.1 Various phenomena predicted by the dilatancy diffusion model and its consecutive stages. [From C.H. Scholz, L.R. Sykes, and Y.P. Aggarwal (1973), *Science* **181**, 803–810, Fig. 1, copyright 1973 by the American Association for the Advancement of Science.]

various phenomena predicted by the dilatancy-diffusion model and its consecutive stages are shown. In stage I, the rate of dilatancy increases with increasing stress. Then the dilatancy rate becomes so high that pore fluid diffusion cannot maintain the pore pressure. This results in dilatancy hardening in stage II, which temporarily strengthens the fault and inhibits further dilatancy. In stage III the pore pressure is reestablished by fluid diffusion, followed by rupture in stage IV. In stage V, the dilatancy recovers after the earthquake.

This sequence of events is most clearly demonstrated in laboratory measurements of the elastic properties of rock as a function of hydrostatic pressure, but its application to earthquakes was motivated by the observed anomalies in seismic velocities premonitory to some earthquakes. For saturated rocks, experiments predict that at low pressure the *P*-wave velocity will be much greater than that for dry rocks. The *S*-wave velocity, on the other hand, will not differ from that for dry rocks. As pressure is increased, the pore fluid may flow away; cracks and pores in the stressed rock will then gradually close, and both the *P*- and *S*-wave velocities will rise and approach the value for the dry rock. In another case, as pressure is increased, the pore fluid remains in the rock and a gradual rise in both the *P*- and *S*-wave velocities may be expected as a result of changes in the intrinsic rock properties and the pore fluid.

Once the dilatancy path has been established from the velocity anomalies, other types of precursory phenomena are predicted by the dilatancy-diffusion model (Fig. 13.1). This model also predicts a relationship between the duration of the precursors and the volume of the dilatant region. If the dilatant volume, in turn, is proportional to the size of the impending earthquake, a precursor time–magnitude relation can be established, such as that shown in Fig. 13.2, reproduced from Scholz *et al.* (1973). This relation was interpreted by Scholz *et al.* (1973) to indicate that the precursor duration is controlled by fluid diffusion. The validity of the dilatancy-diffusion model depends on the assumptions that stress near the earthquake source is sufficient to cause rocks to dilate prior to the earthquake, that water is available in the crust down to considerable depth, and that the permeability of rocks under crustal conditions is sufficient for water to diffuse into the dilatant source area in relatively short time.

Another volume dilatancy model is the dry-dilatancy model described by Mjachkin *et al.* (1975). It is assumed in this model that stages II and III result from dilatancy localization and stress reduction rather then from a pore-pressure interaction as in the dilatancy-diffusion model. Fault zone dilatancy models are reviewed by Rice (1983) and Rudnicki (1988). One objection to dilatancy models is that dilatancy is not observed in laboratory experiments at stresses less than about half the fracture strength, which is greater than the frictional strength. On a geologic scale, however, the fracture strength should

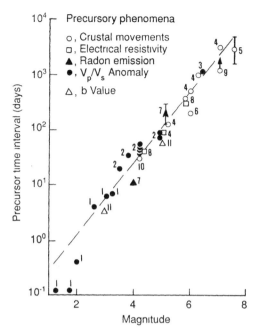

Figure 13.2 An empirical relation between the precursor time and magnitude of earthquakes. The straight line through the data is consistent with the precursor time controlled by diffusion. [From C.H. Scholz, L.R Sykes, and Y.P. Aggarwal (1973), *Science* **181**, 803–810, Fig. 2), copyright 1973 by the American Association for the Advancement of Science.]

be much less than that measured in the laboratory, whereas friction should be about the same. Thus dilatancy will be expected at lower stresses relative to the frictional strength (Scholz, 1990).

13.3 Prevention

Unlike the prevention of natural earthquakes, the prevention of large seismic events in mines, usually associated with rockbursts, or more correctly the mitigation of their severity, is feasible. It can be achieved either by changing the physical properties of the rocks to prevent the storage of strain energy or by changing mining operations to control stress concentration at critical points. Rockmass modification can be approached by destressing during mining and preconditioning during development (e.g., Bolstad, 1990). The most common method of destressing and rock preconditioning in advance of mining is by blasting, using a variety of techniques (e.g., Board and Fairhurst, 1983; Blake, 1984; Rorke and Brummer, 1990). Hydraulic injection of slip-prone features may be applicable in some mines (e.g., Brady, 1990) and destressing drilling in others (e.g., Will, 1984).

Changing the mining procedure to control stress concentrations includes stope sequencing to control mine geometry, blasting at the end of the shift to trigger stress release, controlled excavation to allow a gradual release of strain energy, and active and passive support systems to limit the damage in the event of a rockburst (Bolstad, 1990). The introduction of stabilizing pillars large enough to reduce convergence of the mined-out area has reduced the total seismicity and the occurrence of rockbursts in South Africa gold mines (e.g., McGarr and Wiebols, 1977; Deliac and Gay, 1984; Ortlepp and Spottiswoode, 1984; Lenhardt, 1990; Lenhardt and Hagan, 1990).

The development and improvement of methods for mine design to reduce seismicity is one of the major challenges in mining seismicity studies and rock mechanics. The cumulative seismic moment in a deep-level mine is related to the volume of elastic convergence of the mined-out area (McGarr, 1976; McGarr and Wiebols, 1977). The volumetric closure in turn is closely connected with the spatial rate of energy release. The energy-release rate is at present the most widely used parameter in designing mine layouts to reduce seismicity in deep mines in South Africa (Spottiswoode, 1990). Although volumetric closure and the energy-release rate can be calculated readily in elastic conditions, they provide little information about inelastic effects in the rockmass, such as the distribution of fractures or the behavior of fractured rocks. Mine tremors occurring on geologic faults, therefore, have been modeled in terms of excess shear stress, that is, the shear stress in excess of the frictional resistance present on the fault plane before the slip (Napier, 1987; Ryder, 1987). An alternative model of the excess shear stress type, termed the *volume excess shear stress method*, has recently been proposed by Spottiswoode (1990) as a step toward the development of three-dimensional modeling of inelastic rock deformation around deep-level mines.

References

Abramowitz, M., and Stegun, I. A. (1970). *Handbook of Mathematical Functions*, 9th ed., Dover, New York.

Aggarwal, Y. P., Sykes, L. R., Simpson, D. W , and Richards, P. G. (1973). Spatial and temporal variations of t_s/t_p and in P wave residuals at Blue Mountain Lake, New York: Application to earthquake prediction. *J. Geophys. Res.* **80**, 718–732.

Aki, K. (1965). Maximum likelihood estimate of b in the formula $\log N = a - bM$ and its confidence limits. *Bull. Earthq. Res. Inst. Tokyo Univ.* **43**, 237–239.

Aki, K. (1967). Scaling law of seismic spectrum. *J. Geophys. Res.* **72**, 1217–1231.

Aki, K. (1969). Analysis of the seismic coda of local earthquakes as scattered waves. *J. Geophys. Res.* **74**, 615–631.

Aki, K. (1979). Characterization of barriers on an earthquake fault. *J. Geophys. Res.* **84**, 6140–6148.

Aki, K. (1980a). Attenuation of shear waves in the lithosphere for frequencies from 0.05 to 25 Hz. *Phys. Earth Planet. Interiors* **21**, 50–60.

Aki, K. (1980b). Scattering and attenuation of shear waves in the lithosphere. *J. Geophys. Res.* **85**, 6496–6504.

Aki, K. (1981a). A probabilistic synthesis of precursory phenomena. In *Earthquake Prediction—an International Review* (D. W. Simpson and P. G. Richards, eds.), *Maurice Ewing*, Vol. 4, pp. 566–574, Am. Geophys. Union, Washington, D.C.

Aki, K. (1981b). Source and scattering effects on the spectra of small local earthquakes. *Bull. Seism. Soc. Am.* **71**, 1687–1700.

Aki, K. (1984) Asperities, barriers, characteristic earthquakes and strong motion prediction. *J. Geophys. Res.* **89**, 5867–5872.

Aki, K. (1985). Theory of earthquake prediction with special references to monitoring of the quality factor of lithosphere by coda method. *Earthq. Predict. Res.* **3**, 219–230.

Aki, K. (1987). Magnitude–frequency relation for small earthquakes: A clue to the origin of f_{max} of large earthquakes. *J. Geophys. Res.* **92**, 1349–1355.

Aki, K. (1988). Physical theory of earthquakes. In *Seismic Hazard in Mediterranean Regions* (J. Bonnin, M. Cara, A. Cisternas, and R. Fantechi, eds.), pp. 3–33, Kluwer Academic Publishers, Dordrecht.

Aki, K., and Chouet, B. (1975). Origin of coda waves: Source, attenuation and scattering effects. *J. Geophys. Res.* **80**, 3322–3342.

Aki, K., and Lee, W. H. K. (1976). Determination of three-dimensional velocity anomalies under a seismic array using first P arrival times from local earthquakes. 1. A homogeneous initial model. *J. Geophys. Res.* **81**, 4381–4399.

Aki, K., and Richards, P. G. (1980). *Quantitative Seismology Theory and Methods*, Freeman, San Francisco.

Aki, K., Christofferson, A., and Husebye, E. S. (1977). Determination of the three dimensional seismic structure of the lithosphere. *J. Geophys. Res* **82**, 277–296.

Al-Saigh, N H., and Kusznir, N. J. (1987). Some observations on the influence of faults in mining-induced seismicity. *Eng. Geol.* **23**, 277–289.

Anderson, D. L, Minster, B, and Cole, D (1974). The effect of oriented cracks on seismic velocities. *J. Geophys Res*. **79**. 4011–4015.

Anderson, J. G. (1982). Revised estimates for probabilities of earthquakes following observations of unreliable precursors *Bull. Seism. Soc. Am.* **72**, 879–888.

Anderson, J G. (1986). Seismic strain rates in the central and eastern United States. *Bull. Seism Soc. Am.* **76**, 273–290.

Anderson, J G., and Hough, S. E. (1984). A model for the shape of the Fourier amplitude spectrum of acceleration at high frequencies *Bull. Seism Soc. Am.* **74**, 1969–1993.

Andrews, D. J. (1976). Rupture velocity of plane shear cracks. *J. Geophys. Res*. **81**, 5679–5687.

Andrews, D. J (1985) Dynamic plain-strain shear rupture with a slip-weakening friction law calculated by a boundary integral method. *Bull. Seism. Soc. Am.* **75**, 1–21.

Andrews, D. J. (1986). Objective determination of source parameters and similarity of earthquakes of different size In *Earthquake Source Mechanics* (S. Das, J Boatwright, and C. H Scholz, eds.), *Maurice Ewing*, Vol. 6, pp. 259–267, Am. Geophys. Union, Washington, D.C.

Anonymous (1981). *Catalog of rockbursts in the USSR mines (1973–1980)*, VNIMI, Leningrad (in Russian).

Antsyferov, M. S, ed. (1966) *Seismo-Acoustic Methods in Mining*, Consultants Bureau, New York.

Archuleta, R. J. (1986). Downhole recordings of seismic radiation. In *Earthquake Source Mechanics* (S. Das, J. Boatwright, and C. H. Scholz, eds), *Maurice Ewing*, Vol. 6, pp 319–329, Am. Geophys. Union, Washington, D C.

Archuleta, R. J., and Day, S. M. (1980). Dynamic rupture in a layered medium: The 1966 Parkfield earthquake. *Bull. Seism. Soc. Am.* **70**, 671–689.

Archuleta, R. J., and Frazier, G. A. (1978). Three-dimensional numerical simulations of dynamic faulting in a half-space. *Bull. Seism. Soc. Am* **68**, 573–598.

Archuleta, R. J., Cranswick, E., Mueller, C., and Spudich, P. (1982). Source parameters of the 1980 Mammoth Lakes, California earthquake sequence. *J. Geophys Res*. **87**, 4595–4607.

Aster, R. C., Shearer, P. M., and Berger, J. (1990) Quantitative measurements of shear wave polarizations at the Anza seismic network, southern California. Implications for shear wave splitting and earthquake prediction. *J. Geophys. Res.* **95**, 12, 449–12, 473.

Atkinson, B. K, ed (1987a). *Fracture Mechanics of Rock*, Academic Press, London.

Atkinson, B. K. (1987b). Introduction to fracture mechanics and its geophysical applications. In *Fracture Mechanics* (B K. Atkinson, ed.), pp. 1–26, Academic Press, London.

Aviles, C. A, Scholz, C. H., and Boatwright, J. (1987). Fractal analysis applied to characteristic segments of the San Andreas Fault. *J. Geophys. Res.* **92**, 331–344.

Babich, V. M., and Alekseev, A S. (1958). A ray method of computing wave front intensities. *Izv. Akad. Nauk SSSR, Geofiz*. **1**, 17–31 (in Russian).

Backus, G. E. (1965). Possible forms of seismic anisotropy of the uppermost mantle under oceans *J. Geophys Res*. **70**, 3429–3439.

Backus, G. E. (1977a) Interpreting the seismic glut moments of total degree two or less. *Geophys. J. Roy Astr. Soc.* **51**, 1–25.

Backus, G. E. (1977b) Seismic sources with observable glut moments of spatial degree two. *Geophys. J. Roy. Astr Soc.* **51**, 27–45.

Backus, G. E., and Mulcahy, M (1976) Moment tensors and other phenomenological descriptions of seismic sources. I. Continuous displacements. *Geophys J Roy Astr. Soc.* **46**, 341–371.

Bakun, W. H. (1984). Seismic moments, local magnitudes, and coda-duration magnitudes for earthquakes in central California. *Bull. Seism Soc. Am* **74**, 439–458.

Bakun, W. H., and McEvilly, T. V. (1979). Are foreshocks distinctive? Evidence from the 1966 Parkfield and the 1975 Oroville, California sequences. *Bull. Seism. Soc. Am.* **69**, 1027–1038

Barenblatt, G I. (1962) The mathematical theory of equilibrium cracks in brittle fracture. *Adι Appl. Mech.* **7**, 55–80.

Barker, J. S., and Langston, C. A. (1982). Moment tensor inversion of complex earthquakes *Geophys. J. Roy. Astr. Soc.* **68**, 777–803.

Bates, R. H. J., and McKinnon, G. C. (1979). Towards improving images in ultrasonic transmission tomography. *Austr. Phys. Sci Med* **2**, 134–140.

Båth, M. (1984). Rockburst seismology. In *Rockbursts and Seismicity in Mines* (N. C. Gay and E. H. Wainwright, eds.), Symp. Ser. No. 6, pp. 7–15, S. Afr Inst. Min. Metal., Johannesburg.

Beck, S. L., and Ruff, L. J. (1984). The rupture process of the great 1979 Colombia earthquake: Evidence for the asperity model. *J. Geophys. Res.* **89**, 9281–9291.

Behera, P. K. (1990). Ultradeep mining problems in Kolar gold mines In *Rock at Great Depth* (V. Maury and D. Fourmantraux, eds), pp. 687–694, Balkema, Rotterdam.

Bender, B. (1983). Maximum likelihood estimation of *b* values for magnitude grouped data *Bull. Seism. Soc. Am.* **73**, 831–851.

Bender, B. (1987). Effects of observational errors in relating magnitude scales and fitting the Gutenberg-Richter parameter β *Bull. Seism. Soc Am* **77**. 1400–1428.

Benjamin, J R., and Cornell, C A. (1970). *Probability, Statistics, and Decision for Civil Engineers*, McGraw-Hill, New York

Ben-Menahem, A. (1961). Radiation of seismic surface waves from finite moving sources. *Bull. Seism. Soc. Am.* **51**, 401–435.

Ben-Menahem, A., and Singh, S. J. (1981). *Seismic Waves and Sources*, Springer-Verlag, New York.

Bernard, P., and Madariaga, R (1984) A new asymptotic method for the modeling of near-field accelerograms. *Bull. Seism. Soc. Am.* **74**, 539–557

Beroza, G. C. (1991). Near-source modeling of the Loma Prieta earthquake: Evidence for heterogeneous slip and implications for earthquake hazard. *Bull Seism. Soc. Am.* **81**, 1603–1621.

Bicknell, J, and McGarr, A (1990). Underground recordings of mine tremors—implications for earthquake source scaling. In *Rockbursts and Seismicity in Mines* (C. Fairhurst, ed.), pp. 109–114, Balkema, Rotterdam

Biswas, N. N., and Aki, K. (1984). Characteristics of coda waves: Central and southcentral Alaska. *Bull. Seism. Soc. Am.* **74**, 493–507

Blake, W. (1984). Rock preconditioning as a seismic control measure In *Rockbursts and Seismicity in Mines* (N. C. Gay and E. H. Wainwright, eds.), Symp. Ser. No. 6, pp. 225–229, S. Afr. Inst. Min. Metal., Johannesburg.

Blake, W., Leighton, F., and Duvall, W. I. (1974). *Microseismic Techniques for Monitoring of Behavior of Rock Structures*, Bulletin 665, U.S. Bureau of Mines.

Blakeslee, S., Malin, P., and Alvarez, M. (1989). Fault-zone attenuation of high-frequency seismic waves. *Geophys Res. Lett.* **16**, 1321–1324.

Board, M. P., and Fairhurst, C. (1983). Rockburst control through destressing—a case example *Proc. Symp Rockbursts· Prediction and Control.* pp. 91–101, Inst. Min. Metal., London

Boatwright, J. (1978). Detailed spectral analysis of two small New York State earthquakes. *Bull Seism. Soc. Am.* **68**, 1117–1131.

Boatwright, J. (1980). A spectral theory for circular seismic sources: Simple estimates of source dimension, dynamic stress drop, and radiated seismic energy. *Bull. Seism. Soc. Am* **70**, 1–27

Boatwright, J. (1981) Quasi-dynamic models of simple earthquakes: An application to an aftershock of the 1975 Oroville, California, earthquake. *Bull. Seism. Soc. Am.* **71**, 69–94

Boatwright, J. (1982). A dynamic model for far-field acceleration. *Bull. Seism. Soc. Am.* **72**, 1049–1068.

Boatwright, J. (1984a). The effect of rupture complexity on estimates of source size. *J. Geophys. Res.* **89**, 1132–1146.

Boatwright, J. (1984b). Seismic estimates of stress release. *J. Geophys. Res.* **89**, 6961–6968.

Boatwright, J., and Boore, D. M. (1982). Analysis of the ground accelerations radiated by the 1980 Livermore Valley earthquakes for directivity and dynamic source characteristics. *Bull. Seism. Soc. Am.* **72**, 1843–1865.

Boatwright, J., and Choy, G. L. (1986). Teleseismic estimates of the energy radiated by shallow earthquakes. *J. Geophys. Res.* **91**, 2095–2112.

Boatwright, J., and Fletcher, J. B. (1984). The partition of radiated energy between *P* and *S* waves. *Bull. Seism. Soc. Am.* **74**, 361–376.

Boatwright, J., and Quin, H. (1986). The seismic radiation from a 3-D dynamic model of a complex rupture process. In *Earthquake Source Mechanics* (S. Das, J. Boatwright, and C. H. Scholz, eds.), *Maurice Ewing*, Vol. 6, pp. 97–109, Am. Geophys. Union, Washington, D.C.

Bober, A., and Kazimierczyk, M. (1979). Seismic activity and the roof fall mining in the Lubin mine. *Publ. Inst. Geophys., Pol. Acad. Sci.* **M-2** (123), 271–288.

Bodoky, T., Hermann, L., and Dianiska, L. (1985). Processing of the in-seam seismic transmission measurements. Paper presented at 7th Ann. Meet. Eur. Assoc. Eng. Geol., Budapest, 1985.

Bois, P., La Porte, M., Lavergne, M., and Thomas, G. (1972). Well to well seismic measurements. *Geophysics* **3**, 471–483.

Bollinger, G. A. (1989). Microearthquake activity associated with underground coal-mining in Buchanan County, Virginia, U.S.A. In *Seismicity in Mines* (S. J. Gibowicz, ed.), reprinted from Special Issue, *Pure Appl. Geophys.* **129**, 407–421, Birkhäuser Verlag, Basel.

Bolstad, D. D. (1990). Keynote lecture: Rock burst control research by the U.S. Bureau of Mines. In *Rockbursts and Seismicity in Mines* (C. Fairhurst, ed.), pp. 371–375, Balkema, Rotterdam.

Bolt, B. A., ed. (1987). *Seismic Strong Motion Synthetics*, Academic Press, Orlando, Fla.

Boore, D. M. (1986). The effect of finite bandwidth on seismic scaling relations. In *Earthquake Source Mechanics* (S. Das, J. Boatwright, and C. H. Scholz, eds.), *Maurice Ewing*, Vol. 6, pp. 275–283, Am. Geophys. Union, Washington, D.C.

Boore, D. M., and Boatwright, J. (1984). Average body-wave radiation coefficients. *Bull. Seism. Soc. Am.* **74**, 1615–1621.

Boore, D. M., and Dunbar, W. S. (1977). Effect of the free surface on calculated stress drops. *Bull. Seism. Soc. Am.* **67**, 1661–1664.

Boore, D. M., and Joyner, W. B. (1978). The influence of rupture incoherence on seismic directivity. *Bull. Seism. Soc. Am.* **68**, 283–300.

Boore, D. M., and Joyner, W. B. (1989). The effect of directivity on the stress parameter determined from ground motion observations. *Bull. Seism. Soc. Am.* **79**, 1984–1988.

Booth, D. C., Crampin, S., Evans, R., and Roberts, G. (1985). Shear-wave polarization near the North Anatolian Fault. I. Evidence for anisotropy-induced shear-wave splitting. *Geophys. J. Roy. Astr. Soc.* **83**, 61–73.

Booth, D. C., Crampin, S., Lovell, J. H., and Chiu, J.-M. (1990). Temporal changes in shear wave splitting during an earthquake swarm in Arkansas. *J. Geophys. Res.* **95**, 11, 151–11, 164.

Bording, R. P., Gersztenkorn, A., Lines, L. R., Scales, J. A., and Treitel, S. (1987). Applications of seismic travel-time tomography. *Geophys. J. Roy. Astr. Soc.* **90**, 285–303.

Box, G. E. P. (1957). Use of statistical methods in the elucidation of basic mechanisms *Bull. Internatl. Statist. Inst. Part 3* **36**, 215–225.

Box, G. E. P., and Lucas, H. L. (1959). Design of experiments in non-linear situations. *Biometrika* **46**, 77–90.

Brace, W. F., and Byerlee, J. D. (1966). Stick-slip as a mechanism for earthquakes. *Science* **153**, 990–992.

Brace, W. F., Paulding, B. W., and Scholz, C. H. (1966). Dilatancy in the fracture of crystalline rocks. *J. Geophys. Res.* **71**, 3939–3953.

Brady, B. H. G. (1990). Keynote lecture: Rock stress, structure and mine design. In *Rockbursts and Seismicity in Mines* (C. Fairhurst, ed.), pp. 311–321, Balkema, Rotterdam.

Brady, B. H. G., and Brown, E. T. (1985). *Rock Mechanics*, George Allen and Unwin, London.

Brady, B. T. (1977). Anomalous seismicity prior to rock bursts: Implications for earthquake prediction. In *Stress in the Earth* (M. Wyss, ed.): reprinted from Special Issue, *Pure Appl. Geophys.* **115**, 357–374, Birkhäuser Verlag, Basel.

Brillinger, D. R., Udias, A., and Bolt, B. A. (1980). A probability method for regional focal mechanism solutions. *Bull. Seism. Soc. Am.* **70**, 149–170.

Brink, A. v. Z. (1990). Application of a microseismic system at Western Deep Levels. In *Rockbursts and Seismicity in Mines* (C. Fairhurst, ed.), pp. 355–361, Balkema, Rotterdam.

Brink, A. v. Z., and Mountfort, P. I. (1984). Feasibility studies on the prediction of rockbursts at Western Deep Levels. In *Rockbursts and Seismicity in Mines* (N. C. Gay and E. H. Wainwright, eds.), Symp. Ser. No. 6, pp. 317–325, S. Afr. Inst. Min. Metal., Johannesburg.

Brune, J. N. (1970). Tectonic stress and the spectra of seismic shear waves from earthquakes. *J. Geophys. Res.* **75**, 4997–5009.

Brune, J. N. (1971). Correction. *J. Geophys. Res.* **76**, 5002.

Brune, J. N. (1976). The physics of earthquake strong motion. In *Seismic Risk and Engineering Decisions* (C. Lomnitz and E. Rosenblueth, eds.), pp. 141–177, Elsevier, New York.

Brune, J. N., Archuleta, R., and Hartzell, S. (1979). Far-field *S*-wave spectra, corner frequencies and pulse shapes. *J. Geophys. Res.* **81**, 2262–2272.

Brune, J. N., Fletcher, J., Vernon, F., Haar, L., Hanks, T., and Berger, J. (1986). Low stress-drop earthquakes in the light of new data from the Anza, California telemetered digital array. In *Earthquake Source Mechanics* (S. Das, J. Boatwright, and C. H. Scholz, eds.), *Maurice Ewing*, Vol. 6, pp. 237–245, Am. Geophys. Union, Washington, D.C.

Buchbinder, G. G. R. (1985). Shear wave splitting and anisotropy in the Charlevoix seismic zone, Quebec. *Geophys. Res. Lett.* **12**, 425–428.

Buland, R. (1976). The mechanics of locating earthquakes. *Bull. Seism. Soc. Am.* **66**, 173–187.

Bullen, K. E., and Bolt, B. A. (1985). *An Introduction to the Theory of Seismology*, Cambridge University Press, Cambridge (U.K.).

Burdick, L. J. (1978). t^* for *S* waves with a continental ray path. *Bull. Seism. Soc. Am.* **68**, 1013–1030.

Burdick, L. J., and Mellman, G. R. (1976). Inversion of the body waves from the Borrego Mountain earthquake to the source mechanism. *Bull. Seism. Soc. Am.* **66**, 1485–1499.

Burmin, V. Y. (1986). Optimal placement of seismic stations for registration of near earthquakes. *Izv. Akad. Nauk SSSR, Earth Phys.* **22**, 366–372.

Burridge, R., and Knopoff, L. (1964). Body force equivalents for seismic dislocations. *Bull. Seism. Soc. Am.* **54**, 1875–1888.

Burridge, R., and Willis, J. R. (1969). The self-similar problem of the expanding elliptical crack in an anisotropic solid. *Proc. Camb. Phil. Soc.* **66**, 443–468.

Cagnetti, V., and Pasquale, V. (1979). The earthquake sequence in Friuli, Italy, 1976. *Bull. Seism. Soc. Am.* **69**, 1797–1818.

Calder, P. N., Archibald, J. F., Madsen, D., and Bullock, K. (1990). High frequency precursor analysis prior to a rockburst. In *Rockbursts and Seismicity in Mines* (C. Fairhurst, ed.), pp. 177–181, Balkema, Rotterdam.

Campbell, K. W. (1982). Bayesian analysis of extreme earthquake occurrences. *Bull. Seism. Soc. Am.* **72**, 1689–1705.

352 References

Campillo, M., Bouchon, M., and Massinon, B. (1983). Theoretical study of the excitation, spectral characteristics, and geometrical attenuation of regional seismic phases *Bull Seism Soc Am*. **73**, 79–90.

Campillo, M., Plantet, J L., and Bouchon, M. (1985). Frequency-dependent attenuation in the crust beneath central France from *Lg* waves: Data analysis and numerical modeling. *Bull. Seism. Soc. Am* **75**, 1395–1411.

Casten, U., and Cete, A. (1980) Induzierte Seismizität im Bereich des Steinkohlengergbaus des Ruhrreviers. *Glueckauf-Forschungsh* **41**, 12–16.

Červeny, V. (1987). Ray tracing algorithms in three-dimensional laterally varying structures. In *Seismic Tomography* (G. Nolet, ed.), pp. 99–133, Reidel, Dordrecht.

Červeny, V., Molotkov, I. A., and Psenčik, I. (1977). *Ray Method in Seismology*. Univerzita Karlova, Prague

Chael, E. P. (1987). Spectral scaling of earthquakes in the Miramachi region of New Brunswick. *Bull. Seism. Soc. Am*. **77**, 347–365.

Chael, E P., and Kromer, P. (1988). High-frequency spectral scaling of a main shock/aftershock sequence near the Norwegian coast. *Bull. Seism. Soc Am*. **78**, 561–570.

Chaloner, K., and Larntz, K. (1989) Optimal Bayesian design applied to logistic regression experiments. *J. Stat. Plann. Inf*. **21**, 191–208.

Chander, R. (1977). On tracing seismic rays with specified points in layers of constant velocity and plane interfaces. *Geophys. Prosp*. **25**, 120–124.

Chandra, U. (1971). Combination of *P* and *S* data for the determination of earthquake focal mechanism. *Bull. Seism. Soc. Am*. **61**, 1655–1673

Chang, A. C., Shumway, R. H., Blandford, R. R., and Barker, B. W. (1983) Two methods to improve location estimates—preliminary results. *Bull. Seism. Soc. Am*. **73**, 281–295.

Chapman, M. C., and Rogers, M. J. B. (1989) Coda *Q* in the southern Appalachians. *Geophys. Res. Lett*. **16**, 531–534.

Chen, T.-C., Booth, D. C., and Crampin, S (1987). Shear-wave polarizations near the North Anatolian Fault. III Observations of temporal changes. *Geophys. J. Roy. Astr. Soc*. **91**, 287–311

Chernoff, H. (1953). Locally optimum designs for estimating parameters. *Ann. Math. Stat*. **24**, 586–602.

Chernov, L. (1960). *Wave Propagation in Random Medium*, McGraw-Hill, New York

Chouet, B. (1979). Sources of seismic events in the cooling lava lake of Kilauea Iki, Hawaii. *J Geophys. Res*. **84**, 2315–2330

Chouet, B. (1990). Effect of anelastic and scattering structures of the lithosphere on the shape of local earthquake coda. In *Scattering and Attenuation of Seismic Waves* (R.-S Wu and K. Aki, eds.), Part III, reprinted from Special Issue, *Pure Appl. Geophys*. **132**, 287–310, Birkhauser Verlag, Basel.

Chouet, B., Aki, K., and Tsujiura, M. (1978). Regional variations of the scaling law of earthquake source spectra. *Bull. Seism. Soc. Am*. **68**, 49–79

Chouhan, R. K. S. (1986). Induced seismicity of Indian coal mines. *Phys Earth Planet. Interiors* **44**, 82–86.

Choy, G. L., and Boatwright, J. (1982). Broadband analysis of the extended foreshock sequence of the Miyagi-Oki earthquake of 12 June 1978. *Bull. Seism. Soc. Am* **72**, 2017–2036.

Choy, G. L., and Boatwright, J. (1988). Teleseismic and near-field analysis of the Nahanni earthquakes in the Northwest Territories, Canada *Bull. Seism. Soc. Am*. **78**, 1627–1652.

Cichowicz, A., and Green. R. W. E (1989). Changes in the early part of the seismic coda due to localized scatterers: The estimation of *Q* in a stope environment. In *Seismicity in Mines* (S. J. Gibowicz. ed.), reprinted from Special Issue, *Pure Appl. Geophys* **129**, 497–511, Birkhäuser Verlag, Basel

Cichowicz, A., Green, R. W. E., and Brink, A. v. Z. (1988). Coda polarization properties of high frequency microseismic events. *Bull. Seism. Soc. Am.* **78**, 1297–1318.

Cichowicz, A., Green, R. W. E., Brink, A. v. Z., Grobler, P., and Mountfort, P. I. (1990). The space and time variation of microevent parameters occurring in front of an active stope. In *Rockbursts and Seismicity in Mines* (C. Fairhurst, ed.), pp. 171–175. Balkema, Rotterdam.

Clayton, R. W. and Comer, P. (1984). A tomographic analysis of mantle heterogeneities. *Terra Cognita* **4**, 282–283.

Comer, R. P., and Clayton, R W. (1984). Tomographic reconstruction of velocity heterogeneity in the Earth's mantle (abstract). *EOS Trans. Am. Geophys. Union* **65**, 236.

Console, R., and Di Giovambattista, R. (1987) Local earthquake relative location by digital records. *Phys. Earth Planet Interiors* **47**, 43–49.

Console, R., and Rovelli, A (1981) Attentuation parameters for Friuli region from strong motion accelerograms. *Bull Seism. Soc Am* **71**, 1981–1991.

Cook, J. F, and Bruce, D. (1983). Rockbursts at Macassa mine and the Kirkland Lake mining area. *Proc Symp. Rockbursts: Prediction and Control.* Inst. Min. Metal., London, pp 81–89

Cook, N. G. W. (1963). The seismic location of rockbursts. *Proc. 5th Symp. Rock Mech.*, pp. 493–516, Pergamon Press, Oxford

Cook, N. G. W. (1976). Seismicity associated with mining. *Eng. Geol* **10**, 99–122

Coppersmith, K. J., and Youngs, R. R. (1989). Issues regarding earthquake source characterization and seismic hazard analysis with passive margins and stable continental interiors. In *Earthquakes at North-Atlantic Passive Margins: Neotectonics and Postglacial Rebound* (S. Gregersen and P. W. Basham, eds.), pp. 601–631, Kluwer Academic Publishers, Dordrecht.

Cormier, V. F. (1984). The polarization of S waves in a heterogeneous isotropic Earth's model. *J. Geophys. Res.* **56**, 20–23.

Cornell, C. A. (1968) Engineering seismic risk analysis. *Bull. Seism. Soc. Am.* **58**, 1583–1606.

Cosma, C. (1983). Determination of rock mass quality by the crosshole seismic method. *Bull. Int. Assoc. Eng. Geol.* **20**, 26–27.

Crampin, S. (1977). A review of the effects of anisotropic layering on the propagation of seismic waves. *Geophys. J. Roy. Astr. Soc.* **49**, 9–27.

Crampin, S. (1978). Seismic wave propagation through a cracked solid: Polarization as a possible dilatancy diagnostic. *Geophys. J Roy. Astr Soc.* **53**, 467–496.

Crampin, S. (1981). A review of wave motion in anisotropic and cracked elastic-media. *Wave Motion* **3**, 343–391.

Crampin, S. (1984a). An introduction to wave propagation in anisotropic media. *Geophys. J. Roy. Astr. Soc.* **76**, 17–28.

Crampin, S. (1984b). Effective anisotropic elastic constants for wave propagation through cracked solids. *Geophys. J. Roy. Astr. Soc.* **76**, 135–145

Crampin, S. (1985). Evaluation of anisotropy by shear-wave splitting. *Geophysics* **50**, 142–152.

Crampin, S. (1987a). The basis for earthquake prediction. *Geophys. J Roy Astr Soc.* **91**, 331–347.

Crampin, S. (1987b). Geological and industrial implications of extensive-dilatancy anisotropy. *Nature* **328**, 491–496

Crampin, S. (1990). The scattering of shear-waves in the crust. In *Scattering and Attenuation of Seismic Waves* (R.-S. Wu and K Aki, eds.), Part III, reprinted from Special Issue, *Pure Appl. Geophys.* **132**, 67–91, Birkhäuser Verlag, Basel.

Crampin, S., and Booth, D. C. (1985). Shear-wave polarizations near the North Anatolian Fault. II. Interpretation in terms of crack-induced anisotropy. *Geophys. J. Roy. Astr. Soc.* **83**, 75–92.

Crampin, S., and Booth, D. C. (1989) Shear-wave splitting showing hydraulic dilatation of pre-existing joints in granite. *Sci. Drilling* **1**, 21–26.

Crampin, S., Evans, R., Üçer, B., Doyle, M., Davis, J. P., Yegorkina, G. V., and Miller, A. (1980). Observations of dilatancy-induced polarization anomalies and earthquake prediction. *Nature* **286**, 874–877.

Crampin, S., Chesnokov, E. M., and Hipkin, R. G. (1984a). Seismic anisotropy—the state of the art: II. *Geophys. J. Roy. Astr. Soc.* **76**, 1–16.

Crampin, S., Evans, R., and Atkinson, B. K. (1984b) Earthquake prediction: A new physical basis *Geophys. J. Roy. Astr. Soc.* **76**, 147–156.

Crampin, S., McGonigle, R., and Ando, M. (1986). Extensive-dilatancy anisotropy beneath Mount Hood, Oregon and the effect of aspect ratio on seismic velocities through aligned cracks. *J. Geophys. Res.* **91**, 12,703–12,710.

Crampin, S., Booth, D. C., Evans, R., Peacock, S., and Fletcher, J. B. (1990). Changes in shear wave splitting at Anza near the time of the North Palm Springs earthquake. *J. Geophys. Res.* **95**, 11,197–11,212.

Crampin, S., Booth, D C., Evans, R., Peacock, S., and Fletcher, J. B. (1991). Comment on "Quantitative measurements of shear wave polarizations at the Anza seismic network, southern California: Implications for shear wave splitting and earthquake prediction" by Richard C. Aster, Peter M. Shearer, and Jon Berger. *J. Geophys. Res.* **96**, 6403–6414.

Cranswick, E., Wetmiller, R., and Boatwright, J. (1985). High-frequency observations and source parameters of microearthquakes recorded at hard-rock sites. *Bull. Seism. Soc. Am.* **75**, 1535–1567.

Crosson, R. S. (1976). Crustal structure modeling of earthquake data. 1. Simultaneous least squares estimation of hypocenter and velocity parameters. *J. Geophys. Res.* **81**, 3036–3046.

Dainty, A. M. (1981). A scattering model to explain seismic Q observations in the lithosphere between 1 and 30 Hz. *Geophys. Res. Lett.* **8**, 1126–1128.

Dainty, A. M., and Toksöz, M. N. (1981) Seismic codas on the Earth and the Moon: A comparison. *Phys. Earth Planet. Interiors* **26**, 250–260.

Dainty, A. M., Duckworth, R. M., and Tie, A. (1987). Attenuation and back scattering from local coda. *Bull. Seism. Soc. Am.* **77**, 1728–1747.

Daley, T. M., McEvilly, T. V., and Majer, E. L. (1988). Multiply-polarized shear-wave VSPs from the Cajon Pass drillhole. *Geophys. Res. Lett.* **15**, 1001–1004.

Dan, K., Watanabe, T., Tanaka, T., and Sato, R. (1990). Stability of earthquake ground motion synthesized by using different small-event records as empirical Green's functions. *Bull. Seism. Soc. Am.* **80**, Part A, 1433–1455.

Dargahi-Noubary, G. R. (1983). A procedure for estimation of the upper bound for earthquake magnitudes. *Phys. Earth Planet. Interiors* **33**, 91–93.

Das, S. (1980). A numerical method for determination of source time functions for general three-dimensional rupture propagation. *Geophys. J. Roy. Astr. Soc.* **62**, 591–604.

Das, S. (1981). Three-dimensional spontaneous rupture propagation and implications for the earthquake source mechanism. *Geophys. J. Roy. Astr. Soc.* **67**, 375–393.

Das, S., and Aki, K. (1977). A numerical study of two-dimensional spontaneous rupture propagation. *Geophys. J. Roy. Astr. Soc.* **50**, 643–668.

Das, S., and Boatwright, J. (1985). The breaking of a single asperity: Analysis of an aftershock of the 1975 Oroville, California, earthquake. *Bull Seism. Soc. Am.* **75**, 677–687.

Das, S., and Kostrov, B. V. (1983). Breaking of a single asperity: Rupture propagation and seismic radiation. *J. Geophys. Res.* **88**, 4177–4188.

Das, S., and Scholz, C. H. (1981a). Theory of time-dependent rupture in the earth. *J. Geophys. Res.* **86**, 6039–6051.

Das, S., and Scholz, C. H. (1981b). Off-fault aftershock clusters caused by shear stress increase? *Bull Seism. Soc. Am.* **71**, 1669–1675.

Das, S., Boatwright, J., and Scholz, C H., eds. (1986). *Earthquake Source Mechanics, Maurice Ewing*, Vol. 6, Am. Geophys. Union, Washington, D C.

Davison, F., and Scholz, C. (1985). Frequency-moment distribution of earthquakes in the Aleutian Arc: A test of the characteristic earthquake model. *Bull. Seism. Soc. Am.* **75**, 1349–1362.

Day, S. M. (1982a). Three-dimensional finite difference simulation of fault dynamics: Rectangular faults with fixed rupture velocity. *Bull. Seism. Soc. Am.* **72**, 705–727.

Day, S. M. (1982b). Three-dimensional simulation of spontaneous rupture: The effect of nonuniform prestress. *Bull. Seism. Soc. Am.* **72**, 1881–1902.

Dechelette, O., Josien, J. P., Revalor, R., and Jonis, R. (1984). Seismo-acoustic monitoring in an operational longwall face with a high rate of advance. In *Rockbursts and Seismicity in Mines* (N. C. Gay and E. H. Wainwright, eds.), Symp. Ser. No. 6, pp. 83–87, S. Afr. Inst. Min. Metal., Johannesburg.

Deliac, E. P., and Gay, N. C. (1984). The influence of stabilizing pillars on seismicity at ERPM. In *Rockbursts and Seismicity in Mines* (N. C. Gay and E. H. Wainwright, eds.), Symp. Ser. No. 6, pp. 257–263, S. Afr. Inst. Min. Metal., Johannesburg.

Del Pezzo, E., De Natale, G., Scarcella, G., and Zollo, A. (1985). Q_c of three component seismograms of volcanic microearthquakes at Campari Flegrei volcanic area, southern Italy. *Pure Appl. Geophys.* **123**, 683–696.

Del Pezzo, E., De Natale, G., Martini, M., and Zollo, A. (1987). Source parameters of microearthquakes at Phlegraean Fields (Southern Italy) volcanic area. *Phys. Earth Planet. Interiors* **47**, 25–42.

Dempster, E. L., Tyser, J. A., and Wagner, H. (1983). Regional aspects of mining-induced seismicity: Theoretical and management considerations. *Proc. Symp. Rockbursts: Prediction and Control*, pp. 37–52, Inst. Min. Metal., London.

De Natale, G., Iannaccone, G., Martini, M., and Zollo, A. (1987). Seismic sources and attenuation properties at the Campi Flegrei volcanic area. In *Advances in Volcanic Seismology* (E. A. Okal, ed.); reprinted from Special Issue, *Pure Appl. Geophys.* **125**, 883–917, Birkhäuser Verlag, Basel.

De Natale, G., Ferraro, A., and Virieux, J. (1991). A probability method for local earthquake focal mechanisms. *Geophys. Res. Lett.* **18**, 613–616.

Deschamps, A., Gaudemer, Y., and Cisternas, A. (1982). The El Asnam, Algeria, earthquake of 10 October 1980: Multiple-source mechanism determined from long-period records. *Bull. Seism. Soc. Am.* **72**, 1111–1128.

Dessokey, M. M. (1984). Statistical models of the seismic hazard analysis for mining tremors and natural earthquakes. *Publ. Inst. Geophys., Pol. Acad. Sci.* **A-15** (174), 1–82.

Devaney, A. J. (1984). Geophysical diffraction tomography. *I.E.E.E. Trans. Geosci. Remote Sens.* **22**, 3–13.

Deza, E., and Jaén, H. (1979). Microtemblores de ultrafrecuencia con la explotacion minera a tajo abierto en Cerro de Pasco. *Bol. Soc. Geol. Peru* **63**, 237–247.

Di Bona, M., and Rovelli, A. (1988). Effects of the bandwidth limitation on stress drops estimated from integrals of the ground motion. *Bull. Seism. Soc. Am.* **78**, 1818–1825.

Dieterich, J. H. (1978). Time-dependent friction and the mechanics of stick-slip. *Pure Appl. Geophys.* **116**, 790–806.

Dieterich, J. H. (1979a). Modelling of rock friction: I. Experimental results and constitutive equations. *J. Geophys. Res.* **84**, 2161–2168.

Dieterich, J. H. (1979b). Modelling of rock friction: II. Simulation of preseismic slip. *J. Geophys. Res.* **84**, 2169–2175.

Dillinger, W. H., Harding, S. T., and Pope, A. J. (1972). Determining maximum likelihood body wave focal plane solutions. *Geophys. J. Roy. Astr. Soc.* **30**, 315–329.

Dines, K., and Lytle, J. (1979). Computerized geophysical tomography. *Proc. I.E.E.E.* **67**, 1065–1073.

Dmowska, R., and Li, V. C. (1982). A mechanical model of precursory source processes for some large earthquakes. *Geophys. Res. Lett.* **9**, 393–396.

Dmowska, R., and Rice, J. R. (1986) Fracture theory and its seismological applications. In *Continuum Theories in Solid Earth Physics* (R. Teisseyre. ed.), pp. 187–255, Polish Scientific Publishers, Warsaw; Elsevier, Amsterdam.

Douma, J. (1988). Crack-induced anisotropy and its effect on vertical seismic profiling. *Geologica Ultraiectina*, **54**, 11–162.

Douma, J., and Helbig, K. (1987). What can the polarization of shear waves tell us? *First Break* **5**, 95–104.

Draper, N. R., and Smith, H (1981). *Applied Regression Analysis*, 2nd ed., Wiley, New York.

Droste. Z., and Teisseyre, R (1976) Some cases of the shock mechanism types in the Upper Silesia coal region. *Publ. Inst. Geophys., Pol. Acad Sci* **97**, 141–156

Drzezla, B., Garus, A., and Kijko, A (1986) Energy distribution of largest mining tremors and their connection with geologic structure of rock mass. In *Mining Systems Adjusted to High Rock Pressure Conditions* (A Kidybinski and M Kwasniewski, eds), pp 257–261, Balkema, Rotterdam.

Drzezla, B, Garus, A., and Bialek, J. (1988). Search for quantitative relation between mine-induced stress and seismic activity. In *Modeling of Mine Structures* (A Kidybinski and M. Kwasniewski, eds.), pp. 63–67. Balkema, Rotterdam

Dubinski, J, and Dworak, J (1989) Recognition of the zones of seismic hazard in Polish coal mines by using a seismic method. In *Seismicity in Mines* (S J. Gibowicz, ed); reprinted from Special Issue, *Pure Appl. Geophys*. **129**, 609–617, Birkauser Verlag, Basel

Dugdale, D S (1960). Yielding of steel sheets containing slits. *J. Mech. Phys. Solids* **8**, 100–115.

Dunkin, J W. (1965). Computation of modal solutions in layered elastic media at high frequencies. *Bull. Seism. Soc. Am.* **55**, 335–358.

Duvall, W I, and Stephenson, D. E (1965). Seismic energy available from rockbursts and underground explosions. *Trans Soc. Min Eng, Am Inst Metal Petrol. Eng* **231**, 235–240

Dyer, B, and Worthington, M H. (1988). Some sources of distortion in tomographic velocity images. *Geophys Prosp* **36**, 209–222.

Dysart, P. S., Snoke, J. A., and Sacks, I S. (1988). Source parameters and scaling relations for small earthquakes in the Matsushiro region, southwest Honshu, Japan *Bull. Seism. Soc. Am.* **78**, 571–589.

Dziewonski, A. M., and Woodhouse, J. H. (1983). An experiment in systematic study of global seismicity Centroid-moment tensor solutions for 201 moderate and large earthquakes of 1981 *J Geophys. Res.* **88**, 3247–3271.

Dziewonski, A M., Chou, T.-A., and Woodhouse, J. H. (1981). Determination of earthquake source parameters from waveform data for studies of global and regional seismicity. *J Geophys. Res.* **86**, 2825–2852

Eadie, W T, Drijard, D., James, F E., Sadoulet, B., and Roos, M. (1982). *Statistical Methods in Experimental Physics*, 2nd reprint, North-Holland, Amsterdam

Ebel, J. E., and Bonjer, K.-P (1990) Moment tensor inversion of small earthquakes in southwestern Germany for the fault plane solution. *Geophys. J Internatl* **101**, 133–146.

Ebel, J., and Helmberger, D. V. (1982). P wave complexity and fault asperities: The Borrego Mountain earthquake of 1968. *Bull. Seism Soc. Am* **72**, 413–438.

Eccles, C D, and Ryder, J. A. (1984). Seismic location algorithms: A comparative evaluation. In *Rockburst and Seismicity in Mines* (N. C. Gay and E. H Wainwright, eds.). Symp. Ser. No. 6, pp. 89–92, S. Afr. Inst Min. Metal., Johannesburg.

Engell-Sørensen, L (1991a) Inversion of arrival time of microearthquake sources in the North Sea using a 3-D velocity structure and prior information. Part I. Method. *Bull Seism Soc. Am* **81**, 1183–1194.

Engell-Sørensen, L (1991b). Inversion of arrival times of microearthquake sources in the North Sea using a 3-D velocity structure and prior information Part II Stability, uncertainty analyses, and applications. *Bull. Seism. Soc. Am.* **81**, 1195–1215

Epstein, B , and Lomnitz, C. (1966) A model for the occurrence of large earthquakes. *Nature* **211**, 954–956.

Eshelby, J. D. (1957). The determination of the elastic field of an ellipsoidal inclusion, and related problems. *Proc. Roy. Soc.*, Ser. A, **241**, 376–396.

Evans, J. R. (1984). Effects of the free surface on shear waves. *Geophys. J. Roy Astr. Soc.* **76**, 165–172.

Evans, J. R., and Zucca, J. J. (1988). Active high-resolution seismic tomography of compressional wave velocity and attenuation structure at Medicine Lake Volcano, Northern California Cascade Range. *J. Geophys Res.* **93**, 15,106–15,036.

Evison, F. (1977). Fluctuations of seismicity before major earthquakes. *Nature* **266**, 710–712.

Ewing, W. M., Jardetzky, W. S., and Press, F. (1957). *Elastic Waves in Layered Media*, McGraw-Hill, New York

Fajklewicz, Z., and Jakiel, K. (1989). Induced gravity anomalies and seismic energy as a basis for prediction of mining tremors. In *Seismicity in Mines* (S. J. Gibowicz, ed.), reprinted from Special Issue, *Pure Appl. Geophys.* **129**, 535–552, Birkhäuser Verlag, Basel.

Fedorov, V. V. (1972). *Theory of Optimal Experiments*, Academic Press, New York.

Fedorov, V. V. (1974). Regression problems with controllable variables subject to error *Biometrika* **61**, 49–55.

Fedotov, S. A., Sobolev, G A., Gusev, A A., Kondratenko, A. M., Potapova, O. V., Slavina, L. B., Theophylaktov, V D., Khramov, A. A , and Shirokov, A. V. (1976). Long and short-term earthquake prediction in Kamchatka. *Tectonophysics* **37**, 305–316.

Fehler, M., and Phillips, W. S. (1991). Simultaneous inversion for Q and source parameters of microearthquakes accompanying hydraulic fracturing in granitic rock. *Bull. Seism Soc. Am* **81**, 553–575.

Fernandez, L. M., and Guzman, J. A. (1979). *Seismic History of Southern Africa*, Seism. Ser. No 9. Geol. Surv., Dept. Mines, Pretoria, South Africa.

Fernandez, L. M., and van der Heever, P. K (1984). Ground movement and damage accompanying a large seismic event in the Klerksdorp district. In *Rockbursts and Seismicity in Mines* (N. C. Gay and E. H. Wainwright, eds), Symp. Ser. No. 6, pp 193–198, S Afr Inst Min Metal., Johannesburg.

Fitch, T. J., McCowan, D. W., and Shields, M. W. (1980). Estimation of seismic moment tensor from teleseismic body wave data with application to intraplate and mantle earthquakes. *J. Geophys. Res.* **85**, 3817–3828.

Fletcher, J. B., Haar, L. C., Vernon, F. L., Brune, J N., Hanks, T. C , and Berger, J. (1986). The effects of attenuation on the scaling of source parameters for earthquakes at Anza, California In *Earthquake Source Mechanics* (S. Das, J. Boatwright, and C H. Scholz, eds.), *Maurice Ewing*, Vol 6, pp. 331–338, Am. Geophys. Union, Washington, D.C.

Flinn, E. A. (1965). Confidence regions and error determinations for seismic event location. *Rev. Geophys* **3**, 157–185

Foulger, G. R. (1988). Hengill triple junction, SW Iceland 2 Anomalous earthquake focal mechanisms and implications for process within the geothermal reservoir and at accretionary plate boundaries. *J. Geophys. Res.* **93**, 13,507–13,523.

Frankel, A. (1981). Source parameters and scaling relationships of small earthquakes in the north-eastern Caribbean. *Bull. Seism. Soc. Am* **71**, 1173–1190.

Frankel, A. (1982). The effects of attenuation and site response on the spectra of microearthquakes in the northeastern Caribbean. *Bull Seism. Soc. Am.* **72**, 1379–1402.

Frankel, A. (1989). A review of numerical experiments on seismic wave scattering. In *Scattering and Attenuation of Seismic Waves* (R.-S. Wu and K. Aki, eds.), Part II, reprinted from Special Issue, *Pure Appl. Geophys.* **131**, 639–685, Birkhäuser Verlag, Basel.

Frankel, A., and Clayton, R. W. (1986). Finite difference simulations of seismic scattering: Implications for the propagation of short-period seismic waves in the crust and models of hetrogeneity. *J. Geophys. Res.* **91**, 6465–6489.

Frankel, A., and Wennerberg, L. (1987). Energy-flux model of seismic coda: Separation of scattering and intrinsic attenuation. *Bull. Seism. Soc. Am.* **77**, 1223–1251.

Frankel, A., and Wennerberg, L. (1989). Microearthquake spectra from the Anza, California, seismic network: Site response and source scaling. *Bull. Seism. Soc. Am.* **79**, 581–609.

Frankel, A., Fletcher, J., Vernon, F., Haar, L., Berger, J., Hanks, T., and Brune, J. (1986). Rupture characteristics and tomographic source imaging of $M_L \sim 3$ earthquakes near Anza, southern California. *J. Geophys. Res.* **91**, 12,633–12,650.

Frohlich, C. (1990). Note concerning non-double-couple source components for slip along surfaces of revolution. *J. Geophys. Res.* **95**, 6861–6866.

Frohlich, C., Riedesel, M. A., and Apperson, K. D. (1989). Note concerning possible mechanisms for non-double-couple earthquake sources. *Geophys. Res. Lett.* **16**, 523–526.

Fuchs, K. (1968). The reflection of spherical waves from transition zones with arbitrary depth-dependent elastic moduli and density. *J. Phys. Earth* **16**, 27–41.

Fuchs, K., and Müller, G. (1971). Computation of synthetic seismograms with the reflectivity method and comparison with observations. *Geophys. J. Roy. Astr. Soc.* **23**, 417–433.

Fujii, Y., and Sato, K. (1990). Difference in seismic moment tensors between microseismic events associated with a gas outburst and those induced by longwall mining activity. In *Rockbursts and Seismicity in Mines* (C. Fairhurst, ed.), pp. 71–75, Balkema, Rotterdam.

Futterman, W. I. (1962). Dispersive body waves. *J. Geophys. Res.* **67**, 5279–5291.

Gane, P. G., Hales, A. L., and Oliver, H. A. (1946). A seismic investigation of the Witwatersrand earth tremors. *Bull. Seism. Soc. Am.* **36**, 49–80.

Gao, L.-S., and Li, S. L. (1990). Time domain solution for multiple scattering and the coda envelopes. In *Scattering and Attenuation of Seismic Waves* (R.-S. Wu and K. Aki, eds.), Part III; reprinted from Special Issue, *Pure Appl. Geophys.* **132**, 123–150, Birkhäuser Verlag, Basel.

Gao, L. S., Lee, L. C., Biswas, N. N., and Aki, K. (1983). Comparison of the effects between single and multiple scattering on coda waves for local earthquakes. *Bull. Seism. Soc. Am.* **73**, 377–389.

Garbin, H. D., and Knopoff, L. (1973). The compressional modulus of a material permeated by a random distribution of free circular cracks. *Q. Appl. Math.* **30**, 453–464.

Garbin, H. D., and Knopoff, L. (1975a). The shear modulus of a material permeated by a random distribution of free circular cracks. *Q. Appl. Math.* **33**, 296–300.

Garbin, H. D., and Knopoff, L. (1975b). Elastic moduli of a medium with liquid-filled cracks. *Q. Appl. Math.* **33**, 301–303.

Garcia-Fernandez, M., Kijko, A., Carracedo, J. C., and Soler, V. (1988). Optimum station distribution to monitor seismic activity of Teide volcano, Tenerife, Canary Islands. *J. Volc. Geotherm. Res.* **35**, 195–204.

Gay, N. C., and Ortlepp, W. D. (1979). Anatomy of a mining-induced fault zone. *Bull. Geol. Soc. Am.* **90**, 47–58.

Gay, N. C., Spencer, D., van Wyk, J. J., and van der Heever, P. K. (1984). The control of geological and mining parameters on seismicity in the Klerksdorp gold mining district. In *Rockbursts and Seismicity in Mines* (N. C. Gay and E. H Wainwright, eds.), Symp. Ser. No. 6, pp. 107–120, S. Afr. Inst. Min. Metal., Johannesburg.

Geiger, L. (1912). Probability method for the determination of earthquake epicenters from the arrival time only. *Bull. St. Louis Univ.* **8**, 60–71.

Geller, R. J. (1976). Body force equivalents for stress-drop seismic sources. *Bull. Seism. Soc. Am.* **66**, 1801–1804.

Gendzwill, D. J. (1984). Induced seismicity in Saskatchewan potash mines. In *Rockbursts and Seismicity in Mines* (N. C. Gay and E. H. Wainwright, eds.), Symp. Ser. No. 6, pp. 131–146, S. Afr. Inst. Min. Metal., Johannesburg.

Gerszternkorn, A., Bednar, A., and Lines, L. (1986). Robust iterative inversion for the one-dimensional acoustic wave equation. *Geophysics* **51**, 357–368.

Ghalib, H. A. A., Russel, D. R., and Kijko, A. (1985). Optimal design of regional seismological network for the Arab countries. *Pure Appl. Geophys.* **122**, 694–712.

Giardini, D. (1984). Systematic analysis of deep seismicity: 200 centroid-moment tensor solutions for earthquakes between 1977 and 1980. *Geophys. J. Roy. Astr. Soc.* **77**, 883–914.

Gibowicz, S. (1963). Magnitude and energy of subterranean shocks in Upper Silesia. *Studia Geophys. Geod.* **7**, 1–19.

Gibowicz, S. J. (1973a). Variation of the frequency-magnitude relation during earthquake sequences in New Zealand. *Bull. Seism. Soc. Am.* **63**, 517–528.

Gibowicz, S. J. (1973b). Stress drop and aftershocks. *Bull. Seism. Soc. Am.* **63**, 1443–1446.

Gibowicz, S. J. (1975). Variation of source properties: The Inangahua, New Zealand, aftershocks of 1968. *Bull. Seism. Soc. Am.* **65**, 261–276.

Gibowicz, S. J., (1979). Space and time variations of the frequency–magnitude relation for mining tremors in the Szombierki coal mine in Upper Silesia, Poland. *Acta Geophys. Pol.* **27**, 39–49.

Gibowicz, S. J. (1984). The mechanism of large mining tremors in Poland. In *Rockbursts and Seismicity in Mines* (N. C. Gay and E. H. Wainwright, eds.), Symp. Ser. No. 6, pp. 17–28, S. Afr. Inst. Min. Metal., Johannesburg.

Gibowicz, S. J. (1985). Seismic moment and seismic energy of mining tremors in the Lubin copper basin in Poland. *Acta Geophys. Pol.* **33**, 243–257.

Gibowicz, S. J. (1986). Physics of fracturing and seismic energy release: A review. In *Physics of Fracturing and Seismic Energy Release* (J. Kozák and L. Waniek, eds.), reprinted from Special Issue, *Pure Appl. Geophys.* **124**, 611–658, Birkäuser Verlag, Basel.

Gibowicz, S. J. (1990a). Keynote lecture: The mechanism of seismic events induced by mining—a review. In *Rockbursts and Seismicity in Mines* (C. Fairhurst, ed.), pp. 3–27, Balkema, Rotterdam.

Gibowicz, S. J. (1990b). Seismicity induced by mining. *Adv. Geophys.* **32**, 1–74.

Gibowicz, S. J., and Cichowicz, A. (1986). Source parameters and focal mechanism of mining tremors in the Nowa Ruda coal mine in Poland. *Acta Geophys. Pol.* **34**, 215–232.

Gibowicz, S. J., Cichowicz, A., and Dybel, T. (1977). Seismic moment and source size of mining tremors in Upper Silesia, Poland. *Acta Geophys. Pol.* **25**, 201–218.

Gibowicz, S. J., Bober, A., Cichowicz, A., Droste, Z., Dychtowicz, Z., Hordejuk, J., Kazimierczyk, M., and Kijko, A. (1979). Source study of the Lubin, Poland, tremor of 24 March 1977. *Acta Geophys. Pol.* **27**, 3–38.

Gibowicz, S. J., Droste, Z., Guterch, B., and Hordejuk, J. (1981). The Belchatow, Poland, earthquakes of 1979 and 1980 induced by surface mining. *Eng. Geol.* **17**, 257–271.

Gibowicz, S. J., Guterch, B., Lewandowska-Marciniak, H., and Wysokinski, L. (1982). Seismicity induced by surface mining: The Belchatow, Poland, earthquake of 29 November 1980. *Acta Geophys. Pol.* **30**, 193–219.

Gibowicz, S. J., Niewiadomski, J., Wiejacz, P., and Domanski, B. (1989). Source study of the Lubin, Poland, mine tremor of 20 June 1987. *Acta Geophys. Pol.* **37**, 111–132.

Gibowicz, S. J., Harjes, H.-P., and Schäfer, M. (1990). Source parameters of seismic events at Heinrich Robert mine, Ruhr basin, Federal Republic of Germany: Evidence for nondouble-couple events. *Bull. Seism. Soc. Am.* **80**, 88–109.

Gibowicz, S. J., Young, R. P., Talebi, S., and Rawlence, D. J. (1991). Source parameters of seismic events at the Underground Research Laboratory in Manitoba, Canada. Scaling relations for the events with moment magnitude smaller than −2. *Bull. Seism. Soc. Am.* **81**, 1157–1182.

Gil, H. (1991). *The Theory of Strata Mechanics*, Vol. 63, *Developments in Geotechnical Engineering*, Elsevier, Amsterdam.

Gil, H., and Litwiniszyn, J (1971) Can mining exploitation induce in deeper zones of earth's crust the propagation of fault fissures and consecutive seismic shocks? *Bull. Acad. Pol. Sci.*, *Ser. Sci. Tech.* **19**, 679–684.

Gilbert, F. (1970). Excitation of the normal modes of the earth by earthquake sources. *Geophys. J Roy Astr. Soc.* **22**, 223–226.

Gilbert, F (1973). Derivation of source parameters from low-frequency spectra. *Phil. Trans Roy. Soc A* **274**, 369–371

Gilbert, F., and Buland, R. (1976). An enhanced deconvolution procedure for retrieving the seismic moment tensor from a sparse network. *Geophys. J. Roy. Astr. Soc.* **47**, 251–255.

Glassmoyer, G., and Borcherdt, R. D. (1990). Source parameters and effects of bandwidth and local geology on high-frequency ground motions observed for aftershocks of the northeastern Ohio earthquake of 31 January 1986. *Bull Seism Soc. Am* **80**, 889–912.

Glazer, S. (1991). *Level of Seismic Activity*, Report RS-55-91, Min. Eng. Div., Vaal Reefs Exploration and Mining Company Ltd., Vaal Reefs, South Africa.

Glowacka, E., and Kijko, A. (1989). Continuous evaluation of seismic hazard induced by the deposit extraction in selected coal mines in Poland. In *Seismicity in Mines* (S. J. Gibowicz, ed.); reprinted from Special Issue, *Pure Appl. Geophys* **129**, 523–533, Brikhäuser Verlag, Basel.

Glowacka, E., and Lasocki, S. (1992). Probabilistic synthesis of the seismic hazard evaluation in mines. *Acta Mont.*, **84**, 59–65.

Glowacka, E., Stankiewicz, T., and Holub, K. (1990). Seismic hazard estimate based on the extracted deposit volume and bimodal character of seismic activity *Beitr. Geophys* **99**, 35–43.

Golub, G. H., and Reinisch, C. (1971) Singular value decomposition and least squares solutions. In *Linear Algebra* (J. H. Wilkinson and C. Reinisch, eds.), Springer-Verlag, New York.

Gomberg, J. S, Shedlock, K. M., and Roecker, S. W. (1990). The effect of S-arrival on the accuracy of hypocenter estimation. *Bull. Seism. Soc. Am.* **80**, 1605–1628

Gordon, R., Bender, R., and Herman, G T (1970). Algebraic Reconstruction Techniques (ART) for three-dimensional electron microscopy and X-ray photography *J Theor. Biol.* **29**, 471–481

Gorsky, B. G., and Brodsky, V. Z (1965). Simplexial design of extremal experiments. *Zavodskaya Laboratoriya* **31**, 831–847 (in Russian)

Goszcz, A. (1986). Tectonophysical origin of mining tremors. *Publ Inst Geophys , Pol Acad. Sci.* **M-8** (191), 61–75 (in Polish; English abstract)

Goszcz, A. (1988). The influence of technological factors on seismic and rockburst hazards. *Publ. Inst. Geophys , Pol. Acad Sci* **M-10** (213), 141–153 (in Polish, English abstract).

Green, R W. E. (1990). Keynote lecture: Instrumentation networks for observation of mine-induced seismicity In *Rockbursts and Seismicity in Mines* (C. Fairhurst, ed.), pp. 165–169, Balkema, Rotterdam.

Griffith, A. A. (1920). The phenomena of rupture and flow in solids *Phil. Trans. Roy Soc., Ser. A* **221**, 163–198.

Gu, J.-C., Rice. J., Ruina, A., and Tse, S. T. (1984). Slip motion and stability of a single degree of freedom elastic system with rate and state dependent friction. *J. Mech Phys. Solids* **32**, 167–196

Gumbel, E. J. (1962). *Statistics of Extremes*, Columbia University Press, New York.

Gusev, A. A. (1983). Descriptive statistical model of earthquake source radiation and its application to an estimation of short period strong motion. *Geophys. J. Roy Astr. Soc.* **74**, 787–808.

Gustavsson, M , Ivansson, S., Moren, P., and Pihl, J. (1986). Seismic borehole tomography-measurement system and field studies. *Proc. I.E.E.E.* **74**, 339–346.

Gutenberg, B., and Richter, C. F. (1944). Frequency of earthquakes in California, *Bull. Seism. Soc. Am.* **34**, 185–188.

Gutenberg, B., and Richter, C. F. (1954). *Seismicity of the Earth and Associated Phenomena*, 2nd ed., Princeton University Press, Princeton, N.J.

Gutenberg, B., and Richter, C. F. (1956). Magnitude and energy of earthquakes. *Ann. Geofis.* **9**, 1–15.

Guttorp, P., and Hopkins, D (1986). On estimating varying b values. *Bull Seism. Soc. Am.* **76**, 889–895.

Haar, L. C., Fletcher, J. B., and Mueller, C. S. (1984) The 1982 Enola, Arkansas, swarm and scaling of ground motion in the Eastern United States. *Bull. Seism. Soc. Am.* **74**, 2463–2482.

Habermann, R. E. (1983). Spatial seismicity variations and asperities in the New Hebrides seismic zone. *J. Geophys. Res.* **89**, 5891–5903

Hamano, Y (1974). Dependence of rupture time history on the heterogeneous distribution of strain and strength on the fault plane (abstract). *EOS, Trans. Am. Geophys. Union* **55**, 352.

Hanks, T. C. (1979). b values and $\omega^{-\gamma}$ seismic source models: Implications for tectonic stress variations along active crustal fault zones and the estimation of high-frequency strong ground motion. *J. Geophys. Res.* **84**, 2235–2242.

Hanks, T. C (1981). The corner frequency shift, earthquake source models, and Q. *Bull. Seism Soc Am.* **71**, 597–612.

Hanks, T. C. (1982) f_{max}. *Bull. Seism. Soc. Am.* **72**, Part A, 1867–1880.

Hanks, T. C (1984). a_{rms} and seismic source studies In *Rockbursts and Seismicity in Mines* (N. C. Gay and E. H. Wainwright, eds.), Symp. Ser. No. 6, pp. 39–44, S Afr. Inst. Min. Metal , Johannesburg.

Hanks, T. C., and Boore, D. M. (1984). Moment–magnitude relations in theory and practice. *J. Geophys Res.* **89**, 6229–6235.

Hanks, T. C., and Kanamori, H. (1979). A moment magnitude scale. *J. Geophys Res* **84**, 2348–2350.

Hanks, T. C , and McGuire, R. K. (1981). The character of high frequency strong ground motion. *Bull. Seism. Soc. Am.* **71**, 2071–2096.

Hanks, T. C., and Wyss, M. (1972). The use of body wave spectra in the determination of seismic source parameters. *Bull. Seism. Soc. Am.* **62**, 561–589.

Hanyga, A., ed. (1984). *Seismic Wave Propagation in the Earth*, Polish Scientific Publishers, Warsaw; Elsevier, Amsterdam

Hardy, H. R., Jr. (1981). Application of acoustic emission techniques to rock and rock structures: A state-of-the-art review. In *Acoustic Emissions in Geotechnical Engineering Practice* (V. P. Drnevich and R. E. Gray. eds.), Spec Tech Publ. No. 750, pp. 4–92, Am Soc. Test. Mater., Philadelphia, Pa.

Hardy, H. R., Jr. (1984). Stability monitoring of underground structures using acoustic emission techniques. In *Rockbursts and Seismicity in Mines* (N. C Gay and E H. Wainwright, eds.), Symp. Ser. No. 6, pp. 277–286, S. Afr. Inst. Min. Metal., Johannesburg.

Hartzell, S. (1978). Earthquake aftershocks as Green's functions. *Geophys. Res. Lett.* **5**, 1–4.

Hartzell, S., and Heaton, T. (1983). Inversion of strong ground motion and teleseismic waveform data for the fault rupture history of the 1979 Imperial Valley, California, earthquake *Bull. Seism Soc. Am.* **73**, 1553–1583.

Hartzell, S., and Helmberger, D V. (1982). Strong-motion modeling of the Imperial Valley earthquake of 1979 *Bull. Seism. Soc. Am.* **72**, 571–596.

Hasegawa, H. S., Wetmiller, R. J., and Gendzwill, D. J. (1989). Induced seismicity in mines in Canada–an overview. In *Seismicity in Mines* (S. J. Gibowicz, ed.); reprinted from Special Issue, *Pure Appl. Geophys.* **129**, 423–453, Birkhäuser Verlag, Basel.

Haskell, N. A. (1953). The dispersion of surface waves in multilayered media. *Bull. Seism. Soc. Am.* **43**, 17–34.

Haskell, N. A. (1964). Total energy and energy density of elastic wave radiation from propagating faults. *Bull. Seism. Soc. Am.* **54**, 1811–1841.

Hatton, L., Worthington, M. H., and Makin, J. (1986). *Seismic Data Processing. Theory and Practice*, Blackwell Scientific Publishers, Oxford.

Hauksson, E., Teng, T.-L., and Henyey, T. L. (1987). Results from a 1500 m deep, three-level downhole array: Site response, low Q values, and f_{max}. *Bull. Seism. Soc. Am.* **77**, 1883–1904.

Heaton, T. H., and Helmberger, D. V. (1977). A study of the strong ground motion of the Borrego Mountain, California, earthquake. *Bull. Seism. Soc. Am.* **67**, 315–330.

Heever, P. K. v. D. (1984). Some technical and research aspects of the Klersdorp seismic network. In *Rockbursts and Seismicity in Mines* (N. C. Gay and E. H. Wainwright, eds.), Symp. Ser. No. 6, pp. 349–350. S. Afr. Inst. Min. Metal., Johannesburg.

Herman, G. T. (1980). *Image Reconstruction from Projections*. Academic Press, New York.

Hermann, L., Dianiska, L., and Verboci, J. (1982). Curved ray algebraic reconstruction technique applied in mining geophysics. *Geophys. Trans. Eotvos Lorand Geophys. Inst., Hungary* **28**, 33–46.

Herraiz, M., and Espinosa, A. F. (1987). Coda waves: A review. *Pure Appl. Geophys.* **125**, 499–577.

Herrmann, R. B. (1975). A student's guide to the use of P and S wave data for focal mechanism determination. *Earthq. Notes* **46**(4), 29–39.

Herrmann, R. B. (1979). FASTHYPO–a hypocentre location program. *Earthq. Notes* **50**, 25–37.

Herrmann, R. B. (1980). Q estimates using the coda of local earthquake. *Bull. Seism. Soc. Am.* **70**, 447–468.

Herrmann, R. B., and Wang, C. Y. (1985). A comparison of synthetic seismograms. *Bull. Seism. Soc. Am.* **75**, 41–56.

Hestenes, M., and Stiefel, E. (1952). Method of conjugate gradients for solving linear systems. *Natl. Bur. Standards J. Res.* **49**, 409–436.

Hill, W. J., and Hunter, W. G. (1974). Design of experiments for subsets of parameters. *Technometrics* **16**, 425–434.

Himmelblau, M. (1972). *Applied Nonlinear Programming*, McGraw-Hill, New York.

Hinzen, K.-G. (1982). Source parameters of mine tremors in the eastern part of the Ruhr district (West Germany). *J. Geophys.* **51**, 105–112.

Hirata, T. (1989). A correlation between the b value and the fractal dimension of earthquakes. *J. Geophys. Res.* **94**, 7507–7514.

Hoang-Trong, P., Guéguen, J. F., and Holl, J. M. (1988). Near field seismological observations in the Lorraine coal mine (France): Preliminary results. In *Induced Seismicity and Associated Phenomena* (D. Procházková, ed.), pp. 64–74. Geophys. Inst., Czech. Acad. Sci., Prague.

Holub, K., Knotek, S., and Vajter, Z. (1988). Seismic activity in relation to the coal mining. In *Induced Seismicity and Associated Phenomena* (D. Procházková, ed.), pp. 93–106. Geophys. Inst., Czech. Acad. Sci., Prague.

Honda, S., and Yomogida, K. (1991). Normal fault earthquake off the northern Mozambique: A possible isotropic source. *Geophys. Res. Lett.* **18**, 1381–1384.

Horner, R. B., and Hasegawa, H. S. (1978). The seismotectonics of southern Saskatchewan. *Can. J. Earth Sci.* **15**, 1341–1355.

Hough, S. E., Anderson, J. G., Brune, J., Vernon III, F., Berger, J., Fletcher, J., Haar, L., Hanks, T., and Baker, L. (1988). Attenuation near Anza, California. *Bull. Seism. Soc. Am.* **78**, 672–691.

Hough, S. E., Anderson, J. G., and Patton, H. J. (1989). Attenuation in western Nevada: Preliminary results from earthquake and explosion sources. *Geophys. Res. Lett.* **16**, 207–210.

Hounsfield, G. N. (1972). A method and apparatus for examination of body by radiation such as *X* or gamma radiation. Patent Specification 1283915, Patent Office, London.

Houston, H., and Kanamori, H. (1986). Source spectra of great earthquakes: Teleseismic constraints on rupture process and strong motion. *Bull. Seism. Soc. Am.* **76**, 19–42.

Hsu, V., Helsley, C. E., Berg, E., and Novelo-Casanova, D. A. (1984). Correlation of foreshocks and aftershocks and asperities. *Pure Appl. Geophys.* **122**, 878–893.

Hudson, J. A. (1980). Overall properties of a cracked solid. *Math. Proc. Camb. Phil. Soc.* **88**, 371–384.

Hudson, J. A. (1981). Wave speeds and attenuation of elastic waves in material containing cracks. *Geophys. J. Roy. Astr. Soc.* **64**, 133–150.

Hudson, J. A., Pearce, R. G., and Rogers, R. M. (1989). Source type plot for inversion of the moment tensor. *J. Geophys. Res.* **94**, 765–774.

Hull, S. W. (1983). *The Mechanics of Aftershocks*, MIT Dept. Civil Eng., Res. Report R 83-6, Cambridge, Mass.

Humphreys, E., and Clayton, R. W. (1988). Adaptation of back projection tomography to seismic travel time problems. *J. Geophys. Res.* **93**, 1073–1085.

Humphreys, E., Clayton, R. W., and Hager, B. H. (1984). A tomographic image of mantle structure beneath southern California. *Geophys. Res. Lett.* **11**, 625–627.

Hurtig, E., Bormann, P., Knoll, P., and Tauber, F. (1979). Seismological and geomechanical studies of a strong seismic event in the potash mines of the GDR: Implications for predicting mining tremors. *Internatl. Symp. Earthq. Pred., Paris*, Contrib. Paper II-4, UNESCO, Paris

Hurtig, E., Grosser, H., Knoll, P., and Neunhöfer, H. (1982). Seismologische und geomechanische Untersuchungen des seismischen Ereignisses vom 23.6.1975 im Werragebiet bei Sünna (DDR). *Beitr. Geophys.* **91**, 45–61.

Iannoccone, G., and Deschamps, A. (1989). Evidence of shear wave anisotropy in the upper crust of central Italy. *Bull. Seism. Soc. Am.* **79**, 1905–1912.

Ida, Y. (1972). Cohesive force across the tip of a longitudinal shear crack and Griffith's specific surface energy. *J. Geophys. Res.* **77**, 3796–3805.

Idziak, A., Sagan, G., and Zuberek, W. M. (1991). An analysis of frequency distributions of shocks from the Upper Silesian Coal Basin. *Publ. Inst. Geophys., Pol. Acad. Sci.* **M-15** (235), 163–182.

Imoto, M. (1987). A Bayesian method for estimating earthquake magnitude distribution and changes in the distribution with time and space in New Zealand. *N.Z. J. Geol. Geophys.* **30**, 103–116.

Irwin, G. R. (1960). Fracture mechanics. In *Structural Mechanics* (J. N. Goodier and N. J. Hoff, eds.), pp. 557–591. Pergamon Press, New York.

Ishida, M., and Kanamori, H. (1978). The foreshock activity of the 1971 San Fernando earthquake, California. *Bull Seism. Soc. Am.* **68**, 1265–1279.

Ishida, M., and Kanamori, H. (1980). Temporal variation of seismicity and spectrum of small earthquakes preceding the 1952 Kern County, California, earthquake. *Bull. Seism. Soc Am* **70**, 509–527.

Ivansson, S. (1985). A study of methods for tomographic velocity estimation in the presence of low-velocity zones. *Geophysics* **50**, 969–988.

Ivansson, S. (1986). Seismic borehole tomography—theory and computational methods. *Proc. I.E.E.E.* **74**, 328–338.

Ivansson, S. (1987). Crosshole transmission tomography. In *Seismic Tomography* (G Nolet, ed.), pp. 159–188. D. Reidel Publ. Comp., Dordrecht.

Jackson, D. D., and Matsu'ura, M. (1985) A Bayesian approach to nonlinear inversion. *J. Geophys. Res.* **90**, 581–591.

364   References

Jackson, P. J. (1985). Horizontal seismic in coal seams: Its use by the U.K. coal industry. *First Break* **11**, 15–24.

Jaeger, J. C., and Cook, N. G. W. (1976). *Fundamentals of Rock Mechanics*, 2nd ed., Chapman and Hall, London.

Jech, J. (1989). Seismic tomography in the Ostrava-Karviná mining region. In *Seismicity in Mines* (S. J. Gibowicz, ed.); reprinted from Special Issue, *Pure Appl. Geophys* **129**, 597–608, Birkhauser Verlag, Basel.

Jeffreys, H. (1932). An alternative to the rejection of observations. *Proc. Roy. Soc. London Ser. A* **137**, 78–87.

Jeffreys, H (1958). A modification of Lomnitz's law of creep in rocks *Geophys J Roy. Astr Soc.* **1**, 92–95

Jin, A., and Aki, K. (1986) Temporal change in coda Q before the Tangshan earthquake of 1976 and the Haicheng earthquake of 1975 *J. Geophys. Res.* **91**, 665–673.

Jin, A., Cao, T., and Aki, K. (1985). Regional change of coda Q in the oceanic lithosphere. *J. Geophys. Res.* **90**, 8651–8659.

John, R. C. S., and Draper, N. R. (1975). D-optimality for regression designs: A review *Technometrics* **17**, 15–23.

Johnson, L. R. (1974) Green's function for Lamb's problem. *Geophys. J.* **37**. 99–131.

Johnston, D. E., and Langston, C A. (1984). The effect of assumed source structure on inversion of earthquake source parameters: The eastern Hispaniola earthquake of 14 September 1981. *Bull Seism Soc. Am.* **74**, 2115–2134.

Johnston, J. C., and Einstein, H. H. (1990). A survey of mining associated rockbursts. In *Rockbursts and Seismicity in Mines* (C Fairhurst, ed.), pp. 121–127, Balkema, Rotterdam

Jones, L. M., and Molnar, P. (1979) Some characteristics of foreshocks and their possible relationship to earthquake prediction and premonitory slip on faults. *J. Geophys. Res.* **84**, 3596–3608.

Jones, L. M., Wang, B., Xu, S., and Fitch, T J. (1982). The foreshock sequence of the February 4, 1975 Haicheng earthquake ($M = 7.3$) *J. Geophys. Res.* **87**, 4575–4584.

Jordan, T. H., and Sverdrup, K. A. (1981). Teleseismic location techniques and their application to the earthquake clusters in the south-central Pacific *Bull. Seism. Soc Am.* **71**, 1105–1130.

Jost, M. L, and Herrmann, R. B (1989). A student's guide to and reivew of moment tensors. *Seism. Res. Lett.* **60**, 37–57.

Joughin, N. C., and Jager, A. J. (1983). Fracture of rock at stope faces in South African gold mines. *Proc. Symp. Rockbursts: Prediction and Control*, pp. 53–66, Inst. Min Metal, London.

Joyner, W. B (1991). Directivity for nonuniform ruptures. *Bull. Seism. Soc Am* **81**, 1391–1395.

Julian, B. R., and Gubins, D. (1977). Three-dimensional seismic ray tracing. *J. Geophys.* **43**, 95–113.

Julian, B. R., and Sipkin, S. A. (1985). Earthquake processes in the Long Valley caldera area, California *J. Geophys Res.* **90**, 11.155–11,169.

Justice, J. H., and Vassiliou, A. A. (1990) Diffraction tomography for geophysical monitoring of hydrocarbon reservoirs. *Proc I E.E.E* **78**, 711–722.

Kagan, Y. Y., and Jackson, D. D. (1991). Seismic gap hypothesis Ten years after. *J. Geophys. Res.* **96**. 21,419–21,431

Kagan, Y., and Knopoff, L. (1976) Statistical search for non-random features of the seismicity of strong earthquakes *Phys. Earth Planet. Interiors* **12**, 291–318

Kagan, Y., and Knopoff, L. (1978) Statistical study of the occurrence of shallow earthquakes. *Geophys. J. Roy. Astr. Soc.* **55**, 67–86

Kanamori, H. (1977). The energy release in great earthquakes. *J. Geophys. Res.* **82**, 2981–2987.

Kanamori, H, and Anderson, D L. (1975). Theoretical basis of some empirical relations in seismology. *Bull. Seism. Soc. Am* **65**, 1073–1096.

Kanamori, H., and Given, J. W. (1981). Use of long-period surface waves for rapid determination of earthquake source-parameters. *Phys. Earth Planet Interiors* **27**, 8–31.

Kanamori, H., and Hadley, D. M. (1975). Crustal structure and temporal velocity change in southern California. *Pure Appl. Geophys.* **113**, 257–280.

Kanamori, H., and Stewart, G. S. (1978). Seismological aspects of the Guatemala earthquake of February 4, 1976 *J. Geophys. Res.* **83**, 3427–3434.

Kaneko, K., Sugawara, K., and Obara, Y. (1990). Rock stress and microseismicity in a coal burst district. In *Rockbursts and Seismicity in Mines* (C. Fairhurst, ed.), pp. 183–188. Balkema, Rotterdam

Kaneshima, S. (1990). Origin of crustal anisotropy· Shear wave splitting studies in Japan. *J. Geophys. Res.* **95**, 11,121–11,133.

Kaneshima, S., Ando, M., and Crampin, S. (1987). Shear-wave splitting above small earthquakes in the Kinki district of Japan. *Phys. Earth Planet. Interiors* **45**, 45–58.

Kaneshima, S., Ito, H., and Sugihara, M. (1989). Shear-wave polarization anisotropy observed in a rift zone in Japan. *Tectonophysics* **157**, 281–300

Karczmarz, S. (1937). Solution by approximation of system of linear equations. *Bull. Internatl. Acad. Pol. Sci. Lett.* **35**, 355–357 (in German).

Karpin, T. L., and Thurber, C. H. (1987). The relationship between earthquake swarms and magma transport: Kilauea Volcano, Hawaii. In *Advances in Volcanic Seismology* (E. A. Okal, ed.); reprinted from Special Issue, *Pure Appl Geophys.* **125**, 971–991, Birkhäuser Verlag, Basel.

Kasahara, K. (1981). *Earthquake Mechanics*. Cambridge University Press, Cambridge (U.K.)

Kasina, Z. (1988). Selected aspects of tomography with application of curvilinear seismic ray. *Proc. 33rd Internatl. Geophys. Symp., Prague 1988*, pp. 21–32. Geophys. Inst., Czech. Acad. Sci., Prague.

Kawakatsu, H. (1991a). Insignificant isotropic component in the moment tensor of deep earthquakes. *Nature* **351**, 50–53.

Kawakatsu, H. (1991b). Enigma of earthquakes at ridge-transform-fault plate boundaries: Distribution of non-double couple parameter of Harvard CMT solutions. *Geophys Res. Lett.* **18**, 1103–1106.

Kazimierczyk, M., Kijewski, P., and Szelag, T. (1988). Tectonic and miningaspect of major mining tremors occurring in the Legnica-Glogow Basin. *Publ. Inst. Geophys., Pol. Acad Sci.* **M-10** (213), 187–202 (in Polish; English abstract)

Keilis-Borok, V. L., Pisarenko, V. F., Pyatetskii-Shapiro, I I., and Zhelankina, T. S. (1972). Computer determination of earthquake mechanism. In *Computational Seismology* (V I Keilis-Borok, ed.), pp. 32–45, Consultants Bureau, New York.

Keith, C. M , and Crampin, S. (1977a) Seismic body waves in anisotropic media. Reflection and refraction at a plane interface. *Geophys. J. Roy. Astr. Soc.* **49**, 181–208.

Keith, C. M., and Crampin, S. (1977b). Seismic body waves in anisotropic media: Propagation through a layer *Geophys. J. Roy. Astr. Soc.* **49**, 209–224.

Keith, C. M., and Crampin, S. (1977c). Seismic body waves in anisotropic media: Synthetic seismograms. *Geophys. J. Roy. Astr. Soc.* **49**, 225–243.

Kennett, B. L N. (1983). *Seismic Wave Propagation in Stratified Media*, Cambridge University Press, Cambridge (U.K.)

Kennett, B. L. N. (1988). Radiation from a moment-tensor source. In *Seismological Algorithms* (D. J. Doornbos, ed.), pp. 427–441, Academic Press, London.

Kijko, A. (1975). Some methods and algorithms for locating very near earthquakes with a digital computer. *Publ. Inst. Geophys., Pol. Acad. Sci.* **84**, 7–88

Kijko, A (1977a). An algorithm for the optimum distribution of a regional seismic network *Pure Appl. Geophys.* **115**, 999–1009.

Kijko, A. (1977b). An algorithm for the optimum distribution of a regional seismic network. II. An analysis of the accuracy of location of local earthquakes depending on the number of the seismic stations. *Pure Appl. Geophys.* **115**, 1011–1021.

Kijko, A. (1978). Methods of the optimal planning of regional seismic networks. *Publ. Inst. Geophys., Pol. Acad. Sci.* **A-7** (119), 3–63.

Kijko, A. (1985). Theoretical model for a relationship between mining seismicity and excavation area. *Acta Geophys. Pol.* **33**, 231–241.

Kijko, A. (1988). Application of Bayesian estimation theory for location of seismic events in mines. *Acta Geophys. Pol.* **36**, 343–352.

Kijko, A., and Dessokey, M. (1987). Application of extreme magnitude distribution to incomplete earthquake files. *Bull. Seism. Soc. Am.* **77**, 1429–1436.

Kijko, A., and Sellevoll, M. A. (1989). Estimation of earthquake hazard parameters from incomplete data files. Part I. Utilization of extreme and complete catalogs with different threshold magnitudes. *Bull. Seism. Soc. Am.* **79**, 645–654.

Kijko, A., and Sellevoll, M. A. (1992). Estimation of earthquake hazard parameters from incomplete data files. Part II. Incorporation of magnitude heterogeneity. *Bull. Seism. Soc. Am.* **82**, 120–134.

Kijko, A., and Syrek, B. (1988). Energy and frequency distributions of mining tremors and their relation to rockburst hazard. *Acta Geophys. Pol.* **36**, 189–201.

Kijko, A., Dessokey, M. M., Glowacka, E., and Kazimierczyk, M. (1982). Periodicity of strong mining tremors in the Lubin copper mine. *Acta Geophys. Pol.* **30**, 221–230.

Kijko, A., Stankiewicz, T., and Krol, M. (1986). Relative location of mining events. *Przegl. Gorn.* **10**, 219–223 (in Polish; English abstract).

Kijko, A., Drzezla, B., and Stankiewicz, T. (1987). Bimodal character of the distribution of extreme seismic events in Polish mines. *Acta Geophys. Pol.* **35**, 157–166.

Kikuchi, M., and Kanamori, H. (1982). Inversion of complex body waves. *Bull. Seism. Soc. Am.* **72**, 491–506.

King, G. C. P. (1983). The accommodation of large strains in the upper lithosphere of the earth and other solids by self-similar fault systems: The geometrical origin of *b*-value. *Pure Appl. Geophys.* **121**, 761–815.

King, G. C. P., and Yielding G. (1984). The evolution of a thrust fault system: Processes of rupture initiation, propagation and termination in the 1980 El Asnam (Algeria) earthquake. *Geophys. J. Roy. Astr. Soc.* **77**, 915–933.

Kisslinger, C. (1976). A review of theories of mechanisms of induced seismicity. *Eng. Geol.* **10**, 85–98.

Kisslinger, C. (1980). Evaluation of S to P amplitude ratios for determining focal mechanism from regional network observations. *Bull. Seism. Soc. Am.* **70**, 999–1014.

Kisslinger, C. (1982). Correction. *Bull. Seism. Soc. Am.* **72**, 344.

Kisslinger, C., Bowman, J. R., and Koch, K. (1981) Procedure for computing focal mechanisms from local $(SV/P)_z$ data. *Bull. Seism. Soc. Am.* **71**, 1719–1729.

Kjartansson, E. (1979). Constant Q-wave propagation and attenuation. *J. Geophys. Res.* **84**, 4737–4748.

Knoll, P. (1990). The fluid-induced tectonic rockburst of March 13, 1989 in the "Werra" potash mining district of the GDR (first results). *Beitr. Geophys.* **99**, 239–245.

Knoll, P., and Kuhnt, W. (1990). Seismological and geotechnical investigations of the mechanics of rock bursts. In *Rockbursts and Seismicity in Mines* (C. Fairhurst, ed.), pp. 129–138, Balkema, Rotterdam.

Knoll, P., Kuhnt, W., Behrens, H.-J., and Sievers, J. (1989). Experience in controlling the mining-induced seismicity in potash mining in the DDR. *Beitr. Geophys.* **98**, 461–473.

Knopoff, L. (1958) Energy release in earthquakes. *Geophys. J. Roy. Astr. Soc.* **1**, 44–52

Knopoff, L. (1961a). Analytical calculation of the fault-plane problem. *Publ. Dominion Obs. (Ottawa)* 24, 309–315.

Knopoff, L. (1961b). Statistical accuracy of the fault-plane problem. *Publ. Dominion Obs. (Ottawa)* 24, 317–319.

Knopoff, L. (1964). A matrix method for elastic wave problems. *Bull. Seism. Soc. Am.* 54, 431–438.

Knopoff, L., and Gilbert, F. (1959). Radiation from a strike-slip fault. *Bull. Seism. Soc. Am* 49, 163–178.

Knopoff, L., and Kagan, Y. (1977). Analysis of the theory of extremes as applied to earthquake problems. *J. Geophys. Res.* 82, 5647–5657.

Knopoff, L., and Randall, M. J. (1970). The compensated linear-vector dipole: A possible mechanism for deep earthquakes. *J. Geophys. Res.* 75, 1957–1963.

Koch, K. (1991). Moment tensor inversion of local earthquake data. I. Investigation of the method and its numerical stability with model calculations. *Geophys. J. Internatl.* 106, 305–319.

Koch, M. (1985a). Nonlinear inversion of local seismic travel times for the simultaneous determination of the 3D-velocity structure and hypocenters—application to the seismic zone Vrancea. *J. Geophys.* 56, 160–173.

Koch, M. (1985b). A numerical study on the determination of 3-D structure of the lithosphere by linear and non-linear inversion of teleseismic travel times. *Geophys. J. Roy. Astr. Soc.* 80, 73–93.

Koehler, F., and Taner, M. (1985). The use of conjugate-gradient algorithms in computation of predictive deconvolution operators. *Geophysics* 50, 2752–2758.

Konecny, P. (1989). Mining-induced seismicity (rock bursts) in the Ostrava-Karviná Coal Basin, Czechoslovakia. *Beitr. Geophys.* 98, 525–547.

Kopnichev, Y. F. (1977). The role of multiple scattering in the formation of a seismogram's tail. *Izv. Akad. Nauk, Fiz. Zemli* 13, 394–398 (in Russian).

Kostrov, B. V. (1964). Self-similar problem of propagation of shear cracks. *J. Appl. Math. Mech.* 28, 1077–1087 (in Russian).

Kostrov, B. V. (1966). Unsteady propagation of longitudinal shear cracks. *J. Appl Math. Mech.* 30, 1241–1248 (in Russian).

Kostrov, B. V., and Das, S. (1982). Idealized models of fault behavior prior to dynamic rupture. *Bull. Seism. Soc. Am.* 72, 679–703.

Kostrov, B. V., and Das, S. (1988) *Principles of Earthquake Source Mechanics*, Cambridge University Press, Cambridge (U.K.).

Kremenetskaya, E. O. (1991). Contemporary seismicity of the NW part of the USSR. *NORSAR Sci. Rep.* 1-91 / 92, 134–145.

Krishnamurthy, R., and Shringarputale, S. B. (1990). Rockburst hazards in Kolar Gold Fields. In *Rockbursts and Seismicity in Mines* (C. Fairhurst, ed.), pp. 411–420, Balkema, Rotterdam.

Król, M., and Kijko, A. (1991). Relative location of mining events: Estimation of the method efficiency. *Przegl. Gorn.* (in Polish; English abstract).

Kuhnt, W., Knoll, P., Grosser, H., and Behrens, H.-J. (1989). Seismological models for mining-induced seismic events. In *Seismicity in Mines* (S. J. Gibowicz, ed.); reprinted from Special Issue, *Pure Appl. Geophys.* 129, 513–521, Birkhauser Verlag, Basel.

Kusznir, N. J., Ashwin, D. P., and Bradley, A G. (1980). Mining induced seismicity in the North Staffordshire coal field, England. *Internatl. J Rock Mech. Min. Sci., Geomech. Abstr.* 17, 44–55.

Kusznir, N. J., Al-Saigh, N. H., and Ashwin, D. P. (1984). Induced seismicity generated by longwall coal mining in the North Staffordshire coal-field, U.K. In *Rockbursts and Seismicity in Mines* (N. C. Gay and E. H. Wainwright, eds), Symp. Ser. No. 6, pp. 153–160, S. Afr. Inst. Min. Metal., Johannesburg.

Kvamme, L. B., and Havskov, J. (1989). Q in southern Norway. *Bull. Seism. Soc. Am.* **79**, 1575–1588.

Lanczos, C (1961). *Linear Differential Operators*, Van Nostrand, London

Langston, C. A. (1981). Source transmission of seismic waveforms The Koyna, India earthquakes of September 13, 1967. *Bull Seism Soc. Am.* **71**, 1–24

Lasocki, S (1990). Prediction of strong mining tremors. *Zesz. Naukowe Akad. Gorn.-Hutn., Geofiz. Stosowana* **7**, 1–110 (in Polish; English abstract).

Lawn, B. R. and Wilshaw, T R. (1975). *Fracture of Brittle Solids*, Cambridge University Press, Cambridge (U.K.).

Lawrence, D. A. (1984). Seismicity in the Orange Free State gold mining district. In *Rockbursts and Seismicity in Mines* (N. C. Gay and E. H. Wainwright, eds.), Symp. Ser No 6, pp. 121–130, S. Afr. Inst. Min. Metal , Johannesburg.

Lawson, C. L., and Hanson, R. J (1974) *Solving Least Square Problems*, Prentice-Hall, Englewood Cliffs, N J.

Leach, A. R., and Lenhardt, W. A. (1990). Pillar associated seismicity at Western Deep Levels Mine. In *Static and Dynamic Considerations in Rock Engineering* (R Brummer, ed), pp. 197–205. Balkema, Rotterdam.

Leary, P. C., Crampin, S., and McEvilly, T. V (1990) Seismic fracture anisotropy in the Earth's crust: An overview. *J Geophys. Res.* **95**, 11,105–11,114.

Lee, C. P. (1987) Performance of underground coal mines during the 1976 Tangshan earthquake. *Tunnelling Underground Space Technol.* **2**, 199–202.

Lee, M. F., Beer, G., and Windsor, C. R. (1990) Interaction of stopes, stresses and geologic structure at the Mount Charlotte Mine, Western Australia. In *Rockbursts and Seismicity in Mines* (C. Fairhurst, ed.), pp. 337–343, Balkema, Rotterdam

Lee, W. H. K., and Stewart, S. W. (1981). *Principles and Applications of Microearthquake Networks*, Academic Press, New York.

Lee, W H. K., Aki, K., Chouet, B , Johnson, P., Marks, S., Newberry, J. T., Ryall, A. S., Stewart, S. W., and Tottingham, D. M (1986). A preliminary study of coda Q in California and Nevada. *Bull. Seism. Soc. Am.* **76**, 1143–1150.

Lees, J. M., and Crosson, R. S. (1989) Tomographic inversion for three-dimensional velocity structure at Mount St. Helens using earthquake data. *J. Geophys. Res.* **94**, 5716–5728.

Leighton, F. J. (1984). Microseismic monitoring and warning of rockbursts In *Rockbursts and Seismicity in Mines* (N. C. Gay and E H. Wainwright, eds.), Symp. Ser. No 6, pp. 287–295, S. Afr. Inst. Min. Metal., Johannesburg.

Lenhardt, W. A. (1990). Seismic event characteristics in a deep level mining environment. In *Rock at Great Depth* (V. Maury and D Fourmaintraux, eds.), pp 727–732. Balkema, Rotterdam.

Lenhardt, W A , and Hagan, T. O. (1990). Observations and possible mechanisms of pillar-associated seismicity at great depth. In *Technical Challenges in Deep Level Mining*, pp. 1183–1194, S. Afr. Inst. Min. Metal., Johannesburg.

Li, V C (1987). Mechanics of shear rupture applied to earthquake zones. In *Fracture Mechanics of Rock* (B. K. Atkinson, ed.), pp. 351–428, Academic Press, London

Li, V C., and Fares, N. (1986). Rupture processes in the presence of creep zones. In *Earthquake Source Mechanics* (S. Das, J. Boatwright, and C H. Scholz, eds), *Maurice Ewing*, Vol. 6, pp 71–80, Am. Geophys Union, Washington, D.C.

Li, V. C., and Rice, J. R. (1983). Preseismic rupture progression and great earthquake instabilities at plate boundaries *J Geophys Res* **88**, 4231–4246

Li, Q. L., Chen, L. Yu., and Hao, B L. (1978). Time and space scanning of the b-value: A method for monitoring the development of catastrophic earthquakes. *Acta Geophys. Sin.* **21**, 101–125.

Lienert, B. R., Berg, E., and Frazer, L. N. (1986) Hypocenter: An earthquake location method using centered, scaled, and adaptively damped least squares. *Bull. Seism. Soc. Am.* **76**, 771–783.

Liu, E., and Crampin, S. (1990). Effects of the internal shear wave window· Comparison with anisotropy induced splitting. *J. Geophys. Res.* **95**, 11,275–11,281.

Liu, H. L., and Helmberger, D. V. (1985). The 23·19 aftershock of the 15 October 1979 Imperial Valley earthquake· More evidence for an asperity. *Bull. Seism. Soc. Am.* **75**, 689–708.

Liu, H.-P, Anderson, D. L., and Kanamori, H. (1976). Velocity dispersion due to anelasticity; implications for seismology and mantle composition. *Geophys. J Roy. Astr. Soc.* **47**, 41–58.

Lomakin, V. S., Grigorovich, S. V., Potekhin, R. P, and Khalevin, N. I. (1989). Correlation of volume of focal zone of destruction with seismic energy of rock bump. *Sov. Geol Geophys.* **30**(5), 124–127.

Lomnitz, C. (1966). Statistical prediction of earthquakes. *Rev. Geophys.* **4**, 377–393.

Lomnitz, C. (1974). *Global Tectonics and Earthquake Risk*, Elsevier, Amsterdam.

Lomnitz-Adler, J., and Lomnitz, C. (1979). A modified form of the Gutenberg–Richter magnitude–frequency relation. *Bull. Seism. Soc. Am.* **69**, 1209–1214.

Long, L. T., and Copeland, C. W. (1989). The Alabama, U S A., seismic event and strata collapse of May 7, 1986. In *Seismicity in Mines* (S. J. Gibowicz, ed.); reprinted from Special Issue, *Pure Appl. Geophys.* **129**, 415–421, Birkhäuser Verlag, Basel.

Luco, J. E., and Anderson, J. G. (1983). Steady-state response of an elastic half-space to a moving dislocation of finite width. *Bull. Seism. Soc. Am.* **73**, 1–22.

MacBeth, C. D., and Panza, G. F. (1989). Modal synthesis of high-frequency waves in Scotland. *Geophys. J* **96**, 353–364.

MacBeth, C. D., and Redmayne, D. W. (1989). Source study of local coalfield events using the modal synthesis of shear and surface waves *Geophys. J. Internatl.* **99**, 155–172.

McDonald, A. J. (1984). Radon gas emission and seismicity in deep level gold mines. In *Rockbursts and Seismicity in Mines* (N. C. Gay and E. H. Wainwright, eds.), Symp Ser. No. 6, pp. 309–315, S Afr. Inst. Min Metal., Johannesburg.

McEvilly, T. V., and Johnson, L. R. (1974). Stability of *P* and *S* velocities from central California quarry blasts. *Bull Seism. Soc. Am.* **64**, 343–353.

McGarr, A. (1971). Violent deformation of rock near deep-level tabular excavations–seismic events. *Bull. Seism. Soc. Am.* **61**, 1453–1466.

McGarr, A. (1976). Seismic moments and volume changes *J. Geophys. Res.* **81**, 1487–1494.

McGarr, A. (1981). Analysis of peak ground motion in terms of a model of inhomogeneous faulting. *J. Geophys Res.* **86**, 3901–3912.

McGarr, A. (1984). Some applications of seismic source mechanism studies to assessing underground hazard. In *Rockbursts and Seismicity in Mines* (N. C. Gay and E. H. Wainwright, eds), Symp Ser. No. 6, pp. 199–208, S. Afr. Min. Metal., Johannesburg.

McGarr, A (1986). Some observations indicating complications in the nature of earthquake scaling. In *Earthquake Source Mechanics* (S. Das, J. Boatwright, and C. H. Scholz, eds.), *Maurice Ewing*, Vol. 6, 217–225. Am. Geophys. Union, Washington, D.C.

McGarr, A. (1991). Observations constraining near-source ground motion estimated from locally recorded seismograms *J. Geophys. Res.* **96**, 16,495–16,508.

McGarr, A., and Bicknell, J. (1990). Synthetic seismogram analysis of locally-recorded mine tremors. In *Rock at Great Depth* (V. Maury and D. Fourmaintraux, eds), pp 1407–1413 Balkema, Rotterdam.

McGarr, A., and Green, R. W. E. (1978). Microtremor sequences and tilting in a deep mine. *Bull. Seism. Soc. Am* **68**, 1679–1697

McGarr, A., and Wiebols, G. A. (1977). Influence of mine geometry and closure volume on seismicity in a deep-level mine. *Internatl. J. Rock Mech. Min. Sci.. Geomech. Abstr* **14**, 139–145.

McGarr, A., Spottiswoode, S. M., and Gay, N. C. (1975). Relationship of mine tremors to induced stresses and to rock properties in the focal region. *Bull. Seism. Soc. Am.* **65**, 981–993.

McGarr, A., Spottiswoode, S. M., Gay, N. C., and Ortlepp, W. D. (1979). Observations relevant to seismic driving stress, stress drop, and efficiency. *J. Geophys. Res.* **84**, 2251–2261.

McGarr, A., Green, R. W E., and Spottiswoode, S. M. (1981). Strong ground motion of mine tremors: Some implications for near-source ground motion parameters. *Bull. Seism. Soc. Am.* **71**, 295–319.

McGarr, A., Bicknell, J., Sembera, E., and Green, R. W. E. (1989). Analysis of exceptionally large tremors in two gold mining districts of South Africa. In *Seismicity in Mines* (S. J. Gibowicz, ed.), reprinted from Special Issue, *Pure Appl. Geophys.* **129**, 295–307, Birkhäuser Verlag, Basel.

McGaughey, W. J., Towers, J. J., and Bostock, M. G. (1987). *Introduction to Geotomography*, Research Report, Department of Geological Sciences, Queen's Univesity, Kingston, Ontario.

McLaughlin, K L., Johnson, L., and McEvilly, T. (1983). Two-dimensional array measurements of near source ground accelerations. *Bull. Seism. Soc. Am.* **73**, 349–375.

McMechan, G. (1983). Seismic tomography in boreholes. *Geophys. J. Roy. Astr. Soc.* **74**, 601–612.

McMechan, G A., Luetgert, J. H., and Mooney, W. D. (1985). Imaging of earthquake sources in Long Valley Caldera, California, 1983. *Bull. Seism. Soc. Am.* **75**, 1005–1020.

Madariaga, R. (1976). Dynamics of an expanding circular fault. *Bull. Seism. Soc. Am.* **66**, 639–666.

Madariaga, R. (1977). High-frequency radiation from cracks (stress-drop) models of earthquake faulting. *Geophys. J. Roy. Astr. Soc.* **51**, 625–651.

Madariaga, R. (1979). On the relation between seismic moment and stress drop in the presence of stress and strength heterogeneity. *J. Geophys. Res.* **84**, 2243–2250.

Madariaga, R. (1983a). High frequency radiation from dynamic earthquake fault models. *Ann. Geophys* **1**, 17–23.

Madariaga, R. (1983b). Earthquake source theory: A review. In *Earthquakes. Observation, Theory and Interpretation* (H. Kanamori and E. Boschi, eds.), pp. 1–44. North-Holland, Amsterdam.

Main, I. G. (1987). A characteristic earthquake model of the seismicity preceding the eruption of Mount St. Helens on 18 May 1980. *Phys Earth Planet. Interiors* **49**, 283–293.

Main, I. G., and Burton, P. W. (1984). Physical links between crustal deformation, seismic moment and seismic hazard for regions of varying seismicity. *Geophys. J. Roy. Astr. Soc.* **79**, 469–488.

Main, I. G., Meredith, P G, and Jones, C. (1989) A reinterpretation of the precursory seismic b-value anomaly from fracture mechanics. *Geophys. J.* **96**, 131–138.

Majer, E. L., McEvilly, T. V., Eastwood, F. S., and Myer, L. R. (1988). Fracture detection using P wave and S wave vertical seismic profiling at The Geysers. *Geophysics* **53**, 76–84.

Malin, P. E., Waller, J. A., Borcherdt, R. D., Cranswick, E., Jensen, E. G., and Van Schaack, J. (1988). Vertical seismic profiling of Oroville microearthquakes: Velocity spectra and particle motion as a function of depth. *Bull. Seism. Soc. Am.* **78**, 401–420.

Mandelbrot, B. B. (1967) How long is the coast of Britain? Statistical self-similarity and fractional dimension. *Science* **156**, 636–638.

Mandelbrot, B. B. (1977). *Fractals: Form, Chance and Dimension*, Freeman, San Francisco

Mandelbrot, B. B (1983). *The Fractal Geometry of Nature*, Freeman, San Francisco.

Marquardt, D. W. (1963). An algorithm for least-squares estimation of nonlinear parameters. *J Soc. Appl. Math.* **11**, 431–441.

Martin, C D (1990). Failure observations and in situ stress domains at the Underground Research Laboratory In *Rock at Great Depth* (V. Maury and D. Fourmaintraux, eds.), pp. 719–726, Balkema, Rotterdam.

Mason, I. (1981). Algebraic reconstruction of a two-dimensional inhomogeneity in the High Hazles seam of Thorseby colliery. *Geophysics* **46**, 298–308.

Matsu'ura, M., (1984). Bayesian estimation of hypocenter with origin time eliminated. *J Phys Earth* **32**, 469–483.

Matsu'ura, R. S. (1986). Precursory quiescence and recovery of aftershock activity before some large aftershocks. *Bull. Earthq. Res. Inst. Tokyo Univ.* **61**, 1–65.

Means, W. D. (1979). *Stress and Strain. Basic Concepts of Continuum Mechanics for Geologists*, Springer-Verlag, New York.

Mendecki, A. J. (1981). Methods of the joint hypocentre location for mining tremors and determination of parameters of velocity anisotropy. Ph.D. Thesis, Silesian Technical University of Gliwice, Poland (in Polish).

Mendecki, A. J. (1987). Rock mass anisotropy modelling by inversion of mine tremor data. In *Proceedings of the 6th International Congress on Rock Mechanics* (G. Herget and S. Vongpaisal, eds.), pp. 1141–1144, Balkema, Rotterdam.

Mendecki, A. J. (1990). The Integrated Seismic System (ISS). Paper presented at the Seminar on Monitoring and Safety in Civil and Mining Engineering, Nancy, France, June 1990.

Mendecki, A. J., van Aswegen, G., Brown, J. N. R., and Hewlett, P. (1990). The Welkom seismological network. In *Rockbursts and Seismicity in Mines* (C. Fairhurst, ed.), pp. 237–243. Balkema, Rotterdam.

Mendiguren, J. A. (1977). Inversion of surface wave data in source mechanism studies. *J. Geophys. Res.* **82**, 889–894.

Mendoza, C., and Hartzell, S. H. (1988). Aftershock patterns and main shock faulting. *Bull. Seism. Soc. Am.* **78**, 1438–1449.

Menke, W. (1989). *Geophysical Data Analysis: Discrete Inverse Theory*, Academic Press, Orlando, Fla.

Meredith, P. G., and Atkinson, B. K. (1983). Stress corrosion and acoustic emission during tensile crack propagation in Whin sill dolerite and other basic rocks. *Geophys. J. Roy Astr. Soc.* **75**, 1–21.

Meredith, P. G., Main, I. G., and Jones, C. (1990). Temporal variations in seismicity during quasi-static and dynamic rock failure. *Tectonophysics* **175**, 249–268.

Merz, H. A. and Cornell, C. A. (1974). Aftershocks in engineering risk analysis. *Proc. 5th World Confer. Earthq. Engng, Rome 1974* **2**, 2568–2571.

Meskó, A. (1984). *Digital Filtering: Applications in Geophysical Exploration for Oil*, Akadémiai Kiadó, Budapest.

Michael, A. J., and Eberhart-Phillips, D. (1991). Relations among fault behavior, subsurface geology, and three-dimensional velocity models. *Science* **253**, 651–654.

Michaelson, C. A. (1990). Coda duration magnitudes in central California: An empirical approach. *Bull. Seism. Soc. Am.* **80**, 1190–1204.

Mikumo, T., and Miyatake, T. (1983). Numerical modelling of space and time variations of seismic activity before major earthquakes. *Geophys. J. Roy. Astr. Soc.* **74**, 559–583

Mintrop, L. (1909). Die Erdbebenstation der Westfälischen Berggewerkschaftskasse in Bochum. *Glueckauf* **45**, 357–365.

Mjachkin, V. I., Brace, W. F., Sobolev, G. A., and Dieterich, J. H. (1975). Two models of earthquake forerunners. *Pure Appl. Geophys.* **113**, 169–181.

Mogi, K. (1962). Study of elastic shocks caused by the fracture of heterogeneous materials and their relation to earthquake phenomena. *Bull. Earthq. Res. Inst. Tokyo Univ.* **40**, 125–173.

Mogi, K. (1963). Some discussions on aftershocks, foreshocks and earthquake swarms—the fracture of a semi-infinite body caused by inner stress origin and its relation to the earthquake phenomena (3). *Bull. Earthq. Res. Inst. Tokyo Univ.* **41**, 615–658.

Mogi, K. (1985). *Earthquake Prediction*, Academic Press, Tokyo.

Molchan, G. M (1984). Some remarks on the Markov model due to L. Knopoff for earthquake sequences. *Vychisl Seismol*. **16**, 36–51 (in Russian).

Montalbetti, J R., and Kanasewich, E R. (1970). Enhancement of teleseismic body phases with a polarization filter. *Geophys. J. Roy. Astr. Soc*. **21**, 119–129

Mori, J., and Frankel, A (1990). Source parameters for small events associated with the 1986 North Palm Springs, California, earthquake determined using empirical Green functions. *Bull. Seism. Soc Am*. **80**, 278–295.

Mori, J., and Shimazaki, K. (1984). High stress drop of short-period subevents from the 1968 Tokachi-Oki earthquake as observed on strong-motion records. *Bull. Seism. Soc. Am*. **74**, 1529–1544.

Morrison, D. M (1989) Rockburst research at Falconbridge's Strathcona mine, Sudbury, Canada. In *Seismicity in Mines* (S. J. Gibowicz, ed.); reprinted from Special Issue, *Pure Appl. Geophys*. **129**, 619–645, Birkhäuser Verlag, Basel.

Mueller, C. S. (1985). Source pulse enhancement by deconvolution of an empirical Green's function. *Geophys. Res. Lett*. **12**, 33–36.

Murthy, R K., and Gupta, P. D (1983). Rock mechanics studies on the problem of ground control and rockbursts in the Kolar Gold Fields. *Proc. Symp Rockbursts Prediction and Control*, pp. 67–80. Inst. Min. Metal., London.

Nakanishi, I (1985). Three-dimensional structure beneath the Hokkaido-Tohoku region as derived from a tomographic inversion of P arrival times. *J. Phys Earth* **33**, 241–256

Napier, J. A L. (1987). Application of excess shear stress to the design of mine layouts. *J. S. Afr. Inst. Min Metal*. **87**, 397–405.

Nelder, J., and Mead, R. (1965) A simplex method for function minimization. *Computer J.* **7**, 308–312.

Nemat-Nasser, S., and Horii, H (1982). Compression-induced nonplanar crack extension with application to splitting, exfoliation, and rockburst. *J. Gelphys Res*. **87**, 6805–6821.

Neumann-Denzau, G., and Behrens, J. (1984). Inversion of seismic data using tomographic reconstruction techniques for investigation of laterally inhomogeneous media. *Geophys. J Roy Astr. Soc*. **79**, 305–315.

New, B. M (1985). The seismic investigation of rock properties at the Carwynnen Test Mine. In *U.K. Dept. Envir., Report DOERW* **85**.

Niewiadomski, J. (1989). Application of singular value decomposition method for location of seismic events in mines. In *Seismicity in Mines* (S. J. Gibowicz, ed.), reprinted from Special Issue, *Pure Appl. Geophys*. **129**, 553–570, Birkhauser Verlag, Basel.

Niewiadomski, J., and Meyer, K. (1986). Application of the regularization method for determination of seismic source time functions. *Acta Geophys. Pol*. **34**, 137–144.

Niewiadomski, J., and Rybicki, K (1984). The stress field induced by antiplane shear cracks—application to earthquake study. *Bull. Earthq. Res. Inst. Tokyo Univ*. **59**, 67–81.

Nishizawa, O. (1982). Seismic velocity anisotropy in a medium containing oriented cracks—transversely isotropic case. *J. Phys. Earth* **30**, 331–347.

Nolet, G. (1985). Solving or resolving inadequate and noisy tomographic systems. *J. Comp. Phys* **61**, 463–482.

Nolet, G. (1987) Seismic wave propagation and seismic tomography. In *Seismic Tomography* (G. Nolet, ed.), pp 7–29, Reidel, Dordrecht.

Nordquist, J. M. (1945) Theory of largest values applied to earthquake magnitudes. *Trans. Am. Geophys. Union* **26**, 29–31.

Novelo-Casanova, D A., Berg, E, Hsu, V., and Helsley, C E (1985). Time–space variation of seismic S-wave coda attenuation (Q^{-1}) and magnitude distribution (b-values) for the Petatlan earthquake. *Geophys Res. Lett* **12**, 789–792.

Nur, A. (1972). Dilatancy, pore fluids, and premonitory variations in t_s/t_p travel times. *Bull. Seism. Soc Am* **62**, 1217–1222.

O'Connell, R. J., and Budiansky, B. (1974). Seismic velocities in dry and saturated cracked solids. *J. Geophys. Res.* **79**, 5412–5426.

O'Connell, R. J., and Budiansky, B. (1978). Measures of dissipation in viscoelastic media. *Geophys. Res. Lett.* **5**, 5–8.

O'Connell, D. R. H., and Johnson, L. R (1988) Second-order moment tensors of microearthquakes at The Geysers geothermal field, California. *Bull. Seism. Soc. Am.* **78**, 1674–1692.

Oczkiewicz, J., and Szukalski, S. (1974). Example of application of extreme values theory for account of probability distribution of mining tremors maximum energy. *Cuprum* **5/6**, 28–33 (in Polish).

Ogata, Y., and Yamashina, K (1986). Unbiased estimate of *b*-value of magnitude-frequency. *J. Phys Earth* **34**, 187–194.

Ohtake, M. (1987) Temporal change of Q_p^{-1} in focal area of 1984 western Nagano, Japan, earthquake as derived from pulse width analysis *J. Geophys. Res.* **92**, 4846–4852

Ohtake, M., Matumoto, T., and Latham, G (1977) Seismicity gap near Oaxaca, southern Mexico as a probable precursor to a large earthquake. *Pure Appl. Geophys.* **115**, 375–385.

Ohtake, M., Matumoto, T., and Latham, G. (1981). Evaluation of the forecast of the 1978 Oaxaca, southern Mexico earthquake based on a precursory seismic quiescence. *Earthquake Prediction—an International Review* (D W. Simpson and P. G Richards, eds.), *Maurice Ewing*, Vol. 4, pp. 53–62. Am. Geophys Union, Washington, D.C.

Ohtsu, M. (1991). Simplified moment tensor analysis and unified decomposition of acoustic emission source: Application to in situ hydrofracturing test *J. Geophys. Res.* **96**, 6211–6222.

Okubo, P., and Aki, K. (1987). Fractal geometry in the San Andreas fault system. *J. Geophys. Res* **92**, 345–355.

Okubo, P. G., and Dieterich, J. H. (1984). Effects of physical properties on frictional instabilities produced on simulated faults. *J. Geophys Res.* **89**, 5817–5827.

Okubo, P. G , and Dieterich, J. H. (1986). State variable fault constitutive relations for dynamic slip. In *Earthquake Source Mechanics* (S Das, J. Boatwright, and C. H. Scholz, eds.), *Maurice Ewing*, Vol. 6, pp. 25–36, Am Geophys. Union, Washington, D.C.

Olgaard, D. L., and Brace, W. F. (1983). The microstructure of gouge from a mining-induced seismic shear zone. *Internatl J. Rock Mech. Min. Sci., Geomech. Abstr.* **20**, 11–19.

Oncescu, M. C (1986) Relative seismic moment tensor determination for Vrancea intermediate depth earthquakes. In *Physics of Fracturing and Seismic Energy Release* (J. Kozák and L. Waniek, eds.); reprinted from Special Issue, *Pure Appl. Geophys.* **124**, 931–940, Birkhäuser Verlag, Basel

Oncescu, M. C., and Apolozan, L (1984). The earthquake sequence of Romnicu Sarat, Romania, of 21–22 February 1983. *Acta Geophys. Pol.* **32**, 231–238.

O'Neill, M. E. (1984). Source dimensions and stress drop of small earthquakes near Parkfield, California. *Bull Seism. Soc Am.* **74**, 27–40.

Ortlepp, W. D. (1984). Rockbursts in South African gold mines: A phenomenological view In *Rockbursts and Seismicity in Mines* (N C. Gay and E. H Wainright, eds.), Symp. Ser. No 6, pp. 165–178, S. Afr. Inst. Min. Metal., Johannesburg.

Ortlepp, W. D , and Spottiswoode, S (1984) The design and introduction of stabilizing pillars at Blyvooruitzicht Gold Mining Company Limited. *Proc 12th CMMI Congr.*, pp. 353–362. S Afr. Inst. Min. Metal., Johannesburg.

Osterwald, F. W. (1970). Comments on rockbursts, outbursts, and earthquake prediction. *Bull. Seism. Soc. Am.* **60**, 2083–2088

Ostrihansky, R., and Gerlach, Z. (1982). Assessment of rockburst risk in coal mining regions on the basis of changes in seismic activity. *Publ. Inst. Geophys., Pol. Acad. Sci.* **M-5** (155), 57–72 (in Polish; English abstract).

Page, R. (1968). Aftershocks and microaftershocks of the great Alaska earthquake. *Bull. Seism. Soc. Am.* **58**, 1131–1168.

Page, C. C., and Saunders, M. A. (1982). LSQR: An algorithm for sparse least squares. *ACM Trans. Math. Soft.* **8**, 43–71, 195–209.

Palmer, A. C., and Rice, J. R. (1973). The growth of slip surfaces in the progressive failure of overconsolidated clay slopes. *Proc. Roy. Soc. Lond.* A **332**, 527–548.

Panza, G. F. (1985). Synthetic seismograms: The Rayleigh wave modal summation. *J. Geophys.* **58**, 125–145.

Papageorgiou, A. S (1988). On two characteristic frequencies of acceleration spectra: Patch corner frequency and f_{max}. *Bull. Seism. Soc. Am.* **78**, 509–529.

Papageorgiou, A. S., and Aki, K. (1983a). A specific barrier model for the quantitative description of inhomogeneous faulting and the prediction of strong ground motion, Part I. Description of the model. *Bull. Seism. Soc Am.* **73**, 693–722.

Papageorgiou, A. S., and Aki, K. (1983b). A specific barrier model for the quantitative description of inhomogeneous faulting and the prediction of strong ground motion, Part II. Applications of the model. *Bull. Seism. Soc. Am.* **73**, 953–978.

Papageorgiou, A. S., and Aki, K. (1985). Scaling law of far-field spectra based on observed parameters of the specific barrier model. *Pure Appl. Geophys.* **123**, 353–374.

Papastamatiou, D. (1980). Incorporation of crustal deformation to seismic hazard analysis. *Bull. Seism. Soc. Am.* **70**, 1321–1335

Papazachos, B. (1975). Foreshocks and earthquake prediction. *Tectonophysics* **28**, 213–226.

Park, J., Vernon, F. L., and Lindberg, C. R. (1987). Frequency dependent polarization analysis of high-frequency seismograms. *J. Geophys. Res.* **92**, 664–674.

Parsons, I. D., Hall, J. F., and Lyzenga, G A. (1988). Relationships between the average offset and the stress drop for two- and three-dimensional faults. *Bull. Seism. Soc. Am.* **78**, 931–945.

Parysiewicz, W. (1966). *Rockbursts in Mines*, Slask, Katowice, Poland (in Polish).

Patton, H., and Aki, K (1979). Bias in the estimate of seismic moment tensor by the linear inversion method. *Geophys. J. Roy. Astr. Soc.* **59**, 479–495.

Pavlins, G. L. (1986). Appraising earthquake hypocenter location errors: A complete, practical approach for single event locations. *Bull. Seism. Soc. Am.* **76**, 1699–1717.

Peacock, S., and Crampin, S. (1985). Shear wave vibrator signals in transversely isotropic shale. *Geophysics* **50**, 1285–1293.

Peacock, S. P., Crampin, S., Booth, D. C., and Fletcher, J. B. (1988). Shear wave splitting in the Anza seismic gap, southern California. *J. Geophys. Res* **93**, 3339–3356.

Pearce, R. G., Hudson, J A., and Douglas, A. (1988). On the use of P-wave seismograms to identify a double-couple source. *Bull. Seism. Soc. Am.* **78**, 651–671.

Pechmann, J. C., and Kanamori, H. (1982). Waveforms and spectra of foreshocks and aftershocks of the 1979 Imperial Valley, California, earthquake: Evidence for fault heterogeneity? *J. Geophys. Res.* **87**, 10,579–10,597.

Peng, J. Y., Aki, K, Lee, W. H. K., Chouet, B., Johnson, P., Marks, S., Newberry, J. T, Ryal, A. S., Stewart, S. W., and Tottingham, D. M. (1987). Temporal change in coda Q associated with 1984 Round Valley earthquake in California. *J. Geophys. Res.* **92**, 3507–3536.

Peters, D. C., and Crosson, R. S (1972). Application of prediction analysis to hypocenter determination using a local array. *Bull. Seism. Soc. Am.* **62**, 775–788.

Peterson, J. E., Paulsson, B. N. P., and McEvilly, T. V. (1985). Applications of algebraic reconstruction techniques to crosshole seismic data. *Geophysics* **50**, 1566–1580.

Petukhov, I. M., Smirnov, V. A., and Rabota, E. N. (1980). The study of seismicity in the mining areas of the Tkibuli-Shaorsk coal basin. In *Detailed Seismic Zonation* (I. E. Gubin, ed.), pp. 161–167, Nauka, Moscow (in Russian).

Phillips, W S., and Aki, K. (1986). Site amplification of coda waves from local earthquakes in central California *Bull. Seism. Soc. Am.* **76**, 627–648

Phillips, W. S. and Fehler, M. C. (1991). Traveltime tomography: A comparison of popular methods. *Geophysics* **56**, 1639–1649.

Phillips, W. S., Lee, W. H. K., and Newberry, J. T. (1988). Spatial variation of crustal coda Q in California. In *Scattering and Attenuation of Seismic Waves* (R.-S. Wu and K. Aki, eds.), Part I: reprinted from Special Issue, *Pure Appl. Geophys.* **128**, 251–260, Birkhäuser Verlag, Basel.

Plesinger, A., Hellweg, M., and Seidl, D. (1986). Interactive high-resolution polarization analysis of broadband seismograms. *J. Geophys.* **59**, 129–139.

Pomeroy, P. W., Simpson, D. W., and Sbar, M. L. (1976). Earthquakes triggered by surface quarrying—the Wappingers Falls, New York sequence of June, 1974. *Bull. Seism. Soc. Am.* **66**, 685–700.

Pope, A. J. (1972). Fiducial regions for body wave focal plane solutions. *Geophys. J. Roy. Astr. Soc.* **30**, 331–342.

Potgieter, G. J., and Roering, C. (1984). The influence of geology on the mechanisms of mining-associated seismicity in the Klerksdorp gold-field. In *Rockbursts and Seismicity in Mines* (N. C. Gay and E. H. Wainwright, eds.), Symp. Ser. No. 6, pp. 45–50, S. Afr. Inst. Min. Metal., Johannesburg.

Press, W. H., Flannery, B. P., Teukolsky, S. A., and Vetterling, W. T. (1990). *Numerical Recipes: The Art of Scientific Computing*, Cambridge University Press, New York.

Prugger, A. F., and Gendzwill, D. J. (1988). Microearthquake location: A non-linear approach that makes use of a simplex stepping procedure. *Bull. Seism. Soc. Am.* **78**, 799–815.

Pujol, J., and Herrmann, R. B. (1990). A student's guide to point sources in homogeneous media. *Seismol. Res. Lett.* **61**, 209–224.

Pulli, J. J. (1984). Attenuation of coda waves in New England. *Bull. Seism. Soc. Am.* **74**, 1149–1166.

Qaisar, M. (1989). Attenuation properties of viscoelastic material. In *Scattering and Attenuation of Seismic Waves* (R.-S. Wu and K. Aki, eds.), Part II, reprinted from Special Issue, *Pure Appl. Geophys.* **131**, 701–713, Birkhäuser Verlag, Basel.

Rabinowitz, N. (1988). Microearthquake location by means of nonlinear simplex procedure. *Bull. Seism. Soc. Am.* **78**, 380–384.

Rabinowitz, N., and Kulhánek, O. (1988). Application of non-linear algorithm to teleseismic locations using *P*-wave readings from the Swedish seismographic network. *Phys. Earth Planet. Interiors* **50**, 111–115.

Rabinowitz, N., and Steinberg, D. M. (1990). Optimal configuration of seismographic network: A statistical approach. *Bull. Seism. Soc. Am.* **80**, 187–196.

Ramirez, A. L (1986). Reconstruction of simulated lineaments using geophysical tomography. *Internatl. J. Rock Mech. Min. Sci., Geomech. Abstr.* **23**, 157–163.

Randall, M. J. (1971). Shear invariant and seismic moment for deep-focus earthquakes. *J. Geophys. Res.* **76**, 4991–4992.

Randall, M. J. (1973). The spectral theory of seismic sources. *Bull. Seism. Soc. Am.* **63**, 1133–1144.

Rautian, T. G., and Khalturin, V. I. (1978). The use of the coda for determination of the earthquake source spectrum. *Bull. Seism. Soc. Am.* **68**, 923–948.

Rebollar, C. J. (1984). Calculation of Q_β using the spectral ratio method in Northern Baja California. *Bull. Seism. Soc. Am.* **74**, 91–96.

Rebollar, C. J., Traslosheros, C., and Alvarez, R. (1985). Estimates of seismic wave attenuation in northern Baja California. *Bull. Seism. Soc. Am.* **75**, 1371–1382.

Redmayne, D. W. (1988). Mining induced seismicity in UK coalfields identified on the BGS National Seismograph Network. In *Engineering Geology of Underground Movements* (F. G. Bell, M. G. Culshaw, J. C. Cripps, and M. A. Lovell, eds.), Geol. Soc. Eng. Geol. Spec. Publ. No. 5, pp. 405–413.

Reid, H. F. (1911). The elastic-rebound theory of earthquakes. *Bull. Dept. Geol. Univ. Calif.* **6**, 412-444.

Revalor, R., Josien, J. P., Besson, J. L., and Magron, A. (1990). Seismic and seismoacoustic experiments applied to the prediction of rockbursts in French coal mines. In *Rockbursts and Seismicity in Mines* (C. Fairhurst, ed.), pp. 301-306, Balkema, Rotterdam.

Rice, J. R. (1983). Constitutive relations for fault slip and earthquake instabilities. In *Instabilities in Continuous Media* (L. Knopoff, V. F. Keilis-Borok, and G. Puppi, eds.); reprinted from Special Issue, *Pure Appl. Geophys.* **121**, 443-475, Birkhauser Verlag, Basel

Rice, J. R., and Gu, J. C (1983). Earthquake aftereffects and triggered seismic phenomena. *Pure Appl. Geophys* **121**, 187-219

Richards, P. G. (1976). Dynamic motions near an earthquake fault: A three-dimensional solution. *Bull. Seism. Soc. Am.* **66**, 1-32.

Richter, C. F (1935). An instrumental earthquake magnitude scale. *Bull. Seism. Soc. Am.* **25**, 1-32.

Riedesel, M. A., and Jordan, T. H. (1989). Display and assessment of seismic moment tensors. *Bull. Seism Soc. Am.* **79**, 85-100.

Rikitake, T. (1982). Do foreshock epicenters move toward the main shock epicenter? *Earthq. Predict. Res.* **1**, 95-114.

Roberts, J. D. M., Belchamber, R. M., Lilley, T., Betteridge, D., Bishop, I., and Styles, P. (1989). An evaluation of computerized tomography for near-surface geophysical exploration. *Computers Geosci.* **15**, 727-737.

Roček, V., and Skorepova, J. (1982). Convergence and its relation to sudden changes of stress within the rock mass. *Publ. Inst. Geophys. Pol. Acad. Sci.* **M-5** (155), 37-46.

Rodriguez, M., Havskov, J., and Singh, S K. (1983). *Q* from coda waves near Petatlan, Guerrero, Mexico *Bull. Seism. Soc. Am.* **73**, 321-326.

Roecker, S. W., Tucker, B., King, J., and Hatzfeld, D. (1982). Estimates of *Q* in central Asia as a function of frequency and depth using the coda of locally recorded earthquakes *Bull Seism Soc Am.* **72**, 129-149.

Roeloffs, E. (1988). Hydrological precursors to earthquakes. A review. In *Intermediate-Term Earthquake Prediction* (W. D. Stewart and K. Aki, eds.); reprinted from Special Issue, *Pure Appl. Geophys.* **126**, 177-209.

Rogers, P. G., Edwards, S. A., Young, J. A., and Downey, M. (1987) Geotomography for the delineation of coal seam. *Geoexploration* **24**, 301-328

Rorke, A. J., and Brummer, R. K. (1990). The use of explosives in rockburst control techniques. In *Rockbursts and Seismicity in Mines* (C. Fairhurst, ed.), pp. 377-385, Balkema, Rotterdam.

Rovelli, A (1983) Time-frequency analysis of seismic excitation and estimates of attenuation parameters for the Friuli (Italy) local earthquakes. *Phys. Earth Planet. Interiors* **33**, 94-110

Rovelli, A., Cocco, M, Console, R., Alessandrini, B., and Mazza, S. (1991). Ground motion waveforms and source spectral scaling from close-distance accelerograms in a compressional regime area (Friuli, northeastern Italy). *Bull. Seism. Soc. Am.* **81**, 57-80.

Rudajev, V., and Bucha, V., Jr (1988). Some experience with the research of mining induced seismicity in the CSSR. In *Induced Seismicity and Associated Phenomena* (D. Procházková, ed), pp 75-79, Geophys. Inst., Czech. Acad. Sci , Prague.

Rudajev, V , and Šileny, J. (1985). Seismic events with non-shear components: II. Rock bursts with implosive source component. *Pure Appl. Geophys.* **123**, 17-25.

Rudajev, V., Dragan, V., and Kacák, K. (1985). Correlation of seismoacoustic and seismic data for the purpose of prediction of rockbursts. *Publ Inst. Geophys., Pol. Acad. Sci.* **M-6** (176), 249-262.

Rudajev, V., Teisseyre, R., Kozák, J , and Šileny, J. (1986). Possible mechanism of rockbursts in coal mines. In *Physics of Fracturing and Seismic Energy Release* (J. Kozák and L. Waniek, eds); reprinted from Special Issue, *Pure Appl. Geophys.* **124**, 841-855, Birkhäuser Verlag, Basel.

Rudnicki, J. W. (1988). Physical models of earthquake instability and precursory processes. In *Intermediate-Term Earthquake Prediction* (W. D. Stewart and K. Aki, eds.); reprinted from Special Issue, *Pure Appl. Geophys.* 126, 531–554, Birkhäuser Verlag, Basel.

Rudnicki, J., and Kanamori, H. (1981). Effects of fault interaction on moment, stress drop and strain energy release. *J. Geophys. Res.* 86, 1785–1793.

Rudzki, M. P. (1911). Parametrische Darstellung der elastischen Welle in anisotropic Medien. *Bulletin International de l'Academie des Sciences de Cracovie, Ser.* A, 503–536.

Ruff, L. J. (1984). Tomographic imaging of the earthquake rupture process. *Geophys. Res. Lett.* 11, 629–632.

Ruff, L. J. (1987). Tomographic imaging of seismic sources In *Seismic Tomography* (G. Nolet, ed.), pp. 339–369, Reidel, Dordrecht.

Ruina, A. L. (1983). Slip instability and state variable friction laws. *J. Geophys. Res.* 88, 10,359–10,370.

Rundle, J. B. (1989). Derivation of complete Gutenberg-Richter magnitude–frequency relation using the principle of scale invariance. *J. Geophys. Res.* 94, 12,337–12,342.

Rybicki, K. (1971). The elastic residual field of a very long strike-slip fault in the presence of a discontinuity. *Bull. Seism. Soc. Am.* 61, 79–92.

Rybicki, K. (1986). Dislocations and their geophysical application. In *Continuum Theories in Solid Earth Physics* (R. Teisseyre, ed.), pp. 18–186, Polish Scientific Publishers, Warsaw; Elsevier, Amsterdam.

Ryder, J. A. (1987). Excess shear stress (ESS): An engineering criterion for assessing unstable slip and associated rockburst hazards. *Proc. 6th Internatl. Soc. Rock Mech. Congr.*, Montreal 1987, pp. 294–298.

Saikia, C. K., and Herrmann, R. B. (1985). Application of waveform modeling to determine focal mechanisms of four 1982 Miramichi aftershocks. *Bull. Seism. Soc. Am.* 75, 1021–1040.

Saikia, C. K., and Herrmann, R. B. (1986). Moment–tensor solutions for three 1982 Arkansas swarm earthquakes by waveform modeling. *Bull. Seism. Soc. Am.* 76, 709–723.

Salamon, M. D. G. (1983). Rockburst hazard and the fight for its alleviation in South Africa. *Proc. Symp. Rockbursts: Prediction and Control*, pp. 11–36. Inst. Min. Metal., London.

Salamon, M. D. G., and Wiebols, G. A. (1972). Location of seismic events. *S. Afr. Chamber of Mines*, unpublished report.

Satake, K. (1985). Effects of station coverage on moment tensor inversion. *Bull. Seism. Soc. Am* 75, 1657–1667.

Sato, H. (1977a). Energy propagation including scattering effects: Single isotropic scattering approximation. *J. Phys. Earth* 25, 27–41.

Sato, H. (1977b). Single isotropic scattering model including wave conversions: Simple theoretical model of the short-period body wave propagation. *J. Phys. Earth* 25, 163–176.

Sato, H. (1978). Mean free path of *S* waves under the Kanto District of Japan *J. Phys Earth* 26, 185–198.

Sato, H. (1982). Coda wave excitation due to nonisotropic scattering and nonspherical source radiation. *J. Geophys. Res.* 87, 8665–8674.

Sato, H. (1984). Attenuation and envelope formation of three component seismograms of small local earthquakes in randomly inhomogeneous lithosphere. *J. Geophys. Res.* 89, 1221–1241.

Sato, H. (1988). Temporal change in scattering and attenuation associated with the earthquake occurrence: A review of recent studies on coda waves. *Pure Appl. Geophys.* 126, 465–497.

Sato, H. (1990). Unified approach to amplitude attenuation and coda excitation in the randomly inhomogeneous lithosphere. In *Scattering and Attenuation of Seismic Waves* (R.-S. Wu and K. Aki, eds.), Part III; reprinted from Special Issue, *Pure Appl. Geophys.* 132, 93–121, Birhäuser Verlag, Basel.

Sato, K., and Fujii, Y. (1989). Source mechanism of a large scale gas outburst at Sunagawa coal mine in Japan. In *Seismicity in Mines* (S. J. Gibowicz, ed.); reprinted from Special Issue, *Pure Appl. Geophys.* 129, 325–343, Birkhäuser Verlag, Basel.

Sato, T., and Hirasawa, T. (1973). Body wave spectra from propagating shear cracks. *J. Phys. Earth* **21**, 415–431.

Sato, Y., and Skoko, D. (1965). Optimum distribution of seismic observation points, II. *Bull. Earthq. Res. Inst. Tokyo Univ.* **43**, 451–457.

Savage, J. C. (1972). Relation of corner frequency to fault dimensions. *J. Geophys. Res.* **77**, 3788–3795.

Savage, M. K., Peppin, W. A., and Vetter, U. R. (1990). Shear wave anisotropy and stress direction in and near Long Valley Caldera, California, 1979–88. *J. Geophys. Res.* **95**, 11,165–11,177.

Savarenskiy, Y. F., Sofranov, V. V., Peshkov, A. B., Verbova, L. F., and Peshkova, I. V. (1979). Optimum distribution of seismic stations for minimizing errors of epicenter determination. *Izv. Akad. Nauk SSSR, Earth Phys.* **15**, 572–577.

Scales, J. A. (1987). Tomographic inversion via the conjugate gradient method. *Geophysics* **52**, 179–182.

Scheepers, J. B. (1984). The Klerksdorp seismic network: Monitoring of seismic events and system layout. In *Rockbursts and Seismicity in Mines* (N. C. Gay and E. H. Wainwright, eds.), Symp. Ser. No. 6, pp. 341–345, S. Afric. Inst. Min. Metal., Johannesburg.

Scherbaum, F. (1990). Combined inversion for the three-dimensional Q structure and source parameters using microearthquake spectra. *J. Geophys. Res.* **95**, 12,423–12,438.

Scherbaum, F., and Wyss, M. (1990). Distribution of attenuation in the Kaoiki, Hawaii, source volume estimated by inversion of P wave spectra. *J. Geophys. Res.* **95**, 12,439–12,448

Scholz, C. (1968). The frequency–magnitude relation of microfracturing in rock and its relation to earthquakes. *Bull. Seism. Soc. Am.* **58**, 399–415.

Scholz, C. (1972). Crustal movements in tectonic areas. *Tectonophysics* **14**, 201–217.

Scholz, C. H. (1982). Scaling laws for large earthquakes, consequences for physical models. *Bull. Seism. Soc. Am.* **72**, 1–14.

Scholz, C. H. (1990). *The Mechanics of Earthquakes and Faulting*, Cambridge University Press, Cambridge (U.K.).

Scholz, C. H., and Aviles, C. A. (1986). The fractal geometry of faults and faulting. In *Earthquake Source Mechanics* (S. Das, J. Boatwright, and C. H. Scholz, eds.), *Maurice Ewing*, Vol. 6, pp. 147–155, Am. Geophys. Union, Washington, D.C.

Scholz, C. H., and Mandelbrot, B. B., eds. (1989). *Fractals in Geophysics*, Birkhäuser Verlag, Basel; reprinted from Special Issue, *Pure Appl. Geophys.* **131**(1/2).

Scholz, C. H., Sykes, L. R., and Aggarwal, Y. P. (1973). Earthquake prediction: A physical basis. *Science* **181**, 803–810.

Schwartz, D. P., and Coppersmith, K. J. (1984). Fault behavior and characteristic earthquakes: Examples from the Wasatch and San Andreas fault zones. *J. Geophys. Res.* **89**, 5681–5698.

Scott, D. F. (1990). Relationship of geological features to seismic events, Lucky Friday Mine, Mullan, Idaho. In *Rockbursts and Seismicity in Mines* (C. Fairhurst, ed.), pp. 401–405, Balkema, Rotterdam.

Semenov, A. N. (1969). Variations of the travel time of transverse and longitudinal waves before violent earthquakes. *Izv. Acad. Nauk USSR, Phys. Solid Earth* (English translation) **3**, 245–258.

Shi, Y., and Bolt, B. A. (1982). The standard error of the magnitude–frequency b value. *Bull. Seism. Soc. Am.* **72**, 1677–1687.

Shibutani, T., and Oike, K. (1989). On features of spatial and temporal variation of seismicity before and after moderate earthquakes. *J. Phys. Earth* **37**, 201–224.

Shih, X. R., and Meyer, R. P. (1990). Observation of shear wave splitting from natural events: South moat of Long Valley Caldera, California, June 29 to August 12, 1982. *J. Geophys. Res.* **95**, 11,179–11,195.

Shimamoto, T., and Logan, J. M. (1984). Laboratory friction experiments and natural earthquakes: An argument for long-term tests. *Tectonophysics* **109**, 165–175.

Shimamoto, T., and Logan, J. M. (1986). Velocity-dependent behavior in halite simulated fault gauge: An analog for silicates. In *Earthquake Source Mechanics* (S. Das, J. Boatwright, and C H. Scholz, eds.), *Maurice Ewing*, Vol. 6, pp. 49–64, Am. Geophys. Union, Washington, D.C.

Shimazaki, K. (1986). Small and large earthquakes: The effects of the thickness of seismogenic layer and the free surface. In *Earthquake Source Mechanics* (S. Das, J. Boatwright, and C. H. Scholz, eds.), *Maurice Ewing*, Vol. 6, pp. 209–216, Am. Geophys. Union, Washington, D.C.

Siewierski, S., Bugajski, W., Bachowski, C., and Orzepowski, S. (1989a). Vertical and horizontal deformation measurements of boreholes at Rudna copper mine in Poland. *Acta Geophys. Pol.* **37**, 263–270.

Siewierski, S., Bugajski, W., Bachowski, C., and Orzepowski, S. (1989b). The evaluation of rockmass state from borehole vertical and horizontal deformation measurements at a copper mine. *Acta Geophys. Pol.* **37**, 271–276.

Šileny, J. (1986). Inversion of first-motion amplitudes recorded by local seismic network for rockburst study in the Kladno, Czechoslovakia, mining area. *Acta Geophys. Pol.* **34**, 201–213.

Šileny, J. (1989). The mechanism of small mining tremors from amplitude inversion. In *Seismicity in Mines* (S. J. Gibowicz, ed.); reprinted from Special Issue, *Pure Appl. Geophys.* **129**, 309–324, Birkhäuser Verlag, Basel.

Silver, P. G., and Jordan, T. H. (1982). Optimal estimation of scalar seismic moment. *Geophys. J. Roy. Astr. Soc.* **70**, 755–787.

Simpson, D. W. (1986). Triggered earthquakes. *Ann. Rev. Earth Planet. Sci.* **14**, 21–42.

Singh, S. K., Bazan, E., and Esteva, L. (1980). Expected earthquake magnitude from a fault. *Bull. Seism. Soc. Am.* **70**, 903–914.

Sipkin, S. A. (1982). Estimation of earthquake source parameters by the inversion of waveform data: Synthetic waveforms. *Phys. Earth Planet. Interiors* **30**, 242–259.

Skala, V., and Roček, V. (1985). Facing the rockburst phenomena in the bitominuous coal district Kladno. *Publ. Inst. Geophys., Pol. Acad. Sci.* **M-6** (176), 345–356.

Šklenar, J., and Rudajev, V. (1975). Application of some seismic methods for the evaluation of the stress-strain condition in the rock mass. *Acta Mont.* **32**, 211–230.

Skoko, D., and Sato, Y. (1966). Optimum distribution of seismic observation points, III. *Bull. Earthq. Res. Inst. Tokyo Univ.* **44**, 13–22.

Slunga, R., Norman, P., and Glans, A. C. (1984). Seismicity of Southern Sweden. *FOA Report C20543-T1*, National Deference Research Institute, Stockholm.

Smalley, R., Turcotte, D., and Solla, S. (1985). A renormalization group approach to the stick-slip behavior of faults. *J. Geophys. Res.* **90**, 1894–1900.

Smith, B. D., and Ward, S. H. (1974). Short note: On the computation of polarization ellipse parameters. *Geophysics* **39**, 867–869.

Smith, R. B., Winkler, P. L., Anderson, J. G., and Scholz, C. H. (1974). Source mechanism of microearthquakes associated with underground mines in eastern Utah. *Bull. Seism. Soc. Am.* **64**, 1295–1317.

Snoke, J. A. (1987). Stable determination of (Brune) stress drops. *Bull. Seism. Soc. Am.* **77**, 530–538.

Snoke, J. A., Linde, A. T., and Sacks, I. S. (1983). Apparent stress: An estimate of the stress drop. *Bull. Seism. Soc. Am.* **73**, 339–348.

Soloviev, S. L. (1965). Seismicity of Sakhalin. *Bull. Earthq. Res. Inst., Tokyo Univ.* **43**, 95–102.

Spakman, W. (1985). A tomographic image of the upper mantle in the Eurasian-African-Arabian collision zone. *EOS Trans. Am. Geophys Union* **66**, 975.

Spakman, W. (1986). The upper mantle structure in the Central European-Mediterranean region. In *European Geotraverse (EGT) Project, the Central Segment* (R. Freeman, S Mueller, and P. Giese, eds.), pp. 215–222. European Science Foundation, Strassburg.

380   References

Spakman, W., and Nolet, G. (1988). Imaging algorithms, accuracy and resolution in delay time tomography. In *Mathematical Geophysics* (N. J. Vlaar, G. Nolet, M. J. R Wortel, and S. A. P. L. Cloetingh, eds.), pp. 155–187, Reidel, Dordrecht.

Spence, W. (1980). Relative epicenter determination using *P*-wave arrival-time differences. *Bull. Seism. Soc Am.* **70**, 171–183.

Spencer, C., and Gubbins, D. (1980). Travel-time inversion for simultaneous earthquake location and velocity structure determination in laterally varying media. *Geophys. J. Roy. Astr. Soc.* **63**, 95–116.

Spendly, W., and Hest, G (1962). Sequential application of simplex design in optimization and evolutionary operations. *Technometrics* **4**, 441–461.

Spottiswoode, S. M. (1984). Source mechanisms of mine tremors at Blyvooruitzicht gold mine. In *Rockbursts and Seismicity in Mines* (N. C. Gay and E. H. Wainwright, eds.), Symp. Ser. No. 6, pp. 29–37. S. Afr. Inst. Min. Metal., Johannesburg.

Spottiswoode, S M. (1989). Perspectives on seismic and rockburst research in the South African gold mining industry: 1983–1987. In *Seismicity in Mines* (S. J Gibowicz, ed); reprinted from Special Issue, *Pure Appl. Geophys.* **129**, 673–680, Birkhäuser Verlag, Basel.

Spottiswoode, S. M. (1990). Volume excess shear stress and cumulative seismic moments. In *Rockbursts and Seismicity in Mines* (C. Fairhurst, ed.), pp. 39–43, Balkema, Rotterdam.

Spottiswoode, S. M., and McGarr, A. (1975). Source parameters of tremors in a deep-level gold mine. *Bull. Seism. Soc. Am.* **65**, 93–112.

Sprenke, K. F., Stickney, M. C., Dodge, D. A., and Hammond, W. R. (1991). Seismicity and tectonic stress in the Coeur d'Alene mining district *Bull. Seism. Soc. Am.* **81**, 1145–1156.

Spudich, P., and Cranswick, E. (1984). Direct observation of rupture propagation during the 1979 Imperial Valley earthquake using a short baseline accelerometer array. *Bull. Seism. Soc. Am.* **74**, 2083–2114.

Spudich, P., and Frazer, L. N. (1984). Use of ray theory to calculate high-frequency radiation from earthquake sources having spatially variable rupture velocity and stress drop. *Bull. Seism. Soc. Am.* **74**, 2061–2082.

Stankiewicz, T. (1989). Stochastic model of seismic activity and its application to seismic hazard estimates in mines. Ph.D. thesis, Inst. Geophys., Pol. Acad. Sci., Warsaw (in Polish).

Steensma, G. J., and Biswas, N N. (1988). Frequency dependent characteristics of coda wave quality factor in central and southcentral Alaska. In *Scattering and Attenuation of Seismic Waves* (R.-S. Wu and K. Aki, eds.), Part I; reprinted from Special Issue, *Pure Appl. Geophys.* **128**, 295–307, Birkhäuser Verlag, Basel.

Steidl, J. H., Archuleta, R. J., and Hartzell, S. H. (1991). Rupture history of the 1989 Loma Prieta, California, earthquake. *Bull. Seism. Soc. Am.* **81**, 1573–1602.

Stein, R. S., and Lisowski, M. (1983). The 1979 Homestead Valley earthquake sequence, California. Control of aftershocks and postseismic deformation. *J. Geophys. Res.* **88**, 6477–6490.

Stopinski, W., and Dmowska, R. (1984). Rock resistivity in the Lubin (Poland) copper mine and its relation to variations of strain field and occurrence of rockbursts. In *Rockbursts and Seismicity in Mines* (N. C. Gay and E. H. Wainwright, eds.), Symp. Ser. No. 6, pp. 297–307, S. Afr. Inst. Min. Metal., Johannesburg.

Strelitz, R. A. (1978). Moment tensor inversions and source models. *Geophys. J. Roy. Astr. Soc.* **52**, 359–364

Strelitz, R. A. (1980) The fate of downgoing slabe: A study of the moment tensors from body waves of complex deep-focus earthquakes. *Phys. Earth Planet. Interiors* **21**, 83–96.

Stuart, W. D., and Mavko, G. M. (1979). Earthquake instability on a strike-slip fault. *J. Geophys. Res* **84**, 2153–2160.

Stump, B. W., and Johnson, L. R. (1977). The determination of source properties by the linear inversion of seismograms. *Bull. Seism. Soc. Am.* **67**, 1489–1502

Su, F., and Aki, K. (1990). Temporal and spatial variation in coda Q^{-1} associated with the North Palm Springs earthquake of July 8, 1986. *Pure Appl. Geophys* **133**, 23–52.

Suyehiro, S. (1966). Difference between aftershocks and foreshocks in the relationship of magnitude to frequency of occurrence for the great Chilean earthquake of 1960. *Bull. Seism. Soc. Am.* **56**, 185–200.

Sykes, L. R. (1970). Earthquake swarms and sea-floor spreading. *J. Geophys. Res.* **75**, 6598–6611.

Syratt, P. P. (1990). Seismicity associated with the extraction of stressed remnants in the Klerksdorp gold mining district, South Africa In *Rockbursts and Seismicity in Mines* (C. Fairhurst, ed.), pp. 77–80, Balkema, Rotterdam.

Syrek, B., and Kijko, A. (1988). Energy and frequency distribution of mining tremors and their relation to rockburst hazard in the Wujek coal mine, Poland. *Acta Geophys. Pol.* **36**, 189–201.

Tada, H., Paris, P. C., and Irwin, G R. (1973). *The Stress Analysis of Cracks Handbook*, Del Research Corp., Hellertown, Pa.

Takeo, M. (1983). Source mechanisms of Utsu Volcano, Japan, earthquakes and their tectonic implications. *Phys. Earth Planet. Interiors* **32**, 241–264.

Tarantola, A. (1987). *Inverse Problem Theory*, Elsevier, Amsterdam.

Tarantola, A., and Valette, B. (1982). Inverse problems = quest for information. *J. Geophys.* **50**, 159–170.

Taylor, D. W. A., Snoke, J. A., Sacks, I. S., and Takanami, T. (1990). Nonlinear frequency–magnitude relationships for the Hokkaido corner, Japan. *Bull. Seism. Soc. Am.* **80**, 340–353.

Teisseyre, R. (1980). Some remarks on the source mechanism of rock bursts in mines and on the possible source extension. *Acta Mont.* **55**, 7–13.

Teisseyre, R. (1985a). New earthquake rebound theory. *Phys. Earth Planet. Interiors* **39**, 1–4.

Teisseyre, R. (1985b). Creep flow and earthquake rebound: System of the internal stress evolution. *Acta Geophys. Pol.* **33**, 11–23.

Teufel, L. W., and Logan, J. M. (1978). Effect of shortening rate on the real area of contact and temperature generated during frictional sliding. *Pure Appl. Geophys.* **116**, 840–865.

Thomas, D. (1988). Geochemical precursors to seismic activity. In *Intermediate-Term Earthquake Prediction* (W. D. Stewart and K. Aki, eds.); reprinted from Special Issue, *Pure Appl Geophys.* **126**, 241–266, Birkhäuser Verlag, Basel.

Thomsen, L. (1986). Weak elastic anisotropy. *Geophysics* **51**, 1954–1966.

Thorbjarnardottir, B. S., and Pechmann, J. C. (1987). Constraints on relative earthquake locations from cross-correlation of waveforms. *Bull. Seism. Soc. Am.* **77**, 1626–1634.

Thurber, C. H. (1985). Nonlinear earthquake location: Theory and examples. *Bull. Seism. Soc. Am.* **75**, 779–790.

Thurber, C. H., and Aki, K. (1987). Three-dimensional seismic imaging. *Ann. Rev. Earth Planet. Sci.* **15**, 115–139

Tikhonov, A. N., and Arsenin, V. J. (1979). *Solution Methods of Ill-Conditioned Problems*, Nauka, Moscow (in Russian).

Tinti, S., and Mulargia, F. (1985). Effect of the magnitude uncertainties on estimating the parameters in the Gutenberg–Richter frequency–magnitude law. *Bull. Seism. Soc. Am.* **75**, 1681–1697.

Tinti, S., and Mulargia, F. (1987). Confidence intervals of b values for grouped magnitudes. *Bull. Seism. Soc. Am.* **77**, 2125–2134.

Toksöz, M. N., Dainty, A. M., Reiter, E., and Wu, R.-S. (1988). A model for attenuation and scattering in the Earth's crust. In *Scattering and Attenuation of Seismic Waves* (R.-S. Wu and K. Aki, eds.), Part I; reprinted from Special Issue, *Pure Appl. Geophys.* **128**, 81–100, Birkhäuser Verlag, Basel.

Trampert, J., and Leveque, J.-J. (1990). Simultaneous iterative reconstruction technique: Physical interpretation based on generalized least squares solution. *J. Geophys. Res.* **95**, 12,553–12,559.

Trifu, C. I. (1983). Optimal development of a regional seismic network. Examplification for Romania. *Rev. Roum. Phys.* **28**, 81–90

Tse, S. T., and Rice, J. R. (1986). Crustal earthquake instability in relation to the depth variation of frictional slip properties. *J. Geophys. Res.* **91**, 9452–9472.

Tsukuda, T. (1988). Coda Q before and after the 1983 Misasa earthquake of M 6.2, Tottori prefecture, Japan. *Pure Appl. Geophys.* **128**, 261–279.

Tullis, T. E., ed. (1986). *Friction and Faulting*, Birkhäuser Verlag, Basel; reprinted from Special Issue, *Pure Appl. Geophys.* **124**(3).

Tullis, T. E., and Weeks, J. D. (1986). Constitutive behavior and stability of frictional sliding of granite. In *Friction and Faulting* (T. E. Tullis, ed.); reprinted from Special Issue, *Pure Appl. Geophys.* **124**, 383–414, Birkhäuser Verlag, Basel.

Turcotte, D. L. (1989). Fractals in geology and geophysics. In *Fractals in Geophysics* (C. H. Scholz and B. B. Mandelbrot, eds.); reprinted from, *Pure Appl. Geophys.* **131**, 171–196, Birkhäuser Verlag, Basel.

Udias, A. (1989). Development of fault-plane studies for the mechanism of earthquakes. In *Observatory Seismology* (J. J. Litehiser, ed.), pp. 243–356, University of California Press, Berkeley.

Udias, A., and Baumann, D. (1969). A computer program for focal mechanism determination combining P and S data. *Bull. Seism. Soc. Am.* **59**, 503–519.

Uhrhammer, R. A. (1982) The optimal estimation of earthquake parameters *Phys. Earth. Planet. Interiors* **30**, 105–118.

Ulomov, V. I., and Mavashev, B. Z. (1971). *The Tashkent Earthquake of 26 April 1966*, FAN, Akad. Nauk Uzbek. SSR, Tashkent.

Utsu, T (1965). A method for determining the value of b on the formula $\log n = a - bM$ showing the magnitude–frequency relation for earthquakes. *Geophys. Bull., Hokkaido Univ.* **13**, 99–103 (in Japanese; English abstract)

Utsu, T. (1966). A statistical significance test of the difference in b-value between two earth-quakes groups. *J. Phys. Earth* **14**, 37–40.

Utsu, T. (1971). Aftershocks and earthquake statistics. III. *J. Fac Sci., Hokkaido Univ., Ser. VII* **3**, 379–441.

Valdés, C. M., and Novelo-Casanova, D. A. (1989). User manual for QCODA. In *Toolbox for Seismic Acquisition, Processing and Analysis* (W. H. K. Lee, ed.), Vol. 1, pp. 237–255, Internatl. Assoc. Seism. Phys. Earth Interior and Seism. Soc. Am.

Van Aswegen, G. (1990). Fault stability in SA gold mines. In *Mechanics of Jointed and Faulted Rock* (H. P Rossmanith, ed.), pp. 717–725, Balkema, Rotterdam.

Van der Sluis, A., and van der Vorst, H. A. (1987). Numerical solution of large, sparse linear systems arising from tomographic problems. In *Seismic Tomography* (G. Nolet, ed.), pp. 53–87, Reidel, Dordrecht.

Vasco, D. W. (1990) Moment–tensor invariants: Searching for non-double-couple earthquakes. *Bull. Seism. Soc. Am.* **80**, 354–371.

Vasco, D. W., and Johnson, L. R. (1989). Inversion of waveforms for extreme source models with an application to the isotropic moment tensor component. *Geophys. J.* **97**, 1–18.

Vidale, J. E. (1986). Complex polarization analysis of particle motion. *Bull. Seism. Soc. Am.* **76**, 1393–1405.

Vidale, J. E. (1988) Finite-difference calculation of travel times. *Bull. Seismol. Soc. Am.* **78**, 2062–2076.

Vidale, J. E. (1990). Finite-difference calculation of travel times in three dimensions. *Geophysics* **55**, 521–526.

Virieux, J (1991). Fast and accurate ray tracing by Hamiltonian perturbation *J. Geophys. Res.* **96**, 579–594.

Virieux, J., and Madariaga, R. (1982). Dynamic faulting studied by a finite difference method. *Bull. Seism. Soc. Am.* **72**, 345–369.

Voinov, K. A., Krakov, A. S., Lomakin, V. S., and Khalevin, N. I. (1987). Seismological studies of rock bursts at the Northern Ural bauxite deposits. *Izv. Akad. Nauk SSSR, Fiz. Zemlı* (10), 98–104 (in Russian).

Wakita, H., Nakamura, Y., and Sano, Y. (1988). Short-term and intermediate-term geochemical precursors. In *Intermediate-Term Earthquake Prediction* (W. D. Stewart and K. Aki, eds.); reprinted from Special Issue, *Pure Appl. Geophys.* **126**, 267–278, Birkhäuser Verlag, Basel.

Wallace, T. C. (1985). A reexamination of the moment tensor solutions of the 1980 Mammoth Lakes earthquakes. *J. Geophys. Res.* **90**, 11,171–11,176.

Wallace, T. C., Helmberger, D. V., and Mellman, G. R. (1981). A technique for the inversion of regional data in source parameter studies. *J. Geophys. Res.* **86**, 1679–1685.

Waltham, D. A. (1988). Two-point ray tracing using Fermat's principle. *Geophys. J.* **93**, 575–582.

Wang, C.-Y., and Herrmann, R. B. (1988). Synthesis of coda waves in layered medium. In *Scattering and Attenuation of Seismic Waves* (R.-S. Wu and K. Aki, eds.), Part I; reprinted from Special Issue, *Pure Appl. Geophys.* **128**, 7–42, Birkhäuser Verlag, Basel.

Weeks, J., Lockner, D., and Byerlee, J. (1978). Changes in *b*-values during movement on cut surfaces in granite. *Bull. Seism. Soc. Am.* **68**, 333–341.

Weichert, D. H. (1980). Estimation of the earthquake recurrence parameters for unequal observation periods for different magnitudes. *Bull. Seism. Soc. Am.* **70**, 1337–1346.

Wennerberg, L., and Frankel, A. (1989). On the similarity of theories of anelastic and scattering attenuation. *Bull. Seism. Soc. Am.* **79**, 1287–1293.

Westbrook, G. K., Kusznir, N. J., Browitt, C. W. A., and Holdsworth, B. K. (1980). Seismicity induced by coal mining in Stoke-on-Trent (U.K.). *Eng. Geol.* **16**, 225–241.

Whitcomb, J. H., Garmony, J. D., and Anderson, D. L. (1973). Earthquake prediction: Variation of seismic velocities before the San Fernando earthquake. *Science* **180**, 632–641.

White, J. E. (1983). *Underground Sound. Application of Seismic Waves*, Elsevier, Amsterdam.

Wiejacz, P. (1991). Investigation of focal mechanisms of mine tremors by the moment tensor inversion, Ph.D. thesis, Inst. Geophys., Pol. Acad. Sci., Warsaw (in Polish).

Wieland, E. (1987). On the validity of the ray approximation for interpreting delay times. In *Seismic Tomography* (G. Nolet, ed.), pp. 85–98, Reidel, Dordrecht.

Will, M. (1984). Seismic observations during test drilling and destressing operations in German coal mines. In *Rockbursts and Seismicity in Mines* (N. C. Gay and E. H. Wainwright, eds.), Symp. Ser. No. 6, pp. 231–234, S. Afr. Inst. Min. Metal., Johannesburg.

Williams, D. J., and Arabasz, W. J. (1989). Mining-related and tectonic seismicity in the East Mountain area, Wasatch Plateau, Utah, U.S.A. In *Seismicity in Mines* (S. J. Gibowicz, ed.); reprinted from Special Issue, *Pure Appl. Geophys.* **129**, 345–368, Birkhäuser Verlag, Basel.

Wong, I. G. (1984). Mining-induced seismicity in the Colorado Plateau, western United States, and its implications for the sitting of an underground high-level nuclear waste repository. In *Rockbursts and Seismicity in Mines* (N. C. Gay and E. H. Wainwright, eds.), Symp. Ser. No. 6, pp. 147–152, S. Afr. Inst. Min. Metal., Johannesburg.

Wong, I. G., Humprey, J. R., Adams, J. H., and Silva, W. J. (1989). Observations of mine seismicity in the eastern Wasatch Plateau, Utah, U.S.A.: A possible case of implosional failure. In *Seismicity in Mines* (S. J. GIbowicz, ed.); reprinted from Special Issue, *Pure Appl. Geophys.* **129**, 369–405, Birkhäuser Verlag, Basel.

Wong, J., Hurley, P., and West, G. F. (1985). Investigation of subsurface geological structure at the Underground Research Laboratory with crosshole seismic scanning. *Proc. 17th Information Meet. Nucl. Fuel Waste Managem. Progr., Canada* **TR-229**, 593–608.

Wong, T.-F. (1986). On the normal stress dependence of the shear fracture energy. In *Earthquake Source Mechanics* (S. Das, J. Boatwright, and C. H. Scholz, eds.), *Maurice Ewing*, Vol. 6, pp. 1–11, Am. Geophys. Union, Washington, D C.

Woodgold, C. R. D. (1990). Estimation of Q in eastern Canada using coda waves. *Bull. Seism. Soc. Am.* **80**, 411–429.

Wu, K.-T., Yue, M.-S., Wu, H.-Y., Chao, S.-L., Chen, H.-T., Huang, W.-Q., Tien, K.-Y , and Lu, S.-D. (1978). Foreshocks to the Haicheng earthquake of 1975. Certain characteristics of the Haicheng earthquake ($M = 7.3$) sequence. *Chin. Geophys.* **1**, 289–308.

Wu, R. (1982) Attenuation of short period seismic waves due to scattering. *Geophys. Res. Lett.* **9**, 9–12.

Wu, R.-S. (1985). Multiple scattering and energy transfer of seismic waves—separation of scattering effect from intrinsic attenuation, I. Theoretical modeling. *Geophys. J. Roy. Astr Soc.* **82**, 57–80

Wu, R.-S. (1989). The perturbation method in elastic wave scattering. In *Scattering and Attenuation of Seismic Waves* (R.-S. Wu and K. Aki, eds.), Part II; reprinted from Special Issue, *Pure Appl. Geophys.* **131**, 605–637, Birkhäuser Verlag, Basel.

Wu, R.-S., and Aki, K., eds. (1988a). *Scattering and Attenuation of Seismic Waves*, Part I, Birkhäuser Verlag, Basel; reprinted from Special Issue, *Pure Appl. Geophys.* **128** (1/2).

Wu, R.-S., and Aki, K. (1988b). Multiple scattering and energy transfer of seismic waves— separation of scattering effect from intrinsic attenuation. II. Application of the theory to Hindu Kush region In *Scattering and Attenuation of Seismic Waves*, Part I (R.-S. Wu and K. Aki, eds.); reprinted from Special Issue, *Pure Appl. Geophys.* **128**, 49–80, Birkhäuser Verlag, Basel

Wu, R.-S., and Aki, K., eds. (1989). *Scattering and Attenuation of Seismic Waves*, Part II, Birkhäuser Verlag, Basel; reprinted from Special Issue, *Pure Appl. Geophys.* **131** (4).

Wu, R. S., and Aki, K., eds. (1990) *Scattering and Attenuation of Seismic Waves*, Part III, Birkhäuser Verlag, Basel; reprinted from Special Issue, *Pure Appl. Geophys.* **132** (1/2).

Wyss, M. (1979). Estimating maximum expectable magnitude of earthquake from fault dimensions. *Geology* **7**, 336–340.

Wyss, M., ed. (1991). *Evaluation of Proposed Earthquake Precursors*, Am. Geophys. Union, Washington, D C.

Wyss, M., and Brune, J. N. (1968). Seismic moment, stress and source dimensions for earthquakes in the California-Nevada region. *J. Geophys. Res.* **73**, 4681–4694.

Wyss, M , and Lee, W. H. K. (1973). Time variations of the average earthquake magnitude in central California. *Stanford Univ. Publ., Geol. Sci.* **13**, 24–42.

Yamashina, K. (1978). Induced earthquakes in the Izu Peninsula by the Izu-Hanto-Oki earthquake of 1974, Japan *Tectonophysics* **51**, 139–154.

Yegulalp, T. M., and Kuo, J. T. (1974) Statistical prediction of occurrence of maximum magnitude earthquakes. *Bull. Seism. Soc. Am.* **64**, 393–414.

Young, R. P., and Hill, J. J. (1985). Seismic characterisation of rock masses before and after mine blasting. *Proc 26th U.S. Symp. Rock Mec.*, Rapid City (S. Dak.), 1151–1158.

Young, R. P., Hutchins, D. A., McGaughey, W. J., Urbancic, T., Falls, S., and Towers, J. (1987). Current seismic tomographic imaging and acoustic techniques: A new approach to rockburst In *Proceedings of the 6th International Congress on Rock Mechanics* (G. Hergert and S. Vongpaisal, eds.), pp 1333–1338, Balkema, Rotterdam.

Young, R. P., Talebi, S., Hutchins, D. A., and Urbancic, T. I. (1989a). Analysis of mining-induced microseismic events at Strathcona mine, Sudbury, Canada. In *Seismicity in Mines* (S J. Gibowicz, ed.); reprinted from Special Issue, *Pure Appl. Geophys.* **129**, 455–474, Birkhäuser Verlag, Basel.

Young, R. P., Hutchins, D. A., McGaughey, J., Towers, J., Jansen, D., and Bostock, M. (1989b). Geotomographic imaging in the study of mining induced seismicity. In *Seismicity in Mines* (S. J. Gibowicz, ed.); reprinted from Special Issue, *Pure Appl. Geophys.* **129**, 571–596, Birkhäuser Verlag, Basel

Zavyalov, A. D., and Sobolev, G. A. (1988). Analogy in precursors of dynamic events at different scales. *Tectonophysics* **152**, 277–282.

Zeng, Y., Su, F., and Aki, K. (1991). Scattered wave energy propagation in a random isotropic scattering medium. *J. Geophys. Res.* **96**, 607–619

Zhang, J. Z. and Song, L. Y. (1981). On the method of estimating *b*-value and its standard error *Acta Seism. Sin.* **3**, 292–301.

Zollo, A., and Bernard, P. (1991). Fault mechanisms from near-source data: Joint inversion of *S* polarizations and *P* polarities. *Geophys. J. Internatl.* **104**, 441–451.

Zuberek, W. M. (1989). Application of the Gumbel asymptotic extreme value distributions to the description of the AE amplitude distribution. *Proc. 4th Conf. Acoustic Emission/Microseismic Activity in Geological Structures and Materials*, Pennsylvania State University, October 1985, pp. 649–665, Trans. Tech. Publ., Clausthal-Zellerfeld, Germany

Selected Bibliography

Aki, K., and Richards, P. G (1980). "Quantitative Seismology. Theory and Methods." W. H. Freeman, San Francisco.

Atkinson, B. K., ed. (1987). "Fracture Mechanics of Rock." Academic Press, London.

Ben-Menahem, A., and Singh, S. J., (1981). "Seismic Waves and Sources." Springer-Verlag, New York.

Bolt, B. A., ed. (1987). "Sesmic Strong Motion Synthetics." Academic Press, New York.

Brady, B. H. G., and Brown, E. T. (1985). "Rock Mechanics." George Allen and Unwin, London.

Bullen, K. E., and Bolt, B. A (1985) "An Introduction to the Theory of Seismology." Cambridge University Press, Cambridge.

Červeny, V., Molotkov, I. A., and Psenčik, I. (1977). "Ray Method in Seismology." Univerzita Karlova, Prague.

Doornbos, D. J., ed. (1988). "Seismological Algorithms." Academic Press, London.

Ewing, W. M., Jardetzky, W. S., and Press, F (1957) "Elastic Waves in Layered Media." McGraw-Hill, New York.

Fairhurst, C., ed. (1990). "Rockbursts and Seismicity in Mines." Balkema, Rotterdam.

Gay, N. C., and Wainwright, E. H., eds. (1984). "Rockbursts and Seismicity in Mines." Symp. Ser. No. 6, S. Afr. Inst. Min. Metal., Johannesburg.

Gibowicz, S. J., ed. (1989). "Seismicity in Mines." Birkhäuser Verlag, Basel; reprint from *Pure Appl. Geophys.* **129**, *No.* 3/4.

Hanyga, A., ed. (1984). "Seismic Wave Propagation in the Earth. "Polish Scientific Publishers, Warsaw; Elsevier, Amsterdam.

Hatton, L., Worthington, M. H., and Makin, J. (1986). "Seismic Data Processing: Theory and Practice." Blackwell Scientific Publishers, Oxford.

Isaaks, E. H., and Srivastava, R. M. (1989). "An Introduction to Applied Geostatistics." Oxford University Press, New York.

Jaeger, J. C., and Cook, N. G. W. (1976). "Fundamentals of Rock Mechanics", 2nd ed. Chapman and Hall, London.

Kanamori, K., and Boschi, E., eds. (1983). "Earthquakes: Observation, Theory and Interpretation." North-Holland Publ. Comp., Amsterdam.

Kanasewich, E. R. (1981). "Time Sequence Analysis in Geophysics." University of Alberta Press, Edmonton, Alberta.

Kasahara, K. (1981). "Earthquake Mechanics." Cambridge University Press, Cambridge.

Kennett, B. L. N. (1983). "Seismic Wave Propagation in Stratified Media." Cambridge University Press, Cambridge.

Kostrov, B. V., and Das, S. (1988). "Principles of Earthquake Source Mechanics." Cambridge University Press, Cambridge.

Lee, W. H. K., and Stewart, S. W. (1981) "Principles and Applications of Microearthquake Networks." Academic Press, New York.

Litehiser, J. J., ed. (1989). "Observatory Seismology." University of California Press, Berkeley.

Means, W. D. (1979). "Stress and Strain. Basic Concepts of Continuum Mechanics for Geologists." Springer-Verlag, New York.

Menke, W. (1989). "Geophysical Data Analysis: Discrete Inverse Theory." Academic Press, Orlando.

Meskó, A. (1984). "Digital Filtering: Applications in Geophysical Exploration for Oil." Akadémiai Kiadó, Budapest.

Mogi, K. (1985). "Earthquake Prediction." Academic Press, Tokyo.

Nolet, G., Ed. (1987). "Seismic Tomography." D. Reidel Publ. Comp., Dordrecht.

Press, W. H., Flannery, B. P., Teukolsky, S. A., and Vetterling, W. T. (1990). "Numerical Recipes: The Art of Scientific Computing." Cambridge University Press, New York.

Richter, C. F (1958). "Elementary Seismology." W. H. Freeman, San Francisco.

Robinson, E. (1988). "Seismic Inversion and Deconvolution." Pergamon Press, Oxford.

Scholz, C. H. (1990). "The Mechanics of Earthquakes and Faulting." Cambridge University Press, Cambridge.

Scholz, C. H., and Mandelbrot, B. B., eds. (1989). "Fractals in Geophysics." Birkhäuser Verlag, Basel; reprint from *Pure Appl. Geophys.* **131**, No. 1/2.

Tarantola, A. (1987). "Inverse Problem Theory." Elsevier, Amsterdam.

Tullis, T. E., ed. (1986). "Friction and Faulting." Birkhäuser, Verlag, Basel; reprint from *Pure Appl. Geophys.* **124**, No. 3.

White, J E. (1983). "Underground Sound: Application of Seismic Waves." Elsevier, Amsterdam.

Wu, R.-S., and Aki, K., eds. (1988). "Scattering and Attenuation of Seismic Waves", Part I. Birkhäuser Verlag, Basel; reprint from *Pure Appl. Geophys.* **128**, No. 1/2.

Wu, R.-S., and Aki, K., eds. (1989). "Scattering and Attenuation of Seismic Waves", Part II. Birkhäuser Verlag, Basel; reprint from *Pure Appl. Geophys.* **131**, No. 4.

Wu, R.-S., and Aki, K., eds. (1990). "Scattering and Attenuation of Seismic Waves", Part III. Birkhäuser Verlag, Basel; reprint from *Pure Appl. Geophys.* **132**, No. 1/2.

Index

Absorption coefficient, 158
Acceleration spectrum, 169
Acoustic emission, 338
Active tomography, 93
Aftershocks, 244, 249, 250, 251–253, 255, 277, 337
Albedo, seismic, 160
Amonton's law, 229
Amplitude spectrum, 150
Anelasticity, 128, 138, 140
  constitutive laws, 131
  internal friction coefficient, 140
Angle
  critical, 44, 124
  emergence, 40
  incidence, 40, 265
  loss, 143
  polarization, 124, 203–204
  rake, 194
  reflection, 40
  shear, 25
  slip, 194
  take-off, 195, 198, 266
Angular frequency, 37
Anisotropy, 105
  aspect ratio, 112
  azimuthal, 105, 107
  complex transmission coefficients, 125
  crack density, 112, 118, 340, 341
  extensive-dilatancy (EDA), 105, 122, 340
  sagittal plane, 107
  shear-wave splitting, 105, 115–122, 123–124
  symmetry systems, 109
  time delay, 121
Apparent stress, 291
Asperities, 230, 243–245, 247–248, 250, 292
Attenuation, 128, 141–146
  absorption coefficient, 158
  coefficient (factor), 140, 141–142
  intrinsic, 128, 137, 156, 162
  operator, 143, 170–171, 268, 275
  scattering, 128, 147–148, 152, 153, 156, 162

Auxiliary plane, 193
Average displacement, source, 293
Azimuth, 195, 198

Backscattering, 152, 153
  coefficient, 165
Barriers, 243–245, 247–248, 252, 292
Bayesian
  estimation, 59
  location procedure, 48, 58–59, 63
  theorem, 59
Best double couple, 212
Biaxial stress, 28
Bimodal distribution, 22–23, 314–315, 317–318, 321
Body forces, 26
  double couple, 177, 189–192
  equivalent, 176, 192, 209, 224
  single, 178, 181, 185, 258
  single couple, 185–186
Body wave magnitude, 279
Bolzmann's equation, 139
Born's approximation, 152
b parameter (value), 261, 300, 302–304, 308, 326, 330, 338
Breakdown of self-similarity, 262, 295, 300
Brune's model, source, 236, 247, 269, 284–285, 287
Brune stress drop, 291
Bulk modulus, 30

Cauchy equations, 31
Cauchy strain tensor, 25
Causality principle, 138
Centroid moment tensor, 217
Characteristic earthquake, 297
Coda-duration magnitude, 280
Coda waves, 151
  energy-flux model, 160
  energy transport, 158

Coda waves (cont'd)
envelope, 154
power spectrum, 173
quality factor, 155, 156, 173, 340
source factor, 155, 158, 167, 173, 174
Compensated linear vector dipole (CLVD),
212–213, 218, 220
Complex
density, 143
elastic moduli, 137, 139, 146
propagation functions, 138
spectral density function, 149
wave velocity, 138, 144
Composite fault-plane solution, 205
Compressional axis, 191, 199–200
Compressional waves, 33
Constitutive equations, 29
Convolution, 179–181
Corner frequency, 264, 268–273, 274–275,
284–285, 287, 295
Correspondence principle, 138
Couple moment, force, 185
Couple without moment, force, 186
Crack modes, 225
Crack source models, 233, 238–239
Creep, 129
functions, 134, 135–137
Critical reflection, 44
Cumulative probability, 303, 306, 310–311, 324,
330, 333

Decomposition theorem, matrix, 52–53, 211
Delta function, 178, 180, 193, 256
Deviatoric strain, 130
Deviatoric stress, 130
Diffraction (diffracted) waves, 45, 98
Dilatancy, 342–344
Dilatancy-diffusion theory, 339, 342
Dilatation, 26
Dipole force, 186
Dip-slip fault, 201, 294
Dirac delta, 81, 178, 180, 193, 256
Directivity function (effects), 281, 282, 294
Dislocations, 233, 234, 252, 255
Somigliana, 233
Volterra, 233, 252
Dislocation source models, 233–234, 264
Dispersion, 45, 138
Displacement
far-field, 181–182, 187, 192
intermediate-field, 187, 192

near-field, 181–182, 187, 192
spectrum, 264, 267–268
vectors, 46, 187
Distribution
apparent magnitude, 330
bimodal, 22–23, 314–315, 317–318, 321
χ^2, 80
extreme values, 310–314, 319
frequency-magnitude, 301, 303, 306
Gaussian (normal), 59, 64, 66, 70, 75–77,
308, 325, 330
Gumbel, 310–314
marginal, 60
Poisson, 311, 317, 319–320
Double-couple
force, 177, 189
source, 189–193, 212–213, 277
Doublets, earthquake, 249
Dry-dilatancy model, 344
Duration of records, seismic, 276
Dynamic source models, 233, 238

Earthquake
doublets, 249
faulting theory, 224
sequences, 249, 253–255
swarms, 249, 253
Effective stress, 237
Efficiency, seismic, 284
Eigenvalues, 53
moment tensor, 211–212, 215, 216
Eigenvectors, 53
moment tensor, 211–212
Eikonal equations, 46–47
Elastic afterworking, 131
Elastic-brittle crack model, 225
Elastic constants (moduli), 29–30, 106–107
complex, 137, 138, 139, 146
effective, 111–112, 146
Elasticoviscous material, 132
Elastic limit, 129
Elastic rebound theory, 176, 222, 224
Electric rock resistivity, 341
Empirical Green's function, 214–215, 256
Energy
density, 36
flux, 268, 281, 282
flux model, coda, 160–162
release rate, 322, 346
seismic, 281–282, 284, 301, 305–306, 310,
312–313, 315–318, 322, 324–326, 328
transport, coda, 158

Equal-area projection, 198
Equivalent body forces, 176, 192, 209, 224
Excess shear stress, 346
Extensive-dilatancy anisotropy (EDA), 105, 122
Extinction coefficient, 158
Extinction length, 160
Extreme values distribution, 310–314, 319

Far-field displacement, 181–182, 187, 192
Fault
  dip, 193–194
  dip-slip, 201, 294
  left-lateral, 201
  normal, 202, 258
  oblique-slip, 201
  plane, 193, 266, 285
  reverse, 202
  right-lateral, 201
  strike, 193
  strike-slip, 201, 258, 294
  thrust, 202
Fault-plane solution, 198, 199–201, 203–206, 266
Fedorov's location procedure, 63–64
Firmoviscous material, 131
Focal sphere, 195, 198–199
Force couple, 185–186
Force dipole, 185–186
Force double couple, 177, 189–192
Foreshocks, 244, 249–251, 337, 338
Forward scattering, 152
Fourier transform, 149–151, 264
  inverse, 149–150
  spatial, 151
Fractal
  dimension, 259–261
  distribution, 260
  tear, 263
Fracture
  criterion, 225–226
  energy, 226, 243
  mechanics, 225
  toughness, 225
Free-surface correction, 277
Frequency-magnitude relation, 261, 300, 302–303, 307, 310–312
Frequency-wavenumber spectrum, 247
Frictional constitutive law, 229–230
Friction coefficient, 229
Futterman's operator, attenuation, 143, 170–171, 268, 275

Geiger location method, 48, 49
Generalized Kelvin-Voigt material, 135
Geometrical spreading, 153, 173
Green's function, 179, 210, 213, 214, 232, 256, 258
  empirical, 214–215, 256
Group velocity, 45, 109
Gumbel distributions, 310–314
Gutenberg-Richter relation, 302–303, 308, 312, 324, 326, 330

Haskell's source model, 234–235
Hazard, seismic, 301, 305, 308, 312–314, 322, 325–331, 334
Head waves, 45
Heaviside step function, 133, 178–179, 193, 234
Hooke's law, 29
Hydrologic precursors, 341–342
Hydrostatic stress, 28

Impulse response, 179
Incompressibily, 30
Information a priori, 58–59, 61–62, 72
Inhomogeneous fault model, 247, 285
Inhomogeneous waves, 125
Intermediate-field displacement, 187, 192
Internal friction, 128
  coefficient, 140
Intrinsic attenuation, 128, 137, 156, 162
  quality factor, 140–143, 156, 163–164, 167–169, 171, 175
Inverse Fourier transform, 149–150
Inversion, moment tensor, 209, 213–215, 216–220
Isochrones, 245
Isotropic scattering, 158
Isotropy, 29
  transverse, 107

Kelvin-Voigt material, 131
Kelvin-Voigt model, 134
Kinematic source models, 233, 264
Kolmagorov equation, 316
Kronecker delta, 30, 178

Lamé parameters (constants), 29
Lapse time, 151, 153, 280

Left-lateral fault, 201
Linear viscoelastic solid, 132, 138
Local magnitude, 279, 323
Location procedure
　arrival time difference (ATD), 65
　Bayesian, 48, 58–59, 63
　damping least squares, 57
　Fedorov's, 63–64
　Geiger, 48, 49
　likelihood function, 59
　linear, 74
　L_1 norm, 75–77
　L_2 norm, 75
　master event, 49, 65
　misfit function, 50, 52, 55, 59, 65, 71, 75, 76
　relative, 49, 65
　simplex, 76
　simultaneous hypocenter and velocity, 49, 69
　singular value decomposition, 53, 58
Logarithmic decrement, 141–142
Love waves, 46

McGarr's fault model, 247, 285
Madariaga's source model, 238–239, 285–286, 287
Magnitude, 276, 278
　body wave, 279
　coda-duration, 280
　local, 279, 323
　maximum, 306, 330, 332–333, 335, 392
　mean, 304, 307
　minimum, 303, 304, 306, 330, 332, 334
　moment, 279
　surface wave, 279
Major double couple, 212
Master event, 49, 65
Matrix decomposition theorem, 52–53, 65, 211
Maximum frequency, seismic spectra, 273, 298–299
Maxwell substance (material), 132, 133
Maxwell model, 132
Mean free path, 155, 164
Mie scattering, 147
Mines
　Belchatow, Poland, lignite, 11–12
　Blyvooruitzicht, South Africa, gold, 3
　Bobrek, Poland, coal, 308–309, 327
　Cerro de Pasco, Peru, limestone, 11
　Doubrava, Czecho-Slovakia, coal, 314, 318, 325–326
　ERPM, South Africa, gold, 3, 16, 253

Fang-Shan, China, coal, 8
Grangesberg, Sweden, iron, 8
Heinrich Robert, Germany, coal, 223, 269, 297
Horonai, Japan, coal, 6, 218
Kirovsk, C I.S., apatite, 7
Kiruna Research Mine, Sweden, 94
Lubin, Poland, copper, 22, 23, 68, 312, 314, 341
Lucky Friday, Idaho, ore, 21
Men-Tou-Gou, China, coal, 8, 63
Miike, Japan, coal, 6
Mount Charlotte, Australia, gold, 8
Nowy Wirek, Poland, coal, 317–318
Otomnaki, Finland, ore, 94
Phalaborwa, South Africa, coal, 10
Polkowice, Poland, copper, 69
Pstrowski, Poland, coal, 339
Restoff, U.S.A., salt, 94
Rozbark, Poland, coal, 2
Rudna, Poland, copper, 69, 219–220, 339
Strathcona, Ontario, Canada, ore, 5
Sunagawa, Japan, coal, 218
Szombierki, Poland, coal, 199, 314–315
Underground Research Laboratory, Manitoba, Canada, 94, 223, 269, 272, 283, 285
Vaal Reefs, South Africa, gold, 3, 259
Wappingers Falls, New York, limestone, 11
Western Deep Levels, South Africa, gold, 20, 157, 296, 340
Wujek, Poland, coal, 313, 328–329
Mine tremor
　bimodal distribution, 22–23, 314–315, 317–318, 321
　complex, 287
　types, 15, 18
Mine tremors
　Belchatow lignite mine, Poland (November 29, 1980), 11–12
　ERPM gold mine, South Africa (May 1973), 253
　Grängesberg ore mine, Sweden (August 30, 1974), 8
　Heinrich Robert coal mine, Germany (March 23, 1987), 269
　Kirovsk apatite mine, C.I.S. (April 16, 1989), 7
　Klerksdorp gold district, South Africa (April 7, 1977), 2
　Lubin copper basin, Poland (March 24, 1977) 2, 252, 255, 339 (June 20, 1987) 255, 259

Orange Free State gold district, South Africa (April 9, 1991), 218
Ostrava-Karvina coal basin, Czecho-Slovakia (April 27, 1983), 4
Wappingers Falls limestone quarry, New York (June 7, 1974), 11
Werra potash district, Germany (June 23, 1975), 2 (March 13, 1989), 2
Minor double couple, 212
Moment-magnitude relations, 261, 279, 323
Moment tensor, 177, 186, 193, 195, 196, 209, 210–213
   centroid, 217
   determinant, 215
   deviatoric component, 211
   eigenvalues, 211–212, 215, 216
   eigenvectors, 211–212
   inversion, 209, 213–215, 216–220
   isotropic component, 211–213
   trace, 215
Multiple scattering, 152, 157–159

Near-field displacement, 181–182, 187, 192
Network design, 78, 82
   classic planning criterion, 81
   D-optimum planning, 83
   optimum distribution, 79, 84
   Ostrava regional network, 89
   quality index, 79, 88
   simplexial designs, 83
Nodal planes, 193, 198, 199, 204
Non-double-couple sources, 216, 218–223, 283
Normal
   distribution (Gaussian), 59, 64, 66, 70, 75–77, 308, 325, 330
   equations, 52, 56–58, 65, 70
   fault, 202, 258
   force, 26
   stress component, 27
Null axis, 191

Oblique-slip fault, 201
Omori law, 252, 337
Orthogonal polarization, 108
Orthorhombic symmetry, 106

Partial stress drop, 288–290
Particle motion, 35, 45, 109, 115–117, 122, 126, 182, 187, 265

Passive tomography, 93
Periodic convolution, 181
Phase spectrum, 150
Phase velocity, 45, 109, 139, 140
Plane of incidence, 203
Plane waves, 34, 38, 40, 108, 118
Plastic flow, 129
Point source, 177, 210
Poisson
   distribution, 311, 317, 319–320
   occurrence, 330
   process, 310–311, 319
   ratio, 30, 106
Polarity, 121, 198, 199, 205
Polarization
   angle, 124, 203–204, 206
   diagrams, 115, 117, 120
   orthogonal, 108
   P waves, 108–109, 265
   S waves, 108–109, 115–116, 120, 203–204, 205, 207
   vector, 121
Power spectral density, 150
Precursory phenomena, 303, 337–342
Principal strain, 26
Principal stress, 28
Probability
   a priori location, 95
   cumulative, 303, 306, 310–311, 324, 330, 333
   magnitude distribution, 301, 303, 306
   occurrence, 301, 309, 316, 320, 326–327
Pulse
   duration, 171, 275–276, 286
   rise time, 171–172, 275
P waves, 33
   coda, 151
   quality factor, 143, 171–172, 175
   quasi, 108–109, 145

Quality factor Q, 95–96, 128, 140–144, 153, 156–157, 166, 167, 268, 274–275
   bulk dissipation parameter, 143
   coda, 155–157, 172–174, 340
   frequency dependence, 163
   intrinsic, 141–142, 156, 157–158, 162
   P waves, 143, 171–172, 175
   scattering, 156, 157–158, 162
   spatial, 142
   S waves, 143, 171–172, 175
   temporal, 142
Quasi P waves, 108–109, 145
Quasi S waves, 108–109, 145

Radiation efficiency, 283-284
Radiation pattern (coefficient) 182-184,
    188-189, 190-191, 192, 195-197, 203,
    214, 232-233, 235-236, 266, 277,
    281-282, 292
Radon emission, 341-342
Rake angle, 194, 200
Ramp function, 234, 256
Ray, seismic, 46-47, 94
    tracing, 47, 93-94, 97
Rayleigh scattering, 147, 159
Rayleigh waves, 45
Rectangular fault model, 234-235, 287
Reflection, 40
    angle, 40
    coefficient, 41
    critical, 44
Refraction, 41
Relative location, 49, 65
Relaxation
    function, 134, 136
    time, 135-136
Residuals, time, 48-52, 59, 61, 65, 70, 71, 75
Return period, 313, 335
Reverse fault, 202
Right-lateral fault, 201
Rise time, pulse, 171-172, 234, 275-276, 286
Rms acceleration, 276, 292
Rupture velocity, 177, 234, 238, 240, 284,
    286-287, 292, 294

Sagittal symmetry, anisotropy, 111
Scalar moment, 192, 216
Scale invariance, 259, 295
Scaling relations, 262, 295, 300
Scattered waves, 128, 146, 151
Scattering, 128, 144, 147
    attenuation, 128, 147-148, 152, 153, 156,
        162
    beckward, 152
    coefficient, 156, 158, 164
    cross-section, 156
    forward, 152
    isotropic, 153, 159
    Mie, 147
    multiple, 151, 152, 157-160
    quality factor, 156, 161-162
    Rayleigh, 147, 158
    single, 151, 152, 153, 155, 157, 159, 160, 175
Schmidt net, projection, 198
Seismic
    albedo, 160

cycle, 336
    efficiency, 284
    energy, 281-282, 284, 301, 305, 310,
        312-318, 322, 324-326, 328
    gap, 336
    hazard, 301, 305, 308, 312-314, 322,
        325-331, 334, 336
    moment, 216, 236, 261, 277-279, 295-297,
        322-323
    quiescence, 337
    ray, 46, 94
        bending, 97-98
        tracing, 47, 93-94, 97
Seismicity
    activity rate, 309, 330, 331, 333
    fractal structure, 262
    pattern, 337
Seismic spectra, 264, 267-268, 273-275
    decay coefficient, 264, 268
    intermediate-frequency part, 287
    low-frequency level, 264, 268-269, 271-273,
        277
    maximum frequency, 273, 298-299
    site, 273-274, 278
    source, 273-275
Self-similar crack, 238
Self-similarity breakdown, 262, 295-296, 300
Sequences, earthquake, 249
Shear, 25
    cracks, 225
    force, 26
Shear-tensional source, 222
SH waves, 35-36, 107, 109, 119, 124-126, 214,
    267
Simplex location procedure, 76
Single body force, 178, 181, 185, 258
Single-couple body force, 185-186
Single scattering, 151, 152, 153, 155, 157, 159,
    160, 175
Singular value decomposition (SVD), 53, 58,
    99, 274
Site
    corrections, 167, 277-278
    effects, 167-168, 170, 174, 280
    spectra, 273-274, 278
Slip angle, 194
Slip function, 232, 234, 237, 239
Slip weakening, 226, 228-229, 240
Slowness, 47, 95
Snell's law, 40, 94, 98
Somigliana dislocation, 233
Source, 176-177
    complexity, 293

point, 177, 209–210
radius, 284, 286, 297
shear-tensional, 222
spectra, 273–274
synchronous, 210
time functions, 210, 213, 237, 242, 256, 257, 258–259
Source model
Brune's, 236–237, 247, 269, 284–285, 288, 290
crack, 233, 237–241
dislocation, 233–237, 264
double couple, 189–193, 212–213, 215, 277
Haskell's, 233, 234, 235
McGarr's, 247, 285
Madariaga's, 238–239, 285–286, 287
stochastic, 243
Spatial Fourier transform, 151
Spectral analysis, 266
Hamming tapers, 266–267
Hanning tapers, 266–267
Nyquist frequency, 268
Spectrum
amplitude, 150, 167, 264
energy density, 150
frequency wavenumber, 247
phase, 150
power density, 150, 154
Spherical waves, 37–38
Splitting, S waves, 105, 115–122, 123–124
Standard linear solid, 135–136
State variables, 229–231
Static stress drop, 291
Statistical prediction, 301
Step function, 133, 179, 193, 234
Stereographic projection, 198
Stick-slip instability, 225, 231
Stochastic fault models, 243
Stopping phases, 239, 242, 245
Strain, 25
energy function, 31, 107
recovery, 129
tensor, Cauchy, 25
Strength degradation, 228
Stress, 26–27
field, 28
glut, 209
intensity factor, 225–227, 240
singularity, 226, 227
tensor, 27, 209
Stress drop, 228, 237, 243, 250, 291, 293, 295–297, 316
Brune, 291

dynamic, 238, 275, 283, 291, 292
effective, 291
partial, 288–290
static, 291
Stress intensity factor, 225–227, 240
Strike-slip fault, 201, 258, 294
Surface waves, 45–46
Love, 46
magnitude, 279
Rayleigh, 45
Stoneley, 45
SV waves, 35–36, 107, 109, 119, 124–126, 214, 267
Swarms, 249, 253
S waves, 33
coda, 151
polarization, 108–109, 115–116, 120, 203–204, 205, 207
quality factor, 143, 171–172, 175
quasi, 108–109, 145
splitting, 105, 115–122, 123–124
Synchronous source, 210
Synthetic seismograms, 118, 120, 255–257, 259

Take-off angle, 195, 198, 266
Tapers, spectral analysis, 266–267
Tensile cracks, 225
Tensional axis, 191, 199–200
Thrust fault, 202
Time functions, source, 210, 213, 237, 242, 256, 258–259
Time residuals, 48–52, 59, 61, 65, 70, 71, 75
Tomography, 93
active, 93
algebraic reconstruction technique (ART), 100
attenuation (amplitude), 95
conjugate gradient method, 101–102
crosshole, 94
diffracted waves, 98
inversion techniques, 99
LSQR procedure, 102
passive, 93
ray tracing, 93, 94, 97
simultaneous iteration reconstruction technique (SIRT), 101, 102
travel time perturbation, 95
velocity (travel time), 94
Traction, 26
Transverse isotropy, 107
Transverse waves, 33–34

Triaxial stress, 28
Turbidity coefficient, 156

Uniaxial stress, 28

Vector dipoles, 212
Velocity-weakening model, 231
Viscoelastic media, 129
  linear solid, 132, 138
Viscous flow, 129
Volcanic tremors (events), 23, 217, 220–221,
  295
Volterra dislocation 233, 252
Volume of excavated (mined) rock, 322–326
Volume sources, 211, 215

Wave
  equations, 33
  front, 47
  number, 35, 37, 109, 137–138, 139
  polarization, 108–109
  reflection 40–44
  refraction, 42, 44
  scattering, 128, 144, 147

Waves, 33
  compressional, 33
  conical, 45
  diffraction (diffracted), 45, 98, 99
  head, 45
  inhomogeneous, 125
  longitudinal, 34
  Love, 46
  plane, 34, 38, 108, 118
  quasi P, 108–109, 145
  quasi S, 108–109, 145
  Rayleigh, 45
  reflected, 40
  refracted, 41
  scattered, 128, 146, 151
  SH, 35–36, 107, 109, 119, 124–126, 214, 267
  spherical, 37, 118
  Stoneley, 45
  surface, 45–46
  SV, 35–36, 107, 109, 119, 124–126, 214, 267
  transverse, 33–34
Wieland's effect, 99
Wulff net, projection, 198

Young's modulus, 30, 106

International Geophysics Series

EDITED BY

RENATA DMOWSKA
Division of Applied Sciences
Harvard University
Cambridge, Massachusetts

JAMES R. HOLTON
Department of Atmospheric Sciences
University of Washington
Seattle, Washington

Volume 1 BENO GUTENBERG. Physics of the Earth's Interior. 1959*

Volume 2 JOSEPH W. CHAMBERLAIN. Physics of the Aurora and Airglow. 1961*

Volume 3 S. K. RUNCORN (ed.). Continental Drift. 1962*

Volume 4 C. E. JUNGE. Air Chemistry and Radioactivity. 1963*

Volume 5 ROBERT G. FLEAGLE AND JOOST A. BUSINGER. An Introduction to Atmospheric Physics. 1963*

Volume 6 L. DUFOUR AND R. DEFAY. Thermodynamics of Clouds. 1963*

Volume 7 H. U. ROLL. Physics of the Marine Atmosphere. 1965*

Volume 8 RICHARD A. CRAIG. The Upper Atmosphere: Meteorology and Physics. 1965*

Volume 9 WILLIS L. WEBB. Structure of the Stratosphere and Mesosphere. 1966*

Volume 10 MICHELE CAPUTO. The Gravity Field of the Earth from Classical and Modern Methods. 1967*

Volume 11 S. MATSUSHITA AND WALLACE H. CAMPBELL (eds.). Physics of Geomagnetic Phenomena. (In two volumes.) 1967*

Volume 12 K. YA. KONDRATYEV. Radiation in the Atmosphere. 1969*

*Out of print.

Volume 13   E. PALMÉN AND C. W. NEWTON. Atmospheric Circulation Systems: Their Structure and Physical Interpretation. 1969

Volume 14   HENRY RISHBETH AND OWEN K. GARRIOTT. Introduction to Ionospheric Physics. 1969*

Volume 15   C. S. RAMAGE. Monsoon Meteorology. 1971*

Volume 16   JAMES R. HOLTON. An Introduction to Dynamic Meteorology. 1972*

Volume 17   K. C. YEH AND C. H. LIU. Theory of Ionospheric Waves 1972*

Volume 18   M. I. BUDYKO. Climate and Life. 1974*

Volume 19   MELVIN E. STERN. Ocean Circulation Physics. 1975

Volume 20   J A. JACOBS The Earth's Core. 1975*

Volume 21   DAVID H. MILLER. Water at the Surface of the Earth: An Introduction to Ecosystem Hydrodynamics. 1977

Volume 22   JOSEPH W. CHAMBERLAIN. Theory of Planetary Atmospheres: An Introduction to Their Physics and Chemistry. 1978*

Volume 23   JAMES R. HOLTON. An Introduction to Dynamic Meteorology, Second Edition. 1979*

Volume 24   ARNETT S. DENNIS. Weather Modification by Cloud Seeding. 1980

Volume 25   ROBERT G. FLEAGLE AND JOOST A. BUSINGER. An Introduction to Atmospheric Physics, Second Edition. 1980

Volume 26   KUO-NAN LIOU. An Introduction to Atmospheric Radiation. 1980

Volume 27   DAVID H. MILLER. Energy at the Surface of the Earth: An Introduction to the Energetics of Ecosystems. 1981

Volume 28   HELMUT E. LANDSBERG. The Urban Climate. 1981

Volume 29   M. I. BUDYKO. The Earth's Climate: Past and Future. 1982

Volume 30   ADRIAN E. GILL. Atmosphere–Ocean Dynamics. 1982

Volume 31   PAOLO LANZANO. Deformations of an Elastic Earth. 1982*

Volume 32   RONALD T. MERRILL AND MICHAEL W. McELHINNY The Earth's Magnetic Field: Its History, Origin, and Planetary Perspective. 1983

Volume 33   JOHN S. LEWIS AND RONALD G. PRINN. Planets and Their Atmospheres: Origin and Evolution. 1983

Volume 34   ROLF MEISSNER. The Continental Crust: A Geophysical Approach. 1986

Volume 35   M. U. SAGITOV, B. BODRI, V. S. NAZARENKO, AND KH. G. TADZHIDINOV. Lunar Gravimetry. 1986

Volume 36   JOSEPH W. CHAMBERLAIN AND DONALD M. HUNTEN. Theory of Planetary Atmospheres: An Introduction to Their Physics and Chemistry, Second Edition. 1987

Volume 37   J. A. JACOBS. The Earth's Core, Second Edition. 1987

Volume 38   J. R. APEL. Principles of Ocean Physics. 1987

Volume 39   MARTIN A. UMAN. The Lightning Discharge. 1987

Volume 40   DAVID G. ANDREWS, JAMES R. HOLTON, AND CONWAY B. LEOVY. Middle Atmosphere Dynamics. 1987

Volume 41   PETER WARNECK. Chemistry of the Natural Atmosphere. 1988

Volume 42   S. PAL ARYA. Introduction to Micrometeorology. 1988

Volume 43   MICHAEL C. KELLEY. The Earth's Ionosphere. 1989

Volume 44   WILLIAM R. COTTON AND RICHARD A. ANTHES. Storm and Cloud Dynamics. 1989

Volume 45   WILLIAM MENKE Geophysical Data Analysis: Discrete Inverse Theory, Revised Edition. 1989

Volume 46   S. GEORGE PHILANDER. El Niño, La Niña, and the Southern Oscillation. 1990

Volume 47   ROBERT A BROWN. Fluid Mechanics of the Atmosphere. 1991

Volume 48   JAMES R. HOLTON. An Introduction to Dynamic Meteorology, Third Edition, 1992

Volume 49   ALEXANDER A. KAUFMAN. Geophysical Field Theory and Method, Part A: Gravitational, Electric, and Magnetic Fields. 1992. Part B: Electromagnetic Fields I. 1994. Part C: Electromagnetic Fields II. 1994

Volume 50   SAMUEL S. BUTCHER, GORDON H. ORIANS, ROBERT J. CHARLSON, AND GORDON V. WOLFE. Global Biogeochemical Cycles 1992

Volume 51   BRIAN EVANS AND TENG-FONG WONG. Fault Mechanics and Transport Properties in Rock. 1992

Volume 52   ROBERT E. HUFFMAN. Atomspheric Ultraviolet Remote Sensing. 1992

Volume 53   ROBERT A. HOUZE, JR. Cloud Dynamics, 1993

Volume 54   PETER V. HOBBS. Aerosol–Cloud–Climate Interactions. 1993

Printed and bound by CPI Group (UK) Ltd, Croydon, CR0 4YY

03/10/2024

01040410-0010